Quantum Mechanics

Quantum Mechanics

Mohammad Saleem

*Centre for High Energy Physics,
Punjab University, Pakistan*

IOP Publishing, Bristol, UK

ISBN 978-0-7503-1206-6 (ebook)
ISBN 978-0-7503-1207-3 (print)
ISBN 978-0-7503-1208-0 (mobi)

DOI 10.1088/978-0-7503-1206-6

Version: 20151001

IOP Expanding Physics
ISSN 2053-2563 (online)
ISSN 2054-7315 (print)

British Library Cataloguing-in-Publication Data: A catalogue record for this book is available from the British Library.

Published by IOP Publishing, wholly owned by The Institute of Physics, London

IOP Publishing, Temple Circus, Temple Way, Bristol, BS1 6HG, UK

US Office: IOP Publishing, Inc., 190 North Independence Mall West, Suite 601, Philadelphia, PA 19106, USA

Contents

Preface

Quantum mechanics is one of the most brilliant, stimulating, elegant and exciting theories of the twentieth century. It has not only explained a wide range of phenomena but has also brought revolutionary changes in the conceptual foundations of the entire field of physics. It is therefore imperative that an attempt should be made to evolve the subject so that it keeps the reader absorbed. The new ideas must be presented such that they become an integral part of the entire analysis. The physical situation and not the mathematical computations should be the main focus during the development of the discipline. Above all, the subject should be presented and its major themes described in a lucid style. Although a large number of books on quantum mechanics have been written, we strongly feel that there is still a need for a book written in the light of the above remarks. In this book we have made such an effort, illustrating how quantum mechanics replaces the structure of classical physics.

As quantum mechanics involves the introduction of a new conceptual framework, the new ideas are explicitly mentioned and explained in detail. Wherever possible, the various aspects of the original thinking of eminent physicists are reflected. We hope that by the time a reader has completed their study of this book, he/she will have developed a taste for realising the significance of the underlying principles and for understanding the ways in which the new concepts were introduced. For instance, Bohr's theory of the hydrogen atom, which does not give any justification for the violation of classical laws, was appreciated in general but was severely criticised by such eminent physicists as Ehrenfest, Laue and Stern. The last two physicists are said to have vowed in a conversation that they would give up physics if and when Bohr's new-fangled ideas about the hydrogen atom should turn out to be correct. Fortunately, they did not keep their vow: both of them later won the Nobel Prize!

The reader is encouraged to develop a critical attitude while learning quantum mechanics. It is pointed out in the first chapter that de Broglie obtained marvellous results by extending the idea of the duality of light to the duality of matter. Einstein, one of the greatest physicists of the twentieth century, had earlier extended the concept of covariance of Newton's second law of motion in inertial frames of reference to the covariance of every law of physics in inertial frames of reference. In fact, Newton noted the covariance of his classical law of motion in inertial frames only as a characteristic of this law, but Einstein adopted this covariance as a *criterion* for the validity of a law in physics. This, combined with the assumption of the constancy of the velocity of light in a vacuum, not only changed the conceptual framework of the entire field of physics but also caused the emergence of a new world of science and philosophy.

We have also focussed attention on some common bases in the solutions of various differential equations for different physical systems, such as harmonic oscillators and the hydrogen atom. Some interesting situations, such as the degeneracy in the solution of the hydrogen atom, are examined in detail to demonstrate the hidden content and bring it to the attention of the reader.

An entire chapter on the Dirac formalism and its relation to wave mechanics is an additional feature of this book. We have not only explained the notation but have exploited this technique to derive certain results. Papers in research journals normally use this technique; anyone interested in stepping into the relevant domain of research has to become familiar with it.

The reasons for solving some mathematical problems which apparently have no relevance to physical situations have been given at appropriate places. The mathematical problems for the decay of alpha particles, the normal and anomalous Zeeman effects, and the Stark effect, have been given appropriate attention. The problems have not only been given at the end of each chapter but are also scattered throughout the text. These problems are an essential ingredient of the subject.

We hope that this book opens new vistas and may prove a trend-setter in the teaching of quantum mechanics.

Mohammad Saleem

Author biography

Mohammad Saleem

Professor Dr Mohammad Saleem obtained his BA (Hons), MA (mathematics), MSc (physics) and PhD degrees from Punjab University and his BSc (Special Hons) from London University. He has been Professor and Chairman of the Department of Physics, Founder Director of the Centre for High Energy Physics and Dean of the Faculty of Science of Punjab University. At present, he is Professor Emeritus, Centre for High Energy Physics, Punjab University, Lahore, Pakistan. He is also a professor at the Institute for Basic Research, Palm Harbor, FL, USA. He has written more than 150 research papers on high energy physics, most of which have either been published in standard international journals or contributed to international conferences. His books *Special Relativity, Group Theory for High Energy Physicists* and on high energy physics, written in collaboration with his colleagues, have been published in the UK and USA. He is also an editor of the *Hadronic Journal*, USA.

IOP Publishing

Quantum Mechanics

Mohammad Saleem

Chapter 1

The failure of classical physics and the advent of quantum mechanics

Quantum mechanics has played a significant role in the development of various disciplines of physics. It was propounded in 1925 and has reigned supreme ever since, extending its domain over the years; of course, in its own domain it has always been in excellent agreement with experiments. It has long since become the language of physics and anyone who tries to understand the basic principles of physics without having a grasp of this subject is doomed to fade away in the darkness of ignorance. In this book an attempt has been made to provide a logical, lucid and user-friendly treatment of this elegant and fascinating subject.

1.1 A challenge for classical physics

Looking through the lattice of history, we observe that the first quarter of the twentieth century was a challenging period for classical physics. The interference, diffraction and polarisation phenomena could *only* be explained by assuming that light had a wave nature. But some other phenomena, such as black body radiation, the photoelectric effect and Compton scattering, defied the wave concept of electromagnetic radiation. The black body radiation spectrum was explained by Planck who assumed that the atoms of the walls of the black body act as electromagnetic harmonic oscillators. An oscillator can radiate energy only in quanta with $E = nh\nu$, where n is a positive integer or zero, ν is the frequency of the oscillator and h is a constant now called *Planck's constant*. The remaining two phenomena were explained by Einstein and Compton by assuming that radiation itself, in particular light, consists of particles, called *photons*, each photon possessing the energy $h\nu$. Thus, light has a dual nature, sometimes exhibiting the behaviour of waves and at other times showing the characteristics of particles. In 1923, a French PhD scholar, de Broglie (pronounced de Broy), extended the idea of the duality of light to the duality of matter. This extension of a concept to cover new realms is not

doi:10.1088/978-0-7503-1206-6ch1

something new in physics. We remember that Newton had shown that the *laws of mechanics* had the same form in all inertial frames of reference. Einstein, whose genius, in the history of physics, is almost unparalleled in the twentieth century, *extended* this idea to the entire field of physics by demanding that *laws of physics* should have the same form in all inertial frames of reference. And this became one of the two basic postulates of special relativity. However, it must be emphasised that Newton proposed it only as a characteristic of the second law of motion but Einstein made it a criterion for the validity of any law of physics. de Broglie wrote that 'after long reflection in solitude and meditation, I suddenly had the idea, during the year 1923, that the discovery made by Einstein in 1905 should be generalised by extending it to all material particles and notably to electrons'. He *assumed* that if p is the magnitude of the three-momentum of a particle of energy E, and λ is the wavelength and ν the frequency of the associated wave, then, in addition to $E = h\nu$, we must have

$$p = \frac{h}{\nu}. \tag{1.1}$$

According to Einstein, this hypothesis of de Broglie's about the dual nature of matter was 'a first feeble ray of light on this worst of our physics enigmas'. In 1927 the experiment of Davisson and Germer, in which electrons were scattered by a crystal surface with typical diffraction effects, confirmed this daring hypothesis which ultimately demolished the classical picture of physics.

To get a taste of quantum theory, we analyse the photoelectric effect and Compton scattering.

1.2 The photoelectric effect

The *photoelectric effect*, the emission of electrons by a metal when light falls on it, was discovered by Hertz in 1887. Experiments showed the following characteristics of this effect. When light falls on a metal surface in a vacuum, the emission of electrons depends upon the frequency of the incident light. There is a minimum frequency of light which is required for the emission of electrons from a metal. The value of this threshold frequency varies from metal to metal. The emission of electrons as well as the energy of the emitted electrons, *photoelectrons*, does not depend upon the intensity of the light source. However, if electrons are emitted, then the magnitude of their current is proportional to the intensity of the incident light. Finally, the energy of the photoelectrons varies linearly with the frequency of the light.

The classical theory of electromagnetic radiation can explain some of these characteristics but not all of them. Credit for solving this problem goes to Einstein who, in 1905, refined and extended the ideas Planck used to explain the black body radiation spectrum and assumed that 'light consists of quanta of energy, called photons'. In fact, Planck had introduced the concept of material resonators possessing quanta of energy $nh\nu$, where n is an integer, while Einstein assumed that each quantum of light possesses the energy $h\nu$. The absorption of a single photon by an electron increases the energy of the electron by $h\nu$. Part of this energy is used to remove the electron from the metal. This is called the *work function*. The remaining part of the energy imparted to the electron increases its velocity and consequently its kinetic energy. Thus if $h\nu$, the energy of a photon incident on a metal is greater than the energy E required to separate the electron from the metal, and v is the velocity of the emitted electron, then the following relation must hold:

$$h\nu = E + \frac{1}{2}mv^2. \tag{1.2}$$

All the characteristics of this effect are easily explained by the concept that light consists of photons. The above formula shows that if the energy of the incident photon is less than the work function, the electrons cannot be separated from the surface of the metal and therefore will not be emitted. For a particular metal, the work function E being constant, the relationship between the energy of the incident photon and the kinetic energy of the emitted electron is linear. It is also clear that a more intense source of light will cause photons to be emitted at a greater speed and this will produce a stronger electron current. Thus Einstein was able to provide a completely satisfactory picture of the photoelectric effect by using the concept of the quantum nature of light.

In fact, the dual nature of light is brilliantly reflected by the very assumption Einstein made about the energy of a photon. The frequency is determined by the wave nature of light and is used to define the energy of the particles constituting the light.

It is interesting to note that, in 1921, Einstein was awarded the Nobel Prize in physics 'for his services to Theoretical Physics and especially for his discovery of the law of the photoelectric effect' and *not* for propounding special relativity in 1905 and general relativity in 1915. His extraordinarily remarkable work on relativity changed the complexion of the entire field of physics and ensured him a seat among the immortals of the subject, but surprisingly this magnificent contribution to the pool of knowledge was never considered *specifically* for that enviable prize!

1.3 The Compton effect

Compton was an American physicist who in 1923 performed a crucial experiment which strongly confirmed the corpuscular nature of light. The results of this experiment were explained by Compton himself and independently by Debye. The experiment not only confirmed the law of the conservation of energy, which was previously verified by the photoelectric effect, but also the law of conservation of linear momentum. It was noticed that when electromagnetic radiation of high frequency is incident on electrons of a light element in which the electrons are loosely bound to the nucleus and can be treated as free, the scattered radiation is found to have a smaller frequency than the radiation of the original frequency. This is known as the *Compton effect*. The experiment exhibits that the change in the frequency of incident radiation is independent of its initial frequency and depends only upon the angle of scattering. This can be satisfactorily explained by the quantum theory of light by making use of relativistic expressions for various quantities.

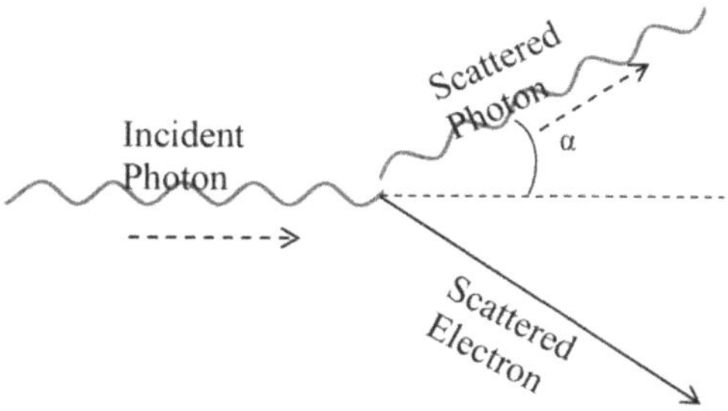

Figure 1.1. The Compton effect.

Consider the scattering of a photon of frequency ν falling on an electron of rest mass m_0, in a frame of reference in which the electron is at rest. Let ν' be the frequency of the scattered photon. Let $\mathbf{p}$ be the linear momentum of the electron after its collision with the photon and α the angle between the final and initial directions of the photon. This is shown in figure 1.1. Since the incident radiation is of high frequency, the incoming photon, by virtue of the relation $E = h\nu$, is very energetic. Therefore the electron which is lightly bound to its atom may be considered, to a good approximation, as free. Thus the Compton effect may be treated as a problem of the collision between a photon and a stationary free electron. The initial energy of the photon is $h\nu$ and its initial momentum is $h\nu/c$, while its final energy and momentum are $h\nu'$ and $h\nu'/c$, respectively. According to the laws of conservation of energy and momentum, we have

$$h\nu + m_0c^2 = h\nu' + E \text{ (conservation of energy)}, \tag{1.3}$$

$$\frac{h\nu}{c}\mathbf{a} = \frac{h\nu'}{c}\mathbf{b} + \mathbf{p} \text{ (conservation of momentum)}, \tag{1.4}$$

where **a** and **b** are unit vectors in the directions of the incident and scattered photons, respectively, and E is the energy of the electron after the collision. Rearranging the terms in (1.3), we have

$$h\nu - h\nu' = E - m_0c^2.$$

Squaring both sides of the above equation, we obtain

$$h^2\nu^2 + h^2\nu'^2 - 2h^2\nu\nu' = E^2 + m_0^2c^4 - 2Em_0c^2. \tag{1.5}$$

Rearranging the terms in (1.4) and multiplying through by c, we have

$$h\nu\,\mathbf{a} - h\nu'\,\mathbf{b} = c\mathbf{p}.$$

Squaring the two sides of the above equation, we obtain

$$h^2\nu^2 + h^2\nu'^2 - 2h^2\nu\nu'\cos\alpha = c^2p^2 \tag{1.6}$$

where α is the angle between **a** and **b**. Substituting the expression for c^2p^2 from the equation $E^2 = m_0^2c^4 + c^2p^2$ into (1.6), we obtain

$$h^2\nu^2 + h^2\nu'^2 - 2h^2\nu\nu' \cos\alpha = E^2 - m_0^2c^4. \tag{1.7}$$

Subtracting (1.5) from (1.7), we have

$$2h^2\nu\nu'(1 - \cos\alpha) = 2m_0c^2\left(E - m_0c^2\right) = 2m_0c^2(h\nu - h\nu')$$

where, in obtaining the expression on the extreme right, we have made use of (1.3). The above equation can be written as

$$\frac{h}{m_0c^2}(1 - \cos\alpha) = \frac{\nu - \nu'}{\nu\nu'} = \frac{1}{\nu'} - \frac{1}{\nu} = \frac{\lambda'}{c} - \frac{\lambda}{c},$$

or

$$\lambda' - \lambda = \frac{h}{m_0c}(1 - \cos\alpha). \tag{1.8}$$

The quantity h/m_0c is called the *Compton wavelength* and has the value 0.02426×10^{-8} cm. Equation (1.8) has been found to be consistent with experiments. For his contribution, Compton shared the Nobel Prize in physics in 1927.

Problem

1.1. Comment on the statement that in the photoelectric effect the photon transfers all of its energy while in the Compton effect only part of the energy is transferred to the electron.

1.4 Heisenberg's uncertainty principle

In classical physics, it is *tacitly assumed* that the operation of observation does not appreciably disturb a system and, at least in principle, the disturbance caused by the measurement process can be rectified exactly. It required the ingenuity of Heisenberg, one of the most brilliant even among the Nobel laureates, to show that wave–particle duality imposes restraints on simultaneous precise measurements of position and momentum. The measurement process in general disturbs a system by an amount which cannot be predicted. For instance, consider the hypothetical experiment shown in figure 1.2, devised for the precise measurement of the position of an electron. The apparatus includes a microscope with 2α as its aperture. The electron beam is moving in the positive direction of the x-axis with a well-defined momentum p_x. In order to measure the position of an electron, it has to be observed. For that purpose, we shine a beam of light along the negative x-axis. Now, it is necessary that at least one photon after falling on it should be scattered into the microscope so that the observer sees it through the microscope. Due to the particle

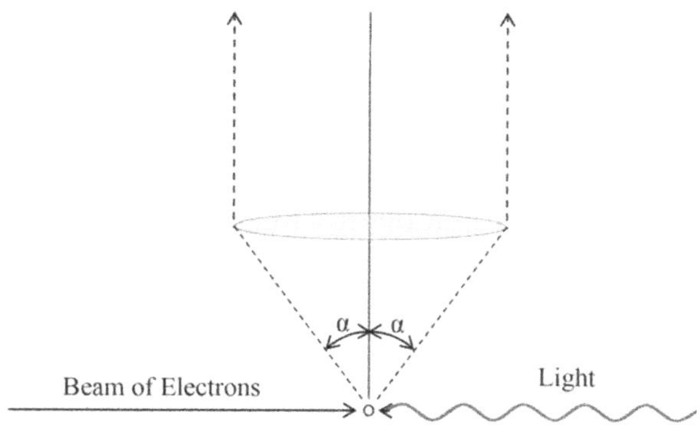

Figure 1.2. Measurement of the position of an electron.

nature of light, as a photon strikes the electron, the latter is disturbed. The momentum of recoil of the electron could be calculated if the initial and final momenta of the electron were known. But because of the finite aperture of the microscope, the photon can enter it along any direction on the illumination cone of the observer. That is, the direction of the photon scattered into the microscope is undetermined within the angle subtended by the aperture. The uncertainty in the measurement of the momentum of the electron is

$$\Delta p_x = 2\frac{h\nu}{c}\sin\alpha. \tag{1.9}$$

We also notice that, using standard optical theory, the resolving power of the microscope is given by

$$\Delta x = \frac{\lambda}{\sin\alpha}, \tag{1.10}$$

where λ is the wavelength of light. This means that Δx gives the uncertainty in the position of the electron. The uncertainty Δx can be made as small as we like by making λ as small as we please and/or making the aperture 2α as large as we desire. But this will enhance the uncertainty in the measurement of momentum. For instance, if we decrease λ, i.e. increase the frequency ν of the incident light, it will certainly decrease the uncertainty Δx in the measurement of the position of the electron, but it will increase the uncertainty Δp_x because the photon striking with greater energy ($h\nu$) will disturb the electron to a greater extent. Multiplying (1.9) and (1.10), we obtain

$$\Delta x \Delta p_x = 2\frac{h\nu}{c}\lambda = 2h \tag{1.11}$$

as $\lambda = c/\nu$. This means that the product of the uncertainties in the simultaneous measurement of the x-components of position and momentum is of the order of Planck's constant. This is just one example. In fact, any experiment designed for a simultaneous precise measurement of position and momentum will encounter the same constraint. This is not an error in experimental measurement. It is inherent in nature in the sense that it is due to the unavoidable interaction between the observer and the observed during the process of observation. The above analysis shows that a simultaneous precise measurement of position and momentum is impossible. This is known as *Heisenberg's uncertainty principle* or *principle of indeterminacy*. It can be shown that similar uncertainty relations exist between *energy and time* and *angular momentum and angle*. In other words, a simultaneous precise measurement of two canonically conjugate variables is impossible.

1.5 The correspondence principle

Although classical mechanics breaks down when applied to determining the behaviour of tiny objects such as electrons, protons, etc, it has been providing correct answers to mechanical phenomena at the macroscopic level. Therefore, at this level, quantum theory should be consistent with classical mechanics. This is known as Bohr's *correspondence principle* and is said to serve as a guide in discovering the correct quantum laws. In fact, under old conditions, a new theory should always yield the same results as the old theory which it is replacing, because the original theory has been explaining the experimental data in its own domain. In the case of quantum mechanics, this correspondence may be specified by claiming that, for large quantum numbers, quantum theory must be consistent with classical physics. Moreover, if a quantum system has a classical analogue, then for the limit $h \rightarrow 0$, it must yield the corresponding classical results. Thus, in the uncertainty principle, as $h \rightarrow 0$ in the classical limit, the product $\Delta x \Delta p_x \rightarrow 0$ and therefore a simultaneous precise measurement of position and momentum at macroscopic level becomes permissible. The importance of the correspondence principle lies not in stating that quantum theory should yield the same results as classical mechanics at the macroscopic level, but in describing the conditions under which it should happen.

1.6 The Schrödinger wave equation

There is no doubt that Planck's quantum theory of black body radiation, Einstein's hypothesis of light quanta for the explanation of the photoelectric effect, Bohr's postulates regarding the interpretation of the spectrum of the hydrogen atom and de Broglie's hypothesis about the dual nature of matter were the milestones in the progress of physics from 1900 to 1923. But physicists desired a differential equation which could govern the behaviour of mechanical phenomena and consequently explain various experimental results. In 1926, Schrödinger set up such a differential equation which is named after him and was supposed to replace Newton's second law of motion as the basic law of nature in mechanics. The mechanics based on this differential equation was called *wave mechanics* and is now known as *quantum mechanics*. Schrödinger's differential equation undoubtedly outshone all the above-mentioned postulates and started commanding immense attention in the physics community immediately after its advent. We will now set up this differential equation. It must be clearly stated that the Schrödinger wave equation *cannot be logically derived*. The historical development may make its presence somewhat plausible. But we will establish it by adopting an operational technique which is to-the-point and simple. We proceed as follows.

In terms of the kinetic energy T and the potential energy V of a particle, the non-relativistic classical expression for its total energy E is given by

$$E = T + V,$$

where V is, in general, a function of space and time coordinates, $V = V(\mathbf{r}, t)$, but

$$T = \frac{1}{2}m\mathbf{v}^2 = \frac{\mathbf{p}^2}{2m},$$

where m is the mass, $\mathbf{v}$ is the velocity and $\mathbf{p}$ is the linear momentum of the particle. This yields

$$E = \frac{\mathbf{p}^2}{2m} + V. \tag{1.12}$$

It is *assumed* that the transition from classical to quantum mechanics is made by interpreting E, $\mathbf{p}$ and V as operators such that

$$E \to \mathrm{i}\hbar\frac{\partial}{\partial t}, \tag{1.13}$$

where $\hbar = \frac{h}{2\pi}$,

$$\mathbf{p} \to -\mathrm{i}\hbar\nabla, \tag{1.14}$$

and

$$V \to V. \tag{1.15}$$

The operator ∇ is given by

$$\nabla = \frac{\partial}{\partial x}\mathbf{i} + \frac{\partial}{\partial y}\mathbf{j} + \frac{\partial}{\partial z}\mathbf{k},$$

where x, y, z are space coordinates and $\mathbf{i}$, $\mathbf{j}$, $\mathbf{k}$ are unit vectors along the x-, y- and z-axes. The time coordinate, wherever it occurs, is denoted by t. Note that while E and $\mathbf{p}$ are interpreted as differential operators, the potential energy is assumed to be only a multiplication operator, its form remaining unchanged when moving from classical to quantum mechanics. Equation (1.12) therefore yields

$$\mathrm{i}\hbar\frac{\partial}{\partial t} = -\frac{\hbar^2}{2m}\nabla^2 + V,$$

where

$$\nabla^2 \equiv \frac{\partial^2}{\partial x^2} + \frac{\partial^2}{\partial y^2} + \frac{\partial^2}{\partial z^2}.$$

If a function $\Psi(x, y, z, t)$, which for convenience we may also write as $\Psi(\mathbf{r}, t)$ or merely as Ψ, represents the particle under consideration, then operating this equation on $\Psi(\mathbf{r}, t) \equiv \Psi$ and interchanging the two sides of the equation thus formed, we obtain

$$\left(-\frac{\hbar^2}{2m}\nabla^2 + V\right)\Psi = \mathrm{i}\hbar\frac{\partial \Psi}{\partial t}, \tag{1.16a}$$

or

$$H\Psi = \mathrm{i}\hbar\frac{\partial \Psi}{\partial t}, \tag{1.16b}$$

where

$$H \equiv \left(-\frac{\hbar^2}{2m}\nabla^2 + V\right) \tag{1.16c}$$

is the Hamiltonian operator for the system. Equation (1.16*a*), equivalently (1.16*b*), is a partial differential equation in four independent variables and is known as the *time-dependent Schrödinger wave equation.* It is called a wave equation as it is similar to a differential equation for waves. It is *assumed* to be the fundamental differential equation governing the behaviour of mechanical phenomena. It replaces Newton's second law of motion in mechanics. However, unlike Newton's law of motion, it is not garbed in words. This law of nature presents itself only as a differential equation. The function $\Psi(\mathbf{r}, t)$, a solution of the time-dependent Schrödinger wave equation for a system, is called a *wave function* and is necessarily a complex function because of the complex nature of the differential equation. It should *not be considered to be a physical wave.* It is actually a mathematical function containing all the information that can be obtained about the system it represents. The time-dependent Schrödinger wave equation is a linear partial differential equation of the first order in the time derivative and of the second order in the spatial derivative. This implies that if its solution at a particular time t_0 is known, it can be calculated at any time t. But just as in the case of Newton's second law of motion, there is *no logical derivation* of the Schrödinger wave equation. It was a brilliant guess of Schrödinger in the perspective of the dual nature of matter. The ultimate test of a theory comes from its confrontation with experimental data. And of course, for non-relativistic mechanical phenomena, the theory based on Schrödinger's wave equation has emerged successful.

Whenever we have to solve a physical problem in quantum mechanics, we resort to this equation, just as in classical mechanics we use the second law of motion. To write the Schrödinger wave equation for a particular system, we have to find the *classical expression for the potential energy V* of the system and substitute it into equation (1.16*a*). This gives us the desired differential equation which may be solved to obtain a complex solution $\Psi(\mathbf{r}, t) \equiv \Psi$. Since the differential equation contains only a first-order time derivative, the wave function is *uniquely* prescribed, once its value at a time $t = t_0$ is known. But how is this function $\Psi(\mathbf{r}, t)$ interpreted so as to relate it to physically measurable characteristics of the system? Certain prescriptions were proposed but after facing insurmountable difficulties, on a suggestion by Born, a consensus was ultimately developed in the physics community. The complex function $\Psi(\mathbf{r}, t)$ representing the particle, being itself not directly observable, is interpreted so that

$$\Psi^*(\mathbf{r}, t)\,\Psi(\mathbf{r}, t)\,\mathrm{d}x\mathrm{d}y\mathrm{d}z$$

represents the *probability of finding the particle in a small volume* $\mathrm{d}x\mathrm{d}y\mathrm{d}z$ ($\equiv \mathrm{d}\tau$) *about the point* $\mathbf{r}$ *and at time* t. It must be emphasised that this interpretation is only a hypothesis. Its validity is established by the success of its predictions. No doubt, the time evolution of the wave function $\Psi(\mathbf{r}, t)$ is inextricably connected to probabilistic concepts. It has been rightly stated, in flowery language, that $\Psi^*\Psi$ is the window through which we can view the world of the atom. Schrödinger himself was shocked when he was told about the statistical interpretation of quantum mechanics. He once told Bohr 'If we are going to stick to this damned quantum jumping, then I regret that I ever had anything to do with quantum theory'. Bohr quipped 'But the rest of us are thankful that you did'. To this interpretation, which demolishes the determinism of classical mechanics, Schrödinger and Einstein could not reconcile themselves, even to the last days of their lives. The total probability of finding the particle in a volume in which it is confined is

$$\int \Psi^*(\mathbf{r}, t)\,\Psi(\mathbf{r}, t)\,\mathrm{d}\tau\left(\equiv \int |\Psi^*(\mathbf{r}, t)|^2\,\mathrm{d}\tau\right)$$

and must be 100%, i.e. 1, because the particle must be somewhere in that volume. Hence we may write

$$\int \Psi^*(\mathbf{r}, t)\,\Psi(\mathbf{r}, t)\,\mathrm{d}\tau = 1. \tag{1.17}$$

The wave function $\Psi(\mathbf{r}, t)$ is then said to be *normalised to unity* or simply *normalised*. The above equation is said to express the *normalisation condition*. This equation exists only if the integral

$$\int \Psi^*(\mathbf{r}, t)\,\Psi(\mathbf{r}, t)\,\mathrm{d}\tau$$

converges, i.e. it is finite; for instance, if the wave function goes to zero sufficiently rapidly as $\mathbf{r}$ tends to infinity. The function $\Psi(\mathbf{r}, t)$ is then said to be *square integrable* or to have an *integrable square*. Symbolically, in this case, $\Psi(\mathbf{r}, t) \to 0$ as $|\mathbf{r}| \to \infty$. This is the boundary condition which is always satisfied when the state is bound. For instance, an electron bound to the hydrogen nucleus, constitutes the hydrogen atom—a bound state. For such a state, the particle can never go to infinity. The integral

$$\int \Psi^*(\mathbf{r}, t)\,\Psi(\mathbf{r}, t)\,\mathrm{d}\tau \equiv \int |\Psi(\mathbf{r}, t)|^2\,\mathrm{d}\tau$$

exists and the function $\Psi(\mathbf{r}, t)$ can be normalised. In certain cases, for instance for a free particle, the above integral may diverge. Then a somewhat different formulation of the normalisation condition is to be given which will be considered at a later stage. $\Psi^*(\mathbf{r}, t)\,\Psi(\mathbf{r}, t)$ is the *probability density*, i.e. the probability per unit volume. Notice that the statement in the above form about the probability is valid only if $\Psi(\mathbf{r}, t)$ is normalised. It may be emphasised that the above prescription *incorporates the statistical interpretation in the basic differential equation*, the Schrödinger wave equation. Since this is a homogeneous, linear differential equation, if $\Psi(\mathbf{r}, t)$ is a solution of this differential equation, then $C\Psi(\mathbf{r}, t)$ is also a solution of the same where C is a constant. In most cases, the value of C can be determined by using the normalisation condition.

In one-dimensional space, $\Psi^*(x, t)\,\Psi(x, t)\,\mathrm{d}x$ is interpreted as the probability of finding the particle in the length $\mathrm{d}x$ between x and $x + \mathrm{d}x$ at time t.

It is important to state at this stage that we have only interpreted Ψ so as to relate it to the probability distribution. Actually, we have to compute various *physical quantities*, representing dynamical variables in classical mechanics, such as *position, linear momentum, energy* and *components of angular momentum*, so that we may compare our theoretical results with experimental data. Another postulate is to be proposed for this purpose. Since new concepts are involved, we will consider it later in this chapter when these concepts have been introduced.

1.7 Constraints on solutions

Every solution of the Schrödinger wave equation is not physically acceptable. The above interpretation of $\Psi^*(\mathbf{r}, t)\,\Psi(\mathbf{r}, t)\,\mathrm{d}\tau$ as the probability of finding the particle in the small volume $\mathrm{d}\tau$ about $\mathbf{r}$ and at time t imposes certain constraints on the solution $\Psi(\mathbf{r}, t)$ of this second-order partial differential equation. In fact, in order for the solution to be physically acceptable, it must be *well-behaved*, i.e. the wave function should be *finite*, *single valued* and *continuous*. Moreover, its *first derivatives* with respect to space co-ordinates must be continuous. This is analysed below.

1. The function $\Psi(\mathbf{r}, t)$ must be finite for all values of x, y, z. In fact, $\Psi(\mathbf{r}, t)$ should be such that it vanishes sufficiently rapidly as infinity is approached so as to give us no trouble; the function remains square integrable. This is so because otherwise the probability of finding the particle in the small region about a point where the function diverges will become infinite which is physically unacceptable.
2. The function $\Psi(\mathbf{r}, t)$ must be single-valued, i.e. for each set of the values of the variables it should have only one value. This is essential because otherwise the probability of finding the particle at a particular point and at a certain time will not be unique; it will have more than one value, each value depending upon the choice of the multivalued function $\Psi(\mathbf{r}, t)$ such as $\tan^{-1}x$. Strictly speaking, according to this argument, it is $\Psi^*(\mathbf{r}, t)\,\Psi(\mathbf{r}, t)$ which should be single-valued. However, successful results for the characteristics of some physical quantities, such as the z-component of orbital angular momentum, require that the wave function be single-valued.
3. The function $\Psi(\mathbf{r}, t)$ and its first derivatives with respect to space co-ordinates should be continuous in all parts of the region.

Before considering the last characteristic in detail, let us recall a mathematical theorem: a function continuous at a point is not necessarily differentiable there *but a function differentiable at a point is necessarily continuous there*. Thus if a function f is differentiable at a point, i.e. if $\mathrm{d}f/\mathrm{d}x$ exists at a point, then f must be continuous there. Similarly, if $\mathrm{d}^2f/\mathrm{d}x^2$ exists, then $\mathrm{d}f/\mathrm{d}x$ must be continuous there. To make things simple, consider the time-dependent Schrödinger wave equation in one-dimension,

$$\left(-\frac{\hbar^2}{2m}\frac{\partial^2}{\partial x^2} + V\right)\Psi = \mathrm{i}\hbar\frac{\partial \Psi}{\partial t}. \tag{1.18}$$

Rearranging the terms, we have

$$V\Psi = \frac{\hbar^2}{2m}\frac{\partial^2 \Psi}{\partial x^2} + \mathrm{i}\hbar\frac{\partial \Psi}{\partial t}.$$

We will consider the case when the potential energy of a physical system is finite, whether continuous or with a number of finite discontinuities, because infinite energies do not occur in nature. Then the left-hand side of the above equation is finite. Consequently, both the terms on the right-hand side should be finite. Thus, as $\partial^2\Psi/\partial x^2$ is finite, i.e. the differential coefficient of $\partial\Psi/\partial x$ exists, the function $\partial\Psi/\partial x$ *must be continuous*. Moreover, as $\partial\Psi/\partial x$ exists, i.e. the differential coefficient of Ψ exists, the function Ψ *must be continuous*. Hence, the condition that the wave function and its first space derivative should be continuous is a requirement imposed by the *finiteness* of Ψ and *consistency* of the Schrödinger wave equation. The analysis can easily be extended to three dimensions.

It can be mentioned that if Ψ is *assumed* to be continuous, then we *need not assume* that it is finite because every continuous function is finite.

Owing to the requirement that the function $\Psi(\mathbf{r}, t)$ must be well-behaved and its first derivative with respect to the space coordinate should be continuous, all the mathematical solutions of the Schrödinger wave equation are not physically acceptable. This in turn means that for a physical system, only those energies will be allowed which correspond to physically acceptable wave functions. We will find that by the requirement of admissible wave functions, energy and some other physical quantities are *quantised*.

The time-dependent Schrödinger wave equation is of first order in the time-derivative. Therefore, if the wave function $\Psi(\mathbf{r}, t_0)$ at any initial time t_0 is known, the wave function $\Psi(\mathbf{r}, t)$ at any time t can be calculated.

Problem

1.2. Comment on the remark that the assumptions that the wave function be finite and single-valued at all points in configuration space may be more rigorous than necessary.

> **Remark**
>
> If Ψ is finite and V is continuous or has finite discontinuities, then the consistency of the Schrödinger wave equation requires that Ψ and its first space derivative should be continuous. It may be emphasised that it is *not* a consequence of the probabilistic interpretation of Ψ. This is a characteristic of certain types of second-order differential equations.

1.8 Eigenfunctions and eigenvalues

Before we proceed further, we will define a few terms and illustrate them with the help of examples. We first define the eigenfunctions and eigenvalues of any operator. Consider an operator A operating on a function ϕ such that

$$A\phi = \lambda\phi, \tag{1.19}$$

that is, it reproduces the function ϕ multiplying it with a constant λ. Then the function ϕ is called an *eigenfunction* (or *characteristic function*) of the operator A with *eigenvalue* (or *characteristic value*) λ or corresponding to the eigenvalue λ. The equation itself is called an *eigenvalue equation* (or characteristic equation). Thus, in the eigenvalue equation

$$\frac{d^2}{dx^2}(\sin 3x) = -9(\sin 3x),$$

-9 is the eigenvalue of the operator d^2/dx^2 corresponding to the eigenfunction $\sin 3x$.

An operator has several eigenfunctions with the corresponding eigenvalues. These eigenvalues may be discrete or continuous. If the eigenvalue spectrum is discrete, the values are written as, say, $\lambda_1, \lambda_2, \lambda_3, \ldots$. If the eigenvalue spectrum is continuous, the values are denoted by $\lambda', \lambda'', \lambda''', \ldots$. The eigenvalue spectrum may be partially discrete and partially continuous. For simplicity, in general analysis, as far as possible, we will be considering only discrete eigenvalues.

A set of functions $\phi_1, \phi_2, \ldots, \phi_n$ is said to be *linearly independent* if its linear combination $a_1\phi_1 + a_2\phi_2 + \cdots + a_n\phi_n$ cannot be made equal to zero for all values of the variables except by taking all a equal to zero. For instance, $\sin x$ and $\cos x$ are two linearly independent functions as their linear combination, $a_1 \sin x + a_2 \cos x$, cannot be made equal to zero for all values of the variable x except by taking $a_1 = a_2 = 0$.

Suppose that $\phi_1, \phi_2, \ldots, \phi_n$ are n linearly independent eigenfunctions of an operator A corresponding to the same eigenvalue λ. Then we may write

$$\begin{aligned} A\phi_1 &= \lambda\phi_1, \\ A\phi_2 &= \lambda\phi_2, \\ &\cdots \\ A\phi_n &= \lambda\phi_n. \end{aligned} \tag{1.20}$$

The number λ is called an *n-fold degenerate eigenvalue* of the operator A, *corresponding to linearly independent eigenfunctions* $\phi_1, \phi_2, \ldots, \phi_n$. These eigenfunctions are called n-fold degenerate, corresponding to the same eigenvalue λ. The number n is called the *degree of degeneracy* of the eigenfunctions.

Problem

1.3. Show that -4 is a two-fold degenerate eigenvalue of the operator $\mathrm{d}^2/\mathrm{d}x^2$ corresponding to linearly independent eigenfunctions $\sin 2x$ and $\cos 2x$.

1.9 The principle of superposition

We have seen that the time-dependent Schrödinger wave equation can be written as

$$H\Psi = \mathrm{i}\hbar\frac{\partial\Psi}{\partial t}. \tag{1.16b'}$$

For convenience, changing the two sides of the above equation, we obtain

$$\mathrm{i}\hbar\frac{\partial\Psi}{\partial t} = H\Psi. \tag{1.21}$$

Let Ψ_1 and Ψ_2 be two solutions (maybe belonging to different values of energy) of this differential equation so that

$$\mathrm{i}\hbar\frac{\partial\Psi_1}{\partial t} = H\Psi_1 \tag{1.22}$$

and

$$\mathrm{i}\hbar\frac{\partial\Psi_2}{\partial t} = H\Psi_2. \tag{1.23}$$

It can easily be verified that a linear combination of these solutions, i.e. $a_1\Psi_1 + a_2\Psi_2$, is also a solution of the Schrödinger wave equation. This is shown below:

$$\begin{aligned}\mathrm{i}\hbar\frac{\partial}{\partial t}(a_1\Psi_1 + a_2\Psi_2) &= a_1\mathrm{i}\hbar\frac{\partial\Psi_1}{\partial t} + a_2\mathrm{i}\hbar\frac{\partial\Psi_2}{\partial t}\\ &= a_1 H\Psi_1 + a_2 H\Psi_2\\ &= H(a_1\Psi_1 + a_2\Psi_2).\end{aligned} \tag{1.24}$$

In fact, we could directly state that as Schrödinger's second order time-dependent partial differential equation is linear (because the function Ψ and its derivatives occur only to the first degree and not as higher powers or products), every linear combination of its solutions is also a solution of this differential equation. This is known as the *principle of superposition* and plays a very important role in quantum mechanics. It can be mentioned that this is a characteristic of every homogeneous linear differential equation.

This is the right moment to point out that in classical mechanics, knowledge about the position and momentum of a particle at any time describes what is called the *state* of the particle. If we know the position and momentum of a particle at any time, then we can compute its position and momentum at any other time by using the second law of motion. This implies that if the initial state of the system is given, the values of all other variables can be determined exactly for all times. In quantum mechanics, according to Heisenberg's uncertainty principle, position and momentum cannot be measured precisely at the same time. Therefore, the state of the particle *in the classical sense* cannot be described. Hence, in quantum mechanics, there is no concept of the trajectory of a particle as it is determined by a simultaneous precise knowledge of position and momentum at every moment; such a trajectory does not exist when quantum effects are important as it is then impossible to keep track of the particle. However, *all the accessible information* about the particle is contained in the wave function Ψ. Naturally, the question arises: how do we define the state of a particle in quantum mechanics? In fact, we *assume* as a fundamental principle of quantum mechanics that every physically acceptable solution of the Schrödinger wave equation represents a *state of the system*. The time-dependent Schrödinger wave equation is a linear partial differential equation of first degree in the time variable. Therefore, if Ψ is known at a particular time t_0, it can be determined at any time t. Suppose that Ψ_1 and Ψ_2 represent two states of a system which correspond to definite energy values, say E_1 and E_2. These states are then called *eigenstates* of the system. Since a linear combination of these solutions, $a_1\Psi_1 + a_2\Psi_2$, where a_1 and a_2 are (complex) numbers, is itself a solution of this differential equation, it also represents a state of the system. However, this state does not correspond to a definite energy. Such a state is called a *quantum state* of the system. It has *no analogue in classical mechanics* where each state, like an eigenstate in quantum mechanics, corresponds to a definite energy. Now suppose that the functions $\Psi_1(t_0)$ and $\Psi_2(t_0)$ are two solutions of the Schrödinger wave equation at time t_0. Let

$$\Psi(t_0) = a_1\Psi_1(t_0) + a_2\Psi_2(t_0)$$

be a linear superposition of $\Psi_1(t_0)$ and $\Psi_2(t_0)$ at time t_0. Suppose that the functions $\Psi_1(t_0)$ and $\Psi_2(t_0)$ develop with time into functions $\Psi_1(t)$ and $\Psi_2(t)$. Then, by virtue of the fact that the Schrödinger wave equation is linear in Ψ, we must have

$$\Psi(t) = a_1\Psi_1(t) + a_2\Psi_2(t),$$

i.e. at any time t, $\Psi(t)$ is the *same* linear combination of the functions $\Psi_1(t)$ and $\Psi_2(t)$. The fact that the first of these equations entails the second equation is a consequence of the linearity of the time development of a state.

An overview of various disciplines of physics shows that the principle of superposition is one of the most significant attributes of the wave concept. It is common to all types of waves. We may transcend a step higher than this and take it as a characteristic of every homogeneous linear differential equation: a linear superposition of the solutions of a homogeneous linear differential equation is also a solution of this differential equation. Quantum mechanics which emphasises this formal aspect was developed by Dirac, a British Nobel laureate. It describes the possible states of a system by abstract quantities called *state vectors* which obey the principle of superposition. Indeed, the essence of this principle as reflected by state vectors is much more attractive than that predicted by wave functions.

A quantity which can be measured is called an *observable*. In classical physics, observables are represented by ordinary variables. In quantum mechanics, however, observables are, *by assumption*, represented by operators. It is customary in the literature to use the same letter for the observable, variable and operator. We will also adopt this convention.

1.10 Complementarity

The statistical interpretation of the wave function and Heisenberg's principle of uncertainty are the concepts with which some eminent physicists of the good old days were finding it difficult to come to terms. Bohr therefore devoted his full attention to detailed analyses of various new concepts which were leading to new trends in scientific attitudes and its philosophical consequences. His analysis that the wave and particle aspects of matter are opposing but complementary modes of its realisation has been named the *principle of complementarity*. This is the essence of his views on the conceptual basis of quantum mechanics but it does not help in making calculations in this field. In fact, classical physics, based upon the knowledge gained from every day experience, i.e. from the behaviour of macroscopic objects, tells us that an object in nature can behave either as a particle or as a wave. It cannot exhibit both characteristics. According to the principle of complementarity, analysis at a microscopic level reveals that an object can behave both as a particle and a wave but the two modes cannot be realised at the same time. A measurement which emphasises one of the wave–particle attributes does so at the expense of the other. An experiment designed to exhibit the particle properties does not give any information on the wave aspect and vice versa. For instance, the observation of cloud chamber tracks does not involve wave aspect at all while interference and diffraction experiments do not contain any information about the particles' demeanour. It is interesting to note that all possibly available knowledge about the characteristics of a microscopic entity, say an electron, is contained in the wave function. We will express it like this. A microscopic entity under one situation shows those properties which at macroscopic level are attributed to particles and we say that it is behaving as a particle. In some other situation, the same entity exhibits characteristics which at macroscopic level are assigned to waves. Then we say that the entity is behaving as a wave. Actually, the behaviour is determined by the same wave function and it is different, as it must be, under different circumstances. We are surprised simply because it is not what we were expecting *from* our experience in everyday life. But nature does behave that way and sometimes it produces results even against those expectations which are based on very careful considerations and analysis!

1.11 Schrödinger's amplitude equation

We will now show that the computations in solving the Schrödinger wave equation are simplified if the *potential energy* V of the system does not depend upon time *explicitly:* $V \equiv V(x, y, z) \equiv V(\mathbf{r})$. Let us *assume* that in this case the wave function $\Psi(\mathbf{r}, t)$ representing the particle can be expressed as a product of two functions, one depending only on space coordinates and the other depending on time alone. If we denote these functions, respectively, by $\psi(x, y, z)$ (also written as $\psi(\mathbf{r})$) and $\phi(t)$, then we may write

$$\Psi(x, y, z, t) = \psi(x, y, z)\phi(t). \tag{1.25}$$

Differentiating twice with respect to x, we obtain

$$\frac{\partial^2 \Psi}{\partial x^2} = \frac{\partial^2 \psi}{\partial x^2}\phi. \tag{1.26}$$

where $\Psi \equiv \Psi(x, y, z, t) \equiv \Psi(\mathbf{r}, t)$, $\psi \equiv \psi(x, y, z) \equiv \psi(\mathbf{r})$ and $\phi \equiv \phi(t)$. We can obtain two similar expressions for the second-order derivatives with respect to space coordinates y and z. Next, differentiating (1.25) with respect to t, we obtain

$$\frac{\partial \Psi}{\partial t} = \psi \frac{\mathrm{d}\phi}{\mathrm{d}t}. \tag{1.27}$$

Notice that for the function ϕ, we have used the ordinary derivative instead of the partial derivative as this function depends upon one variable t only. Substituting the expressions from (1.26), etc, and equation (1.27) in equation (1.16*a*), we obtain

$$-\frac{\hbar^2}{2m}\left[\frac{\partial^2 \psi}{\partial x^2}\phi + \frac{\partial^2 \psi}{\partial y^2}\phi + \frac{\partial^2 \psi}{\partial z^2}\phi\right] + V(x, y, z)\psi\phi = \mathrm{i}\hbar\psi\frac{\mathrm{d}\phi}{\mathrm{d}t},$$

or

$$-\frac{\hbar^2}{2m}(\nabla^2\psi)\phi + V\psi\phi + \mathrm{i}\hbar\psi\frac{\mathrm{d}\phi}{\mathrm{d}t}.$$

Dividing throughout by $\psi\phi$, we obtain

$$-\frac{\hbar^2}{2m}\frac{1}{\psi}(\nabla^2\psi) + V = \mathrm{i}\hbar\frac{1}{\phi}\frac{\mathrm{d}\phi}{\mathrm{d}t}. \tag{1.28}$$

The right-hand side of this equation is a function of time only while, as V depends explicitly only on space coordinates, the left-hand side depends upon space coordinates alone. Therefore, a variation in space coordinates will not affect the right-hand side while a variation in time will not affect the left-hand side. This is possible only if each side is equal to the same constant. We denote this constant by E. Then we may write

$$\mathrm{i}\hbar\frac{1}{\phi}\frac{\mathrm{d}\phi}{\mathrm{d}t} = E \tag{1.29}$$

and

$$-\frac{\hbar^2}{2m}\frac{1}{\psi}(\nabla^2\psi) + V = E,$$

or

$$\left(-\frac{\hbar^2}{2m}\nabla^2 + V\right)\psi = E\psi. \tag{1.30}$$

This is called the *Schrödinger amplitude equation*. The operator in brackets in the above equation is the Hamiltonian operator H of the particle. The amplitude equation therefore can be written as

$$H\psi = E\psi. \tag{1.31}$$

It is called the *Hamiltonian form* of the Schrödinger amplitude equation. This is an eigenvalue equation. The constant E is the eigenvalue of the Hamiltonian operator H corresponding to the eigenfunction ψ. The Schrödinger amplitude equation thus takes the form of an eigenvalue equation for the Hamiltonian H and this simplifies the analysis of the problem.

Let us first solve the differential equation (1.29) involving only time as an independent variable. Transposing $\mathrm{i}\hbar$ to the right-hand side of this differential equation and integrating with respect to time t, we obtain

$$\ln \phi(t) = -\frac{\mathrm{i}Et}{\hbar} + K.$$

Choosing the initial condition that makes $K = 0$, we obtain

$$\ln \phi(t) = -\frac{\mathrm{i}Et}{\hbar},$$

or

$$\phi(t) = \exp\left(-\frac{\mathrm{i}Et}{\hbar}\right). \tag{1.32}$$

The expression in brackets on the right-hand side shows that E has the dimensions of energy. We will assume here but will prove later on that E is the total energy of the system.

Let us next consider the differential equation (1.30) that can be written as

$$\nabla^2\psi + \frac{2m}{\hbar^2}(E - V)\psi = 0. \tag{1.33}$$

This is the *time-independent Schrödinger wave equation* or the *amplitude equation* or the *steady-state Schrödinger equation*. This differential equation for a system can be solved only if the expression for the potential energy of the system is known.

The solution of this differential equation is denoted by $\psi(\mathbf{r})$. Equation (1.25) can now be written as

$$\Psi(\mathbf{r}, t) = \psi(\mathbf{r}) \exp\left(-\frac{\mathrm{i}Et}{\hbar}\right),$$

where for convenience we have written $\mathbf{r}$ for x, y, z. If we are interested in finding the characteristics of a physical system whose *potential energy does not depend explicitly on time*, then instead of using the time-dependent Schrödinger equation, which is relatively much more difficult to solve, we can use the time-independent Schrödinger equation which is easier to solve. We obtain the solution $\psi(\mathbf{r})$ of this differential equation and multiply it by $\phi(t)$ ($\equiv \exp(-\mathrm{i}Et/\hbar)$) so as to obtain $\Psi(\mathbf{r}, t)$ which represents the system. Then a wave function of the system corresponding to a definite energy can be written as

$$\Psi(\mathbf{r}, t) = \psi(\mathbf{r})\phi(t) = \psi(\mathbf{r}) \exp\left(-\frac{\mathrm{i}Et}{\hbar}\right). \tag{1.34}$$

Thus

$$\Psi^*(\mathbf{r}, t)\, \Psi(\mathbf{r}, t) = \Psi^*(\mathbf{r})\, \psi(\mathbf{r}). \tag{1.35}$$

The above analysis shows that the *probability density* for a particle whose potential energy does not depend upon time explicitly *is constant in time*. For this reason, a wave function of the form (1.34) is said to represent a *stationary state* or an *energy eigenstate* of the particle. The energy in a stationary state is said to be *sharp* or well-defined. It can be said that although the wave function $\Psi(\mathbf{r}, t)$ representing the particle is time-dependent, the probability density is independent of time. Therefore, the system would remain in that state indefinitely. Every measurement will always give the same value of energy. This interpretation is consistent with the uncertainty relation

$$\Delta E \Delta t \sim \hbar,$$

as this means that if the system is in an eigenstate with definite energy so that $\Delta E = 0$, then unlimited time should be available to make that measurement. If the energy spectrum is discrete, the lowest energy state is called the *ground state* of the system. The higher energy states are called *excited states* of the system.

For a stationary state, the normalisation condition,

$$\int \Psi^*(\mathbf{r}, t)\, \Psi(\mathbf{r}, t)\, \mathrm{d}\tau = 1,$$

is simplified to

$$\int \Psi^*(\mathbf{r})\mathrm{e}^{\mathrm{i}Et/\hbar}\, \psi(\mathbf{r})\mathrm{e}^{-\mathrm{i}Et/\hbar}\, \mathrm{d}\tau = \int \psi^*(\mathbf{r})\, \psi(\mathbf{r})\, \mathrm{d}\tau = 1. \tag{1.36}$$

In one dimension, the time-independent Schrödinger wave equation reduces to the following form:

$$\frac{d^2\psi}{dx^2} + \frac{2m}{\hbar^2}(E - V)\psi = 0, \tag{1.37}$$

where $\psi \equiv \psi(x)$. It is important to note that the amplitude function $\psi(x)$, a solution of the Schrödinger amplitude equation and related to the wave function $\Psi(x, t)$ by

$$\Psi(x, t) = \psi(x)\phi(t) = \psi(x)\exp\left(-\frac{iEt}{\hbar}\right), \tag{1.33'}$$

should also be well-behaved. This is because if $\psi(x)$ is not finite, then $\Psi(x, t)$ will also be not finite and if $\psi(x)$ is not single-valued, then $\Psi(x, t)$ will also not be single-valued. This is due to the fact that the time function $\phi(t)$ is always finite and does not depend upon space coordinates. As far as the continuity of the function $\psi(x)$ and its first derivative $d\psi/dx$ is concerned, we note from (1.37) that as $\psi(x)$ is always finite and infinite energies cannot be achieved in nature, the first term on the left-hand side of this equation, i.e. the second derivative of $\psi(x)$, should also be finite. Therefore, $d\psi/dx$ must be continuous. Moreover, as $d\psi/dx$ exists, the function $\psi(x)$ should be continuous. Thus, the very consistency of the Schrödinger amplitude equation requires that both $\psi(x)$ and $d\psi(x)/dx$ should be continuous.

Infinite energies do not occur in nature. But let us see what will happen if at some point the potential jumps from a finite to an infinite value. Equation (1.37) shows that in this case $d^2\psi/dx^2$ will be infinite. Therefore, $d\psi/dx$ may or may not be continuous. That is, the condition of continuity of the space derivative of the wave function at the discontinuity of the potential *cannot* be imposed. However, as $d\psi/dx$ exists because otherwise the differential equation will not hold, the function $\psi(x)$ will be continuous. In the solution of the eigenvalue equation $H\psi = E\psi$ for a physical system, we will find that the energy eigenvalues very much depend upon the conditions imposed on the solutions to the eigenvalue equation. We take this opportunity to point out that it is *not* necessary that the integration should always be over the entire space extending from $-\infty$ to $+\infty$.

1.12 The orthonormal set of functions

Consider a set of functions $\phi_1, \phi_2, \phi_3, \ldots$ which are *individually normalised*, i.e.

$$\int \phi_i^* \phi_i \, d\tau = 1, \quad i = 1, 2, 3, \ldots \tag{1.38}$$

and *mutually orthogonal*, i.e.

$$\int \phi_j^* \phi_i \, d\tau = 0, \quad j \neq i, \quad i, j = 1, 2, 3, \ldots \tag{1.39}$$

They are said to form an *orthonormal* set of functions. These two conditions for an orthonormal set can be expressed as

$$\int \phi_i^* \phi_j \, d\tau = \delta_{ij}, \tag{1.40}$$

where the Kronecker delta is defined as

$$\delta_{ij} = 1, \quad \text{for } j = I,$$

$$\delta_{ij} = 0, \quad \text{for } j \neq i.$$

1.13 The equation of continuity

We know that any solution $\Psi(\mathbf{r}, t)$ of the time-dependent Schrödinger wave equation is such that

$$\rho = \Psi^*(\mathbf{r}, t)\, \Psi(\mathbf{r}, t) \tag{1.41}$$

is interpreted as the probability density, i.e. the probability of finding the particle in unit volume about the point $\mathbf{r}$ and at time t. Differentiating with respect to time, we obtain

$$\frac{\partial \rho}{\partial t} = \frac{\partial}{\partial t}(\Psi^* \Psi) = \frac{\partial \Psi^*}{\partial t}\Psi + \Psi^* \frac{\partial \Psi}{\partial t}. \tag{1.42}$$

But by virtue of equation (1.16*a*), i.e.

$$-i\hbar \frac{\partial \Psi}{\partial t} = \left(-\frac{\hbar^2}{2m}\nabla^2 + V \right)\psi$$

and its complex conjugate

$$-i\hbar \frac{\partial \Psi^*}{\partial t} = \left(-\frac{\hbar^2}{2m}\nabla^2 + V \right)\psi^*,$$

(1.42) yields

$$\begin{aligned} -i\hbar \frac{\partial \rho}{\partial t} &= -i\hbar \frac{\partial \Psi^*}{\partial t}\Psi + (-i\hbar)\Psi^* \frac{\partial \Psi}{\partial t} \\ &= \left(-\frac{\hbar^2}{2m}\nabla^2 \psi^* + V\psi^* \right)\Psi - \Psi^* \left(-\frac{\hbar^2}{2m}\nabla^2 \psi + V\psi \right) \\ &= -\frac{\hbar^2}{2m}(\Psi \nabla^2 \psi^* - \Psi^* \nabla^2 \psi), \end{aligned}$$

or

$$\frac{\partial \rho}{\partial t} = -\frac{i\hbar}{2m}\left(\Psi\nabla^2\psi^* - \Psi^*\nabla^2\psi\right)$$
$$= -\frac{i\hbar}{2m}\nabla\left(\Psi\nabla\psi^* - \Psi^*\nabla\psi\right). \tag{1.43}$$

If we write

$$-\frac{i\hbar}{2m}\left(\Psi^*\nabla\Psi - \Psi\nabla\Psi^*\right) = \mathbf{j}, \tag{1.44}$$

then (1.43) can be expressed as

$$\frac{\partial \rho}{\partial t} + \nabla\mathbf{j} = 0,$$

or

$$\frac{\partial \rho}{\partial t} + \operatorname{div}\mathbf{j} = 0. \tag{1.45}$$

This is the well-known *equation of continuity*. This equation also arises in electromagnetic theory and expresses the conservation of charge. If ρ represents the charge density, i.e. the charge per unit volume, and $\mathbf{j} = \rho\mathbf{v}$ is the current density, i.e. the charge passing unit area normal to its direction of motion in one second, then the equation of continuity expresses the law of conservation of charge. The charge which is decreasing with time in a bounded volume is accounted for by the charge which is crossing the surface of the bounded volume. In quantum mechanics, ρ is interpreted as probability density. Therefore, if $\mathbf{j}$ is interpreted as the *probability current density*, then the equation of continuity guarantees the conservation of probability. Thus, if the probability of finding a particle in a certain bounded region decreases with time, it should correspond to the increase in probability of finding it outside that region. The probability current density is also called the *probability flux*. It is given by

$$\mathbf{j} = -\frac{i\hbar}{2m}\left(\Psi^*\nabla\Psi - \Psi\nabla\Psi^*\right) = \frac{\hbar}{m}\operatorname{Im}\left(\Psi^*\nabla\Psi\right). \tag{1.46}$$

Problem

1.4. Prove that

$$\mathbf{j} = \frac{\hbar}{m}\operatorname{Im}\left(\Psi^*\nabla\Psi\right). \tag{1.47}$$

We will now prove that the total probability of finding the particle in space is independent of time. For one-dimensional space, we have

$$\rho(x, t) = \Psi^*(x, t)\, \Psi(x, t). \tag{1.41'}$$

Integrating with respect to x and then differentiating with respect to t, we obtain

$$\frac{\mathrm{d}}{\mathrm{d}t}\int_{-\infty}^{\infty} \rho(x, t)\, \mathrm{d}x = \frac{\mathrm{d}}{\mathrm{d}t}\int_{-\infty}^{\infty} \Psi^*(x, t)\, \Psi(x, t)\, \mathrm{d}x. \tag{1.48}$$

But, in one-dimensional space, we have

$$\left(-\frac{\hbar^2}{2m}\frac{\partial^2}{\partial x^2} + V(x)\right)\Psi(x, t) = \mathrm{i}\hbar\frac{\partial \Psi(x, t)}{\partial t}. \tag{1.16a'}$$

Taking the complex conjugate of both sides of equation (1.16a'), we obtain

$$\left(-\frac{\hbar^2}{2m}\frac{\partial^2}{\partial x^2} + V(x)\right)\Psi^*(x, t) = -\mathrm{i}\hbar\frac{\partial \Psi^*(x, t)}{\partial t}. \tag{1.49}$$

Substituting these expressions for $\partial\Psi^*(x, t)/\partial t$ and $\partial\Psi(x, t)/\partial t$ in (1.48) and simplifying, we obtain

$$\begin{aligned}\frac{\mathrm{d}}{\mathrm{d}t}\int_{-\infty}^{\infty} \rho(x, t)\, \mathrm{d}x &= -\frac{\mathrm{i}\hbar}{2m}\int_{-\infty}^{\infty}\frac{\partial}{\partial x}\left(\Psi(x, t)\frac{\partial \Psi^*(x, t)}{\partial t} + \Psi^*(x, t)\frac{\partial \Psi(x, t)}{\partial t}\right)\mathrm{d}x \\ &= -\frac{\mathrm{i}\hbar}{2m}\left[\left(\Psi(x, t)\frac{\partial \Psi^*(x, t)}{\partial t} + \Psi^*(x, t)\frac{\partial \Psi(x, t)}{\partial t}\right)\right]_{-\infty}^{\infty} = 0. \end{aligned} \tag{1.50}$$

The last result has been obtained because the wave function $\Psi(x, t) \to 0$ as $x \to \infty$ so that the integral

$$\int_{-\infty}^{\infty} \Psi^*(x, t)\, \Psi(x, t)\, \mathrm{d}x$$

may converge.

Equation (1.50) shows that the total probability is conserved. This also ensures that the normalisation is preserved: if the wave function is normalised, it will remain normalised.

1.14 Complete sets of functions

A set of functions $\psi_1(x)$, $\psi_2(x)$, ... in a variable x is said to form a *complete set* if an arbitrary *square integrable* function $\phi(x)$ can be expanded in terms of them:

$$\phi(x) = \sum_i a_i \psi_i(x), \tag{1.51}$$

where a_i are called *expansion coefficients*. The values of a_i can be obtained as follows. Multiplying equation (1.51) by $\psi_j^*(x)$ and integrating with respect to x, we obtain

$$\int \psi_j^*(x)\,\phi(x)\,\mathrm{d}x = \sum_i a_i \int \psi_j^*(x)\,\psi_i(x)\,\mathrm{d}x. \tag{1.52}$$

If the functions $\psi_1(x)$, $\psi_2(x)$, ... form an orthonormal set, then

$$\int \psi_j^*(x)\,\psi_i(x)\,\mathrm{d}x = \delta_{ji}.$$

Therefore, (1.52) reduces to

$$\int \psi_j^*(x)\,\phi(x)\,\mathrm{d}x = \sum_i a_i \delta_{ij}(x).$$

Summing over i, and finally changing j to i throughout the equation, we obtain

$$a_i = \int \psi_i^*(x)\,\phi(x)\,\mathrm{d}x. \tag{1.53}$$

This equation determines all the expansion coefficients.

1.15 The quantum theory of measurement

We will now analyse the process of measurement in quantum mechanics in detail. We will concentrate on how to compute the physical quantities which are to be compared with the experiment. In fact, whenever we want to make an accurate measurement of any quantity, we measure that quantity a large number of times and take the arithmetic mean of the measured values. This is called the *average* or *mean value* of the variable. The average value of a variable x is denoted by $\bar{x}$:

$$\bar{x} = \frac{x_1 + x_2 + \cdots + x_n}{n}. \tag{1.54}$$

In order to have an idea about the precision of the various values we must know the scattering or *dispersion* of these individual values about their average. The individual deviations from the mean are

$$x_1 - \bar{x},\ x_2 - \bar{x},\ \ldots,\ x_n - \bar{x}.$$

But the average of these deviations is equal to

$$\frac{(x_1 - \bar{x}) + (x_2 - \bar{x}) + \cdots + (x_n - \bar{x})}{n} = \frac{x_1 + x_2 + \cdots + x_n}{n} - \frac{n\bar{x}}{n} = \bar{x} - \bar{x} = 0$$

This shows that whatever the deviations of x from its mean value, the average of these deviations is always zero. The average of these deviations is therefore not useful as a standard for measuring dispersion. Its value, being always zero, cannot tell us whether the individual values are close to or far away from the average.

Perhaps a better idea about the dispersion of the values of a variable about their mean can be obtained if we consider the average of the square of deviations of the variable x from its mean value $\bar{x}$. This is called the *variance* of x and is denoted by σ. Thus, we may write:

$$\begin{aligned}
(\text{Variance of } x) &= \sigma \\
&= \frac{(x_1 - \bar{x})^2 + (x_2 - \bar{x})^2 + \cdots + (x_n - \bar{x})^2}{n} \\
&= \frac{x_1^2 + x_2^2 + \cdots + x_n^2}{n} + \frac{n\bar{x}^2}{n} - 2\bar{x}\frac{x_1 + x_2 + \cdots + x_n}{n} \\
&= \bar{x^2} + \bar{x}^2 - 2\bar{x}\bar{x} \\
&= \bar{x^2} + \bar{x}^2 - 2\bar{x}^2 \\
&= \bar{x^2} - \bar{x}^2.
\end{aligned} \tag{1.55}$$

The positive square root of the variance of a variable x is called the *standard deviation* of x or the *root mean square deviation* or the *uncertainty* in the value of x and is denoted by Δx. Thus

$$\Delta x = +\sqrt{\bar{x^2} - \bar{x}^2}. \tag{1.56}$$

1.16 Observables and expectation values

As already mentioned, in quantum mechanics, each physical quantity (or attribute or property) which can be measured experimentally is called an *observable* and is *represented by an operator*. For instance, the energy of a system is an observable and is represented by the Hamiltonian operator H. Linear momentum is another observable and is represented by the linear momentum operator $-i\hbar\nabla$. The *state* of a physical system is a collection of observables and is specified by the wave function representing the system. It is therefore common in the literature to use the term eigenstate to indicate a state represented by an eigenfunction. The *average value*, usually called the *expectation value*, of a sequence of measurements of an observable represented by an operator $A(\mathbf{r}, -i\hbar\nabla, t)$ on a system in a *normalised state* described by Ψ is, *by definition*, given by

$$\langle A \rangle \equiv \bar{A} = \int \Psi^*(\mathbf{r}, t) A - \Psi(\mathbf{r}, t)\, d\tau. \tag{1.57}$$

Frequently, the expectation value of a physical quantity represented by an operator A is denoted by $\langle A \rangle$, i.e. the angle brackets are usually used for expectation values. Moreover, it is customary to use the same symbol for a dynamical variable as well as the operator which represents it. It may again be remarked that in quantum mechanics, a physical system is represented by a wave function such that our entire theoretical knowledge about it is contained in the wave function. Since the interpretation of the wave function is statistical in nature, laws of physics can only make probabilistic predictions. They cannot predict the precise behaviour of a system. In other words, if a physical quantity is measured a large number of times by repeating an experiment under identical conditions or by performing a large number of identical experiments, its average value can be predicted by the above relation. It should be noted that, in general, the result of a single measurement will not be given by $\langle A \rangle$. It is only the average value of a large number of measurements, made in the manner suggested above, which is to be compared with the theoretical value predicted by equation (1.57). It may be emphasised that laws of physics are statistical in nature even when we are dealing with a single particle.

If the wave function is not normalised, then the expectation value $\langle A \rangle$ is defined by the equation

$$\langle A \rangle = \frac{\int \Psi^*(\mathbf{r}, t) A \, \Psi(\mathbf{r}, t) \, d\tau}{\int \Psi^*(\mathbf{r}, t) \, \Psi(\mathbf{r}, t) \, d\tau}$$

which reduces to equation (1.57) for a normalised wave function.

Let us now calculate the uncertainty in the measurement of a physical quantity represented by an operator A when the system is represented by a *normalised eigenfunction* Ψ_n of the operator A corresponding to the eigenvalue λ_n. Then

$$A\Psi_n = \lambda_n \Psi_n \tag{1.58}$$

and the expectation values of A and A^2 are given by

$$\begin{aligned}\langle A \rangle &= \int \Psi^*(\mathbf{r}, t) A \, \Psi(\mathbf{r}, t) \, d\tau = \int \Psi^*(\mathbf{r}, t) \lambda_n \Psi(\mathbf{r}, t) \, d\tau \\ &= \lambda_n \int \Psi^*(\mathbf{r}, t) \, \Psi(\mathbf{r}, t) \, d\tau = \lambda_n\end{aligned}$$

and

$$\langle A^2 \rangle = \int \Psi_n^* A^2 \Psi_n \, d\tau = \lambda_n^2,$$

where we have used (1.57) and (1.58) and the normalisation condition for Ψ_n.

Problem

1.5. The variance in the measurement of a physical quantity represented by an operator A is defined by

$$\langle (A - \langle A \rangle)^2 \rangle.$$

Show that it is equal to

$$\langle A^2 \rangle - \langle A \rangle^2.$$

In the measurement of a physical quantity represented by A, the uncertainty ΔA, defined as the positive square root of variance, is given by

$$\begin{aligned}\Delta A &= +\sqrt{\bar{A}^2 - \bar{A}^2} \\ &= +\sqrt{\bar{\lambda}^2 - \bar{\lambda}^2} \\ &= 0.\end{aligned}$$

This result shows that if a system is represented by an eigenfunction Ψ_n of the operator A, the uncertainty in a measurement of the physical quantity represented by A is zero and it should yield the eigenvalue λ_n of A in an individual measurement.

We may point out that the Schrödinger wave equation written in its Hamiltonian form

$$H\psi(x) = E\psi(x)$$

shows that the constant E which is the eigenvalue of the Hamiltonian operator H must be the *energy of the system*. This was assumed previously and is proved now when we have become familiar with the basic concepts and assumptions of quantum mechanics. The set of all the eigenvalues of the Hamiltonian operator H is called the *energy spectrum* of H.

Let us next see what values of A would be observed if the system happened to be in a quantum state Ψ which is not an eigenfunction of A. We will *assume* that the eigenfunctions of any operator representing a physical quantity form a complete set of functions. We express Ψ as a linear combination of the complete orthonormal set of eigenfunctions Ψ_n of the operator A:

$$\Psi = \sum_n a_n \Psi_n. \tag{1.59}$$

Since Ψ_n are eigenfunctions of the operator A, we must have

$$A\Psi_n = \lambda_n \Psi_n. \tag{1.58'}$$

Then

$$\begin{aligned}\langle A\rangle &= \int \Psi^* A \Psi \, \mathrm{d}\tau = \int \sum_m a_m^* \Psi_m^* A \sum_n a_n \Psi_n \, \mathrm{d}\tau \\ &= \sum_m \sum_n a_m^* a_n \int \Psi_m^* A \Psi_n \, \mathrm{d}\tau \\ &= \sum_m \sum_n a_m^* a_n \lambda_n \int \Psi_m^* \Psi_n \, \mathrm{d}\tau \\ &= \sum_m \sum_n a_m^* a_n \lambda_n \delta_{mn}.\end{aligned}$$

Summing over m, we obtain

$$\begin{aligned}\langle A\rangle &= \sum_n a_n^* a_n \lambda_n \\ &= a_1^* a_1 \lambda_1 + a_2^* a_2 \lambda_2 + \cdots \end{aligned} \tag{1.60}$$

Now according to the normalisation condition, we have

$$\int \Psi^*(\mathbf{r}, t)\, \Psi(\mathbf{r}, t)\tau \, d\tau = 1,$$

or

$$\int \sum_m a_m^* \Psi_m^* \sum_n a_n^* \Psi_n^* \, d\tau = 1,$$

or

$$\sum_m \sum_n a_m^* a_n \int \Psi_m^* \Psi_n \, d\tau = 1,$$

or

$$\sum_m \sum_n a_m^* a_n \delta_{mn} = 1.$$

Summing over m, we obtain

$$\sum_n a_n^* a_n = a_1^* a_1 + a_2^* a_2 + \cdots = 1. \tag{1.61}$$

Equations (1.60) and (1.61) suggest that if we measure the physical quantity, represented by A, on a system in the normalised quantum state Ψ, then $a_n^* a_n$ may be interpreted as the probability that a measurement will yield the eigenvalue λ_n. Of course, the total probability that measurement will yield any one of these eigenvalues is unity. When an eigenvalue, say λ_r, is p-fold degenerate, the probability of finding the value λ_r is $pa_r^* a_r$. The coefficient a_n is sometimes called a *probability amplitude*. Since (1.60) involves only the eigenvalues of the operator A, the probability that a measurement on the system will yield a value which is not an eigenvalue of A is zero. We conclude that the measurement of an observable on a system should always yield one of the eigenvalues of the operator representing the observable. Since the observables are always real, the operators representing the observables should be such that their eigenvalues are always real. Such operators are called *Hermitian operators*. Thus in quantum mechanics, observables are represented by Hermitian operators. It can be emphasised that when the system is in a quantum state, the result of a single measurement on the system is unpredictable. It may also be stated that since (1.61) does not involve time, a wave function which has been normalised will always remain normalised. The constants a_n

occurring in this equation can be evaluated as follows. Multiplying equation (1.59) on the left by $\Psi_m{}^*$ and integrating with respect to τ, we obtain

$$\int \Psi_m^* \Psi \, d\tau = \sum_n a_n^* \int \Psi_m^* \Psi_n \, d\tau = \sum_n a_n^* \delta_{mn} = a_m. \tag{1.62}$$

To summarise: quantum mechanically, the system can be either in an eigenstate or in a quantum state. If it happens to be in an eigenstate, i.e. in a state with a definite eigenvalue, then a measurement will yield that particular value. On the other hand, if the system happens to be in a quantum state, i.e. in a state which in general is a linear combination of various eigenstates, i.e. it is in a state given by

$$a_1\Psi_1 + a_2\Psi_2 + \cdots + a_n\Psi_n,$$

where $\Psi_1, \Psi_2, \ldots, \Psi_n$ represent the eigenstates of the system, then the measurement must yield one of the eigenvalues of the corresponding operator. It may be emphasised that there is no classical analogue of a quantum state.

It may be remarked that if before the measurement of a physical quantity represented by an operator A the system is in a quantum state Ψ, then a measurement can yield any one of the eigenvalues of A. However, when the measurement is made on the system, a *definite* eigenvalue, say λ_n, is obtained. This means that the very act of measurement has so disturbed the system that it has been carried from a quantum state Ψ into an eigenstate Ψ_n.

Problem

1.6. What will happen if a measurement on the system is made immediately after that?

It must have been evident from the above analysis that it is not essential that a system should always be in a quantum state in which a given quantity does not possess a definite value. It can be in an eigenstate corresponding to a definite eigenvalue but this is *not the most general situation.*

1.17 Phases and relative phases

Suppose that $\Psi(\mathbf{r}, t)$ is a solution of the time-dependent Schrödinger wave equation corresponding to a definite value of energy. Since the differential equation is linear, $c\Psi$ will also be its solution, where c is a complex number. The normalisation condition for this function gives

$$c^* c \int \Psi^* \Psi \, d\tau = 1.$$

Now the complex number c may be written as

$$c = |c| e^{i\alpha},$$

where the real α is called the *phase* of c while $e^{i\alpha}$ is called its *phase factor*. Then $c^*c = |c|e^{-i\alpha}\,|c|e^{i\alpha} = |c|^2$. That is, the product c^*c is independent of the phase. Since the probability and expectation values of observables always involve $\Psi^*\Psi$ and therefore the factor c^*c, the corresponding expression is independent of the phase value. We can therefore give it any value we like. It will not affect the physical result. For convenience, we take a phase equal to zero. Then c may be considered as real. This will not affect the physical results.

The situation takes a turn when we consider a system in a quantum state Ψ. Then in terms of eigenstates which correspond to definite energies and are represented by the wave functions Ψ_i, the function Ψ can be represented by

$$\Psi(\mathbf{r}, t) = \sum_i a_i \Psi_i(\mathbf{r}, t).$$

To make the things simple, suppose that Ψ is a linear combination of two eigenstates only:

$$\Psi(\mathbf{r}, t) = a_1\Psi_1(\mathbf{r}, t) + a_2\Psi_2(\mathbf{r}, t). \tag{1.63}$$

Since each complex function can be written as $Re^{i\alpha}$, where R and α are real, we write the functions $a_1\Psi_1$, $a_2\Psi_2$ as

$$a_1\Psi_1 = R_1 e^{i\alpha_1},$$
$$a_2\Psi_2 = R_2 e^{i\alpha_2}.$$

Substituting these expressions in equation (1.63), we obtain

$$\Psi(\mathbf{r}, t) = R_1 e^{i\alpha_1} + R_2 e^{i\alpha_2}.$$

Therefore:

$$\begin{aligned}\Psi^*\Psi &= \left(R_1 e^{-i\alpha_1} + R_2 e^{-i\alpha_2}\right)\left(R_1 e^{i\alpha_1} + R_2 e^{i\alpha_2}\right).\\ &= R_1^2 + R_2^2 + 2R_1R_2\cos(\alpha_1 - \alpha_2).\end{aligned}$$

The above analysis shows that although the overall phase in Ψ can be ignored, the relative phase is important.

1.18 Postulates of quantum mechanics

For our convenience, let us now collect together the various postulates which have been proposed while relating solutions of the Schrödinger wave equation for a physical system with measurable quantities:

1. Every physically acceptable solution of the Schrödinger wave equation for a system represents a state of the system. Since a linear combination of the solutions of the Schrödinger wave equation is a solution of the differential equation, it should also represent a state of the system. This is known as the principle of superposition.
2. To every observable, there corresponds an operator A. In particular, $-i\hbar\nabla$ and $i\hbar\partial/\partial t$ are energy and momentum operators.

3. The nature of the operator should be such that its eigenvalues are always real. Such an operator is called a Hermitian operator.
4. The measurement of a physical quantity represented by a Hermitian operator A must yield one of the eigenvalues of the operator A.
5. The measurement of a physical quantity, represented by an operator A, on a system in an eigenstate always gives the eigenvalue of A corresponding to that eigenstate.
6. The average value of a large number of measurements of a physical quantity on a system in an arbitrary state Ψ, a solution of the Schrödinger wave equation, yields the average or expectation value $\langle A\rangle$ given by

$$\langle A\rangle = \int \Psi^* A\Psi \, \mathrm{d}\tau$$

provided that Ψ is normalised and there exist suitable boundary conditions.

7. The probability of finding a particle represented by the wave function $\Psi(\mathbf{r}, t)$ in a small volume $\mathrm{d}\tau$ about the point $\mathbf{r}$ and at time t is $\Psi^*(\mathbf{r}, t)\ \Psi(\mathbf{r}, t)\ \mathrm{d}\tau$.

1.19 The Schrödinger wave equation under space reflection, space inversion and time reversal

We will now briefly examine the behaviour of the Schrödinger wave equation under space reflection, space inversion and time reversal.

1.19.1 Invariance under space reflection

Consider the time-independent Schrödinger wave equation in one-dimensional space:

$$\frac{\mathrm{d}^2\psi(x)}{\mathrm{d}x^2} + \frac{2m}{\hbar^2}(E - V(x))\psi(x) = 0. \tag{1.64}$$

Changing x to $-x$ throughout the above equation, we obtain

$$\frac{\mathrm{d}^2\psi(-x)}{\mathrm{d}x^2} + \frac{2m}{\hbar^2}(E - V(-x))\psi(-x) = 0. \tag{1.65}$$

If $V(x)$, the potential energy of the system, is an even function of x, i.e. the function does not change by changing the sign of the coordinate x, then $V(-x) = V(x)$ and the above equation reduces to

$$\frac{\mathrm{d}^2\psi(-x)}{\mathrm{d}x^2} + \frac{2m}{\hbar^2}(E - V(x))\psi(-x) = 0. \tag{1.66}$$

Equations (1.64) and (1.66) show that if $V(x)$ is an even function of x, then $\psi(x)$ and $\psi(-x)$ are both solutions of the same amplitude equation. But, as shown in figure 1.3, a change in the sign of x is equivalent to a reflection of x in a plane mirror passing through the origin and placed normal to x. Thus, $\psi(-x)$ is the mirror image of $\psi(x)$

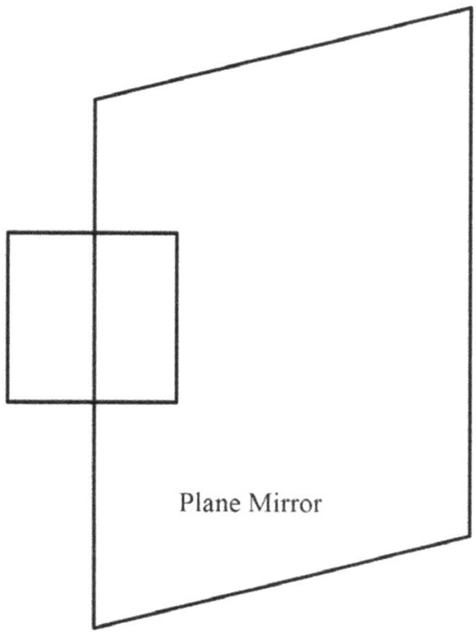

Figure 1.3. Mirror image.

and both are solutions of the *same amplitude equation*, provided $V(-x) = V(x)$. Hence, the amplitude equation with potential energy as an even function of the space coordinate is *invariant under reflection*.

1.19.2 Invariance under space inversion

The analysis can easily be extended to three-dimensional space. Then it is called *invariance under space inversion*.

1.19.3 Invariance under time reversal

Let us next consider the time-dependent Schrödinger wave equation

$$i\hbar\frac{\partial\Psi(\mathbf{r}, t)}{\partial t} = H\Psi(\mathbf{r}, t), \tag{1.16b'}$$

where $H = T + V(\mathbf{r})$ such that the potential energy $V(\mathbf{r})$ is a real function of space coordinates alone. Taking the complex conjugate of equation (1.16b'), we obtain

$$-i\hbar\frac{\partial\Psi^*(\mathbf{r}, t)}{\partial t} = H\Psi^*(\mathbf{r}, t). \tag{1.67}$$

Changing t to $-t$ throughout the equation, we obtain

$$i\hbar\frac{\partial\Psi^*(\mathbf{r}, -t)}{\partial t} = H\Psi^*(\mathbf{r}, -t). \tag{1.68}$$

Equations (1.16b') and (1.68) show that, provided the potential energy, and consequently the Hamiltonian, is real and does not depend upon time explicitly, the wave functions $\Psi(\mathbf{r}, t)$ and $\Psi^*(\mathbf{r}, -t)$ are both solutions of the same time-dependent Schrödinger wave equation. The wave function $\Psi^*(\mathbf{r}, -t)$ is often called the *time-reversed solution* of the time-dependent Schrödinger wave equation with respect to $\Psi(\mathbf{r}, t)$. The time-dependent Schrödinger wave equation (1.16b') is said to be *invariant under time reversal.* This means that if $\Psi(\mathbf{r}, t)$ is a solution of the time-dependent Schrödinger wave equation, then $\Psi^*(\mathbf{r}, -t)$ is also a solution of the same.

If the system is in a stationary state with definite energy E, then we may write

$$\Psi(\mathbf{r}, t) = \psi(\mathbf{r}) \exp\left(-\frac{\mathrm{i}Et}{\hbar}\right). \tag{1.69}$$

Taking the complex conjugate of the above equation, we obtain

$$\Psi^*(\mathbf{r}, t) = \psi^*(\mathbf{r}) \exp\left(\frac{\mathrm{i}Et}{\hbar}\right).$$

Changing t to $-t$ throughout the equation, we obtain

$$\Psi^*(\mathbf{r}, -t) = \psi^*(\mathbf{r}) \exp\left(-\frac{\mathrm{i}Et}{\hbar}\right). \tag{1.70}$$

Thus invariance under time reversal implies that, for a stationary state, if $\psi(\mathbf{r})$ is a solution of the amplitude equation, then $\psi^*(\mathbf{r})$ is also a solution of the same.

1.20 Concluding remarks

We conclude the first chapter by emphasising that the Schrödinger wave equation is *not* a model for the explanation of *some* experimental results. It is a law of nature and has replaced Newton's second law of motion for non-relativistic mechanics. It therefore provides the foundation for the explanation of mechanical phenomena. However, it is in the form of a differential equation and is not garbed in words. Its predictions based on a probabilistic interpretation of its solution are consistent with experiments although sometimes they appear to be at variance with common sense. But what is common sense? It is the knowledge that we gain from our everyday experience with macroscopic objects. And why do we expect that the microscopic world would exhibit the same characteristics and behave the same way? If we define common sense as the knowledge gained from the behaviour manifested by microscopic entities, quantum mechanics becomes fully consistent with common sense! The theory is beautiful and elegant and its achievements are spectacular and fantastic. It has introduced new concepts and changed the pattern of philosophical interpretations. We have established the Schrödinger wave equation and interpreted its solutions so as to relate them to physically measurable quantities. Our next objective is to apply this theory to solve physical problems in mechanics and during this process enhance our understanding of various characteristics of mechanical phenomena.

Additional problems

1.7. Why in the Compton effect is an x-ray photon scattered by an electron with a change in wavelength while in the photoelectric effect an optical photon transfers all its energy to the photoelectron?

1.8. The classical expression for the potential of a linear harmonic oscillator is $1/2kx^2$, where k is a constant and x is the displacement of the oscillator from its mean position. Can the Schrödinger amplitude equation be used for solving the problem of a linear harmonic oscillator?

1.9. A particle is somewhere on a line of about 7 m in length. Calculate the uncertainty in the measurement of its linear momentum. What conclusion can be drawn from this result?

1.10. What is the Compton effect? Give its quantum theory and spell out its significance.

1.11. What is the energy–time uncertainty relation? If an electron in an atom takes about 10^{-8} s to emit radiation in falling from an excited state to a lower energy state, calculate the energy uncertainty of such an excited state.

1.12. State Heisenberg's uncertainty relation for position and momentum. Describe an idealised experiment to illustrate the uncertainty relation.

1.13. Show that the probability current density can be expressed as

$$\mathbf{j} = -\frac{\hbar}{m}\,\mathrm{Re}\left(\mathrm{i}\Psi^*\nabla\Psi\right).$$

1.14. Show that if Ψ is real, then the vector $\mathbf{j}(\mathbf{r}, t)$ vanishes.

IOP Publishing

Quantum Mechanics

Mohammad Saleem

Chapter 2

A particle in a one-dimensional box

As a first application of the Schrödinger wave equation, we will consider a system whose potential energy does not depend upon time explicitly and for which the differential equation can be solved exactly. For further simplicity, we will carry out our mathematical calculations in one dimension only. This is the case of a microscopic entity, say an electron, whose motion is confined to a one-dimensional box of finite length and which is moving with a constant potential with no force acting on it. The analysis can be easily extended to a three-dimensional case.

The solution of the Schrödinger amplitude equation for a microscopic entity moving with constant potential in a one-dimensional box manifests the concept of quantum numbers, shows how, under restraints on a bounded system, discrete energy levels are naturally introduced and exhibits theoretically the existence of zero-point energy in accordance with Heisenberg's principle of uncertainty.

We will proceed in three steps:

(1) Find the classical expression for the potential energy of the physical system.
(2) As the potential energy does not explicitly depend upon the time variable, the Schrödinger amplitude equation is used to find the characteristics of the system. Write the Schrödinger amplitude equation for this potential and solve it.
(3) Compute the quantities which may be compared with experiment.

2.1 Introduction

Consider a *microscopic entity*, usually referred to as a particle, say an electron of mass m and energy E, moving along a straight path of finite length a with a constant potential energy V. The potential energy is assumed to be infinite at the extremities of the length as shown in figure 2.1. This situation is expressed by stating that the particle is moving in a one-dimensional box of length a having perfectly rigid walls at the two extremities. When the electron collides with the perfectly rigid walls of the box, it is reflected back without any loss of energy: its total energy E remains constant. It is not possible for the electron to leave the box as for this purpose it will

doi:10.1088/978-0-7503-1206-6ch2

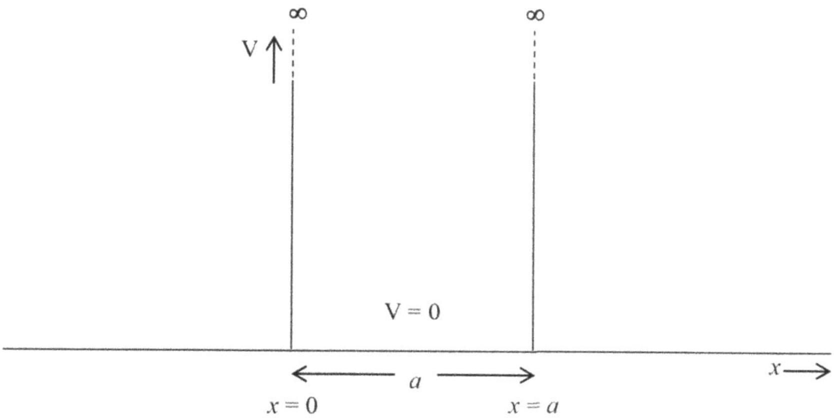

Figure 2.1. Particle in a one-dimensional box.

have to carry out an infinite amount of work. Hence, for $x \leqslant 0$ and for $x \geqslant a$, the wave function representing the electron must vanish. In practice, it is impossible to have such a motion, although it can be approximated, for instance, by a metallic crystal. Although it is a very simple problem, we will study it in detail to show how the concepts of quantum numbers, discrete energy values and zero-point energy arise in a natural way.

Let us take the direction of motion of the electron as the x-axis. Let the origin of the coordinates be at the left corner of the box so that the motion of the electron along the x-axis is confined to $0 < x < a$. The value of the constant potential energy will be determined by the choice of the zero of the energy scale. We choose it so that the potential energy in the box is zero.

2.2 The solution of Schrödinger's amplitude equation

Since the potential energy of the electron does not have explicit time dependence (in fact, it is constant, independent of space as well as time coordinates), it is sufficient to solve the time-independent Schrödinger wave equation for this electron and multiply its solution $\psi(x)$ by $\phi(t) \equiv \exp(-\mathrm{i}Et/\hbar)$ to get the wave function $\Psi(x, t)$ representing the electron. The time-independent Schrödinger wave equation in one-dimension is

$$\frac{\mathrm{d}^2\psi}{\mathrm{d}x^2} + \frac{2m}{\hbar^2}(E - V)\psi = 0, \tag{2.1}$$

where for convenience we have written ψ for $\psi(x)$. For the electron in the one-dimensional box, $V = 0$ and therefore the amplitude equation (2.1) reduces to

$$\frac{\mathrm{d}^2\psi}{\mathrm{d}x^2} + \frac{2m}{\hbar^2}E\psi = 0,$$

or

$$\frac{\mathrm{d}^2\psi}{\mathrm{d}x^2} + k^2\psi = 0, \quad \text{where } k^2 = 2mE/\hbar^2. \tag{2.2}$$

Problem

2.1. Show that for $V = 0$,

$$p = \hbar k,$$

where

$$k^2 = 2mE/\hbar^2.$$

The relation $k^2 = 2mE/\hbar^2$ shows that for constant energy E, the parameter k is also constant.

This is an ordinary second-order linear differential equation and therefore its general solution must have two arbitrary constants. The solution is given by

$$\psi(x) = A \sin kx + B \cos kx, \tag{2.3}$$

where A and B are two constants of integration. The values of these constants can be determined by applying the condition of continuity of wave function at the points where the potential changes so that:

(i) The function $\psi(x) = 0$ at $x = 0$.
(ii) The function $\psi(x) = 0$ at $x = a$.

Applying the first boundary condition to the solution (2.3), we obtain $B = 0$. Equation (2.3) then reduces to

$$\psi(x) = A \sin kx.$$

Applying the second boundary condition, we obtain

$$A \sin ka = 0.$$

There are two possibilities. Either $A = 0$ or $\sin ka = 0$. The first possibility cannot be realised because if $A = 0$, then the function $\psi(x) = A \sin kx$ will vanish for all values of x and the probability of finding the electron anywhere inside the box will also be zero. In other words, the electron will not exist even inside the box! That is, the electron will disappear. This is impossible. The second possibility, i.e. $\sin ka = 0$, yields that

$$ka = n\pi, \quad n = 0, \pm 1, \pm 2, \ldots$$

For $n = 0$, this gives $ka = 0$. As $a \neq 0$, it will yield $k = 0$ and consequently $\sin kx$ will be zero for all values of x. As a result of this, the function $\psi(x)$ $(= A \sin kx)$ will vanish for all x and the electron will disappear from the box! This is impossible as the electron is confined to the box. The solution for $n = 0$ is therefore ruled out. Moreover, as the negative values of n do not provide any information about the characteristics of the system (which always involve $\psi^*\psi$) other than that given by the positive values of n, these are not retained. The imposition of two boundary conditions thus finally leads to the relationship

$$ka = n\pi, \quad n = 1, 2, 3, \ldots, \tag{2.4a}$$

or

$$k^2a^2 = n^2\pi^2, \quad n = 1, 2, 3, \ldots \tag{2.4b}$$

Substituting the expression for k^2, i.e. $2mE/\hbar^2$, from (2.2) in (2.4*b*), we obtain

$$\frac{2mE}{\hbar^2}a^2 = n^2\pi^2, \quad n = 1, 2, 3, \ldots,$$

or

$$E = \frac{n^2\pi^2\hbar^2}{2ma^2}, \quad n = 1, 2, 3, \ldots$$

The number n, which can here take positive integral values, is called a *quantum number*. To emphasise that the energy of the electron is associated with a quantum number and to indicate the value we are dealing with, we write E_n for E. Then, the above equation may be written as

$$E_n = \frac{n^2\pi^2\hbar^2}{2ma^2}, \quad n = 1, 2, 3, \ldots \tag{2.5}$$

This is the formula for energy values of an electron moving in this one-dimensional box of length a. It shows that in this case the energy E, a continuous parameter in classical mechanics, can possess only discrete values in quantum mechanics. Note that:

1. Classically, this is a bounded system, the particle being confined in a box with infinite potential walls. For such a system, it is the imposition of the continuity requirement for ψ at the extremities of the box that makes it imperative for the permissible energy values to be discrete.
2. When the potential changes at the extremities by an infinite amount, the condition of continuity of the derivative of wave function, i.e. $d\psi/dx$, cannot be used because in this case the derivative is *not necessarily continuous*.

The energy level diagram for a particle in a one-dimensional box is shown in figure 2.2.

2.3 Zero-point energy

The *ground state energy*, i.e. the energy in the lowest state of the electron moving in a one-dimensional box, is obtained from (2.5) by choosing $n = 1$. Then, we have

$$E_1 = \frac{\pi^2\hbar^2}{2ma^2}.$$

Thus the electron will have energy, called the *zero-point energy*, even in its lowest state. That is, the electron will never be at rest. This is against the classical concept

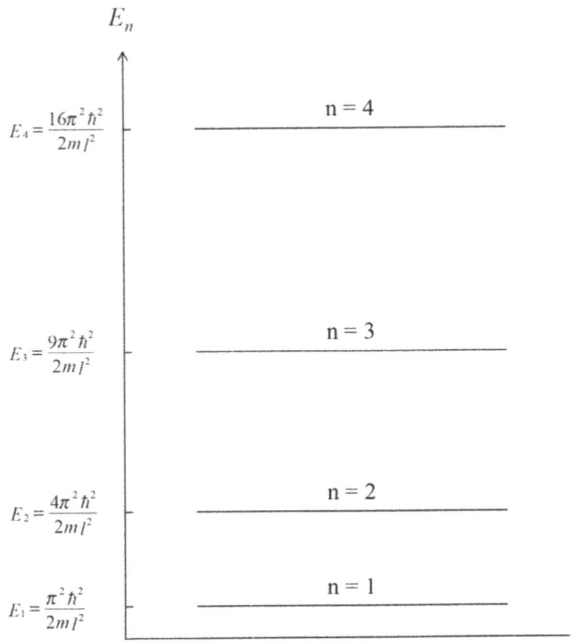

Figure 2.2. Energy level diagram for a particle in a one-dimensional box.

according to which the electron is stationary in its lowest energy state. This energy is a consequence of Heisenberg's uncertainty principle that does not allow the electron to have simultaneous zero values of position and momentum. This means that the particle must possess a minimum non-zero energy; a zero energy would contradict the principle of uncertainty. As the electron is confined within a length a, this is the uncertainty in its position. According to the uncertainty relation, its momentum should be uncertain by $\sim\hbar/a$ so that the product of these two is $\sim\hbar$.

The states with higher energies are called *excited states*. For instance, E_2, the state with $n = 2$, is the first exited state and so on.

The energy of the electron in the nth state is given by equation (2.5), i.e.

$$E_n = \frac{n^2\pi^2\hbar^2}{2ma^2}, \quad n = 1, 2, 3, \ldots \tag{2.5$'$}$$

and the energy of the electron in the $(n + 1)$th state is given by

$$E_{n+1} = \frac{(n+1)^2\pi^2\hbar^2}{2ma^2}.$$

Therefore, the difference between adjacent energy levels is

$$E_{n+1} - E_n = \frac{(2n+1)\pi^2\hbar^2}{2ma^2}. \tag{2.6}$$

This equation shows that for $\hbar \to 0$, the energy difference between adjacent levels disappears and the spectrum becomes continuous. This is in accordance with the correspondence principle which states that for $\hbar \to 0$, quantum mechanics reduces to classical mechanics.

Since for $n = 1$,

$$E_1 = \frac{\pi^2 \hbar^2}{2ma^2},$$

(2.5) and (2.6) can respectively be written as

$$E_n = n^2 E_1$$

and

$$E_{n+1} - E_n = \frac{(2n+1)\pi^2\hbar^2}{2ma^2} = (2n+1)E_1. \tag{2.6'}$$

We have thus found quantum mechanically the expression for discrete energy levels of a particle moving with a constant potential in a one-dimensional box with rigid walls. We have also obtained a formula for the difference between adjacent energy levels.

We will now find an expression for the corresponding Schrödinger wave function.

Problem

2.2. Determine the difference between energy levels E_2 and E_3 for an electron of mass $m = 0.51$ MeV moving in a one-dimensional box of length 10^{-8} cm.

The solution $\psi(x)$ of the amplitude equation has been shown to be given by

$$\psi(x) = A \sin kx.$$

But $ka = n\pi$ (see (2.4*a*)). Substituting the value of k in the above equation, we obtain

$$\psi(x) = A \sin\frac{n\pi x}{a}, \quad n = 1, 2, 3, \ldots$$

To emphasise that the function $\psi(x)$ of the electron is associated with the quantum number n, we write $\psi_n(x)$ for $\psi(x)$. Then the above equation is written as

$$\psi_n(x) = A \sin\frac{n\pi x}{a}, \quad n = 1, 2, 3, \ldots \tag{2.7}$$

This is the wave function representing the nth state of the particle.

Problem

2.3. Show by *explicit calculations* that the functions

$$\psi_n(x) = A \sin\frac{n\pi x}{a}, \quad n = 1, 2, 3, \ldots$$

are the eigenfunctions of the Hamiltonian operator

$$H \equiv -\frac{\hbar^2}{2m}\frac{d^2}{dx^2}$$

with

$$E_n = \frac{n^2\pi^2\hbar^2}{2ma^2}, \quad n = 1, 2, 3, \ldots \tag{2.5}$$

as the corresponding eigenvalues.

2.4 The normalisation constant

To find the value of the constant A occurring in (2.7), called the *normalisation constant*, we make use of the normalisation condition according to which

$$\int_0^a \psi_n^*(x)\,\psi_n(x)\,dx = 1.$$

Substituting the expressions for ψ_n and its complex conjugate ψ_n^*, we obtain

$$\int_0^a A^* \sin\frac{n\pi x}{a} A \sin\frac{n\pi x}{a}\,dx = 1,$$

or

$$A^* A \int_0^a \sin^2\frac{n\pi x}{a}\,dx = 1.$$

Since the physical interpretation of ψ always involves $\psi^*\psi$, the phase of the wave function is immaterial. This is so because only the probability density $\psi^*\psi \equiv |\psi|^2$, and not the wave function ψ, has physical relevance. We therefore take the phase of ψ to be zero. In this case, A^* becomes equal to A and the above equation reduces to

$$A^2 \int_0^a \sin^2\frac{n\pi x}{a}\,dx = 1. \tag{2.8}$$

Problem

2.4. Show that

$$\int_0^a \sin^2\frac{n\pi x}{a}\,dx = \frac{a}{2}. \tag{2.9}$$

Substituting the value of the integral from (2.9) in (2.8) and simplifying, we obtain the value of the constant A:

$$A = \sqrt{\frac{2}{a}}.$$

The normalised wave function can therefore be written as

$$\psi_n(x) = \sqrt{\frac{2}{a}} \sin\frac{n\pi x}{a}, \quad n = 1, 2, 3, \ldots \tag{2.10}$$

The time-dependent wave function $\Psi_n(x, t)$ representing the particle in the nth state is

$$\Psi_n(x, t) = \psi_n(x)\phi(t) = \psi_n(x)\mathrm{e}^{-\mathrm{i}Et/\hbar}. \tag{2.11}$$

Example 2.1.

Problem

Show that the energy eigenfunctions

$$\psi_n(x) = \sqrt{\frac{2}{a}} \sin\frac{n\pi x}{a}, \quad n = 1, 2, 3, \ldots$$

of an electron moving in a one-dimensional box are also eigenfunctions of p^2 where $\mathbf{p}$ is its linear momentum operator.

Solution

We have

$$\begin{aligned} p^2\psi_n(x) &= -\hbar^2\frac{\mathrm{d}^2}{\mathrm{d}x^2}\left(\sqrt{\frac{2}{a}} \sin\frac{n\pi x}{a}\right) \\ &= \hbar^2\left(\sqrt{\frac{2}{a}}\frac{n^2\pi^2}{a^2} \sin\frac{n\pi x}{a}\right) \\ &= \frac{n^2\hbar^2\pi^2}{a^2}\psi_n(x) = p_n^2\psi_n(x), \quad \text{where } p_n^2 = \frac{n^2\hbar^2\pi^2}{a^2}, \\ &= \frac{n^2h^2}{4a^2}\psi_n(x). \end{aligned}$$

This shows that the energy eigenfunction $\psi_n(x)$ is *also* an eigenfunction of the square of the linear momentum operator $p^2 \equiv -\hbar^2\mathrm{d}^2/\mathrm{d}x^2$ with $\frac{n^2\hbar^2\pi^2}{a^2} \equiv n^2h^2/4a^2$ as its eigenvalue.

The relation

$$p_n^2 = \frac{n^2\hbar^2\pi^2}{a^2}$$

shows that

$$p_n = \pm\frac{n\pi\hbar}{a} = \pm\frac{nh}{2a}.$$

The ± sign shows that the electron is moving forward and backward in the potential box.

Problems

2.5. Show that the difference between the momenta corresponding to two consecutive energy levels is equal to $\frac{\hbar}{2a}$.

2.6. Show by explicit calculations that two linearly independent eigenfunctions $\psi_m(x)$ and $\psi_n(x)$ of the Hamiltonian of an electron moving in a one-dimensional box and with different eigenvalues are orthogonal to each other.

The solution of the last problem reflects that the functions ψ_m and ψ_n as described above are orthogonal to each other.

The normalised eigenfunctions obtained from (2.10) by writing n = 1, 2, 3 are given below and are plotted in figure 2.3:

$$\psi_1(x) = \sqrt{\frac{2}{a}}\sin\frac{\pi x}{a},$$
$$\psi_2(x) = \sqrt{\frac{2}{a}}\sin\frac{2\pi x}{a}, \qquad (2.12)$$
$$\psi_3(x) = \sqrt{\frac{2}{a}}\sin\frac{3\pi x}{a}.$$

Figure 2.3 shows that the wave function $\psi_1(x)$ has two nodes, at $x = 0$ and $x = a$; the wave function $\psi_2(x)$ has three nodes, at $x = 0$, $a/2$ and a; the wave function $\psi_3(x)$ has four nodes at $x = 0$, $a/3$, $2a/3$ and a.

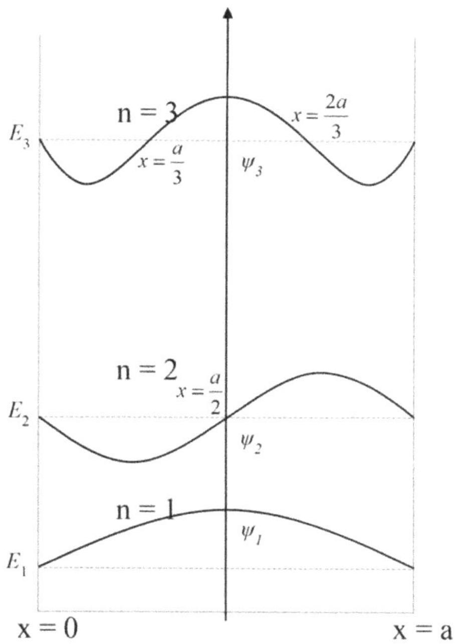

Figure 2.3. Wave functions for $n = 1, 2, 3$ for a particle in a one-dimensional box.

The probability density for the state represented by $\Psi_n(x, t)$ is given by

$$\Psi_n^*(x, t)\,\Psi_n(x, t) = \psi_n^*(x)e^{iEt/\hbar}\,\psi_n(x)e^{-iEt/\hbar}$$

$$= \psi_n^*(x)\,\psi_n(x), \quad n = 1, 2, 3, \ldots, \tag{2.13}$$

$$= \sqrt{\frac{2}{a}}\,\sin\frac{n\pi x}{a}\sqrt{\frac{2}{a}}\,\sin\frac{n\pi x}{a}$$

$$= \frac{2}{a}\,\sin^2\frac{n\pi x}{a}, \quad n = 1, 2, 3, \ldots \tag{2.14}$$

and is independent of time. The particle is then said to be in the stationary state $\Psi_n(x, t)$. The probability of finding the particle at any position x does not depend upon time. Equation (2.14) shows that the probability density varies with x and always has small values near the walls of the box. This equation also shows that the maximum values of the probability density occur for

$$\Psi_n^*\Psi_n = \frac{2}{a}\,\sin^2\frac{n\pi x}{a}.$$

The probability density is therefore maximum when

$$\frac{n\pi x}{a} = \frac{\pi}{2}, \frac{3\pi}{2}, \frac{5\pi}{2}, \ldots,$$

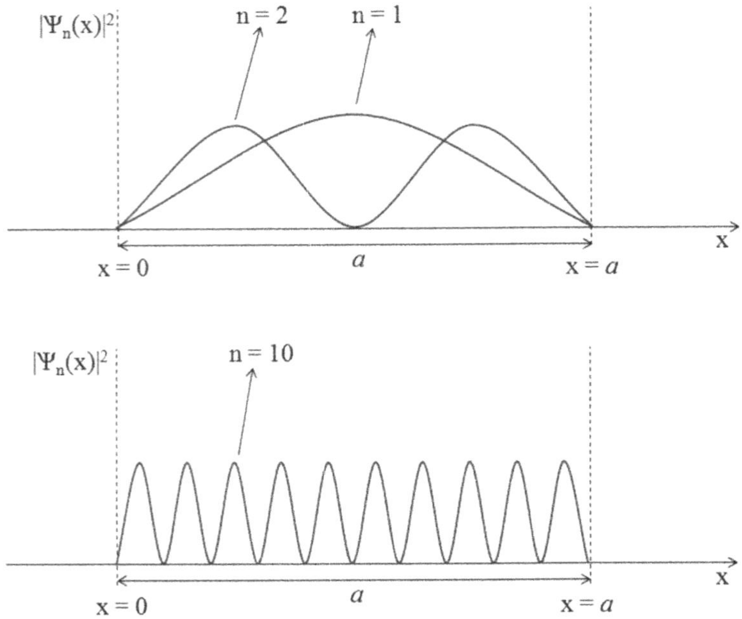

Figure 2.4. Probability distribution curves for $n = 1$, 2 and 10 for a particle moving in a one-dimensional box.

or when

$$x = \frac{a}{2n}, \frac{3a}{2n}, \frac{5a}{2n}, \dots$$

The probability distribution curves for $n = 1$, 2 and 10 are drawn in figure 2.4. It may be noticed that for $n = 1$, i.e. in the lowest energy state, the particle has the maximum chance of being found at the centre of the one-dimensional box, i.e. at $x = a/2$. For $n = 2$, the particle has *no* chance of being found at the centre. The maximum values occur at $x = a/4$ and $x = 3a/4$. For very large values of n, it oscillates very rapidly attaining roughly a uniform limit. An idea of the same can be obtained from the probability distribution curve for $n = 10$.

Note that for $\psi_2(x)$, the probability of finding the electron at $x = a/2$, i.e. at the centre of the box, is zero, while the probability of finding the electron adjacent to $x = a/2$, both on its left as well on its right, is different from zero. The particle can be on the left or on the right of the centre of the one-dimensional box but it can never be at the centre! That is, the particle can go from left of the centre to the right of the centre without passing through the centre. This is a quantum mechanical concept and cannot be visualised. It is this quantum jumping to which Schrödinger, who proposed the differential equation, and Einstein, a great physicist of modern times, could never reconcile themselves. One should become accustomed to such occasional shocks. Bohr, a Nobel laureate in physics and a philosopher of science, has remarked: ‘Those who are not shocked when they first come across quantum theory cannot possibly have understood it’.

Problems

2.7. The ground state and the first excited state of an electron of mass m confined in a one-dimensional infinite potential well of length a are given by two stationary states

$$\psi_1(x) = \sqrt{\frac{2}{a}}\sin\frac{\pi x}{a},$$

$$\psi_2(x) = \sqrt{\frac{2}{a}}\sin\frac{2\pi x}{a}.$$

Consider the time development of these states and find out whether a superposition of two stationary states is a stationary state or not.

2.8. Find the expectation value of the position x of the electron in its nth state in a one-dimensional box.

2.9. A particle is moving in a one-dimensional box of length a. What is its most probable position when it is in the third excited state?

2.5 The parity of eigenfunctions

Equations (2.5) and (2.10) show that, in this one-dimensional case involving only one quantum number n, there is a one-to-one correspondence between the stationary state energies E_n and the corresponding eigenfunctions ψ_n. The energy eigenvalues are therefore *non-degenerate*: two or more than two linearly independent eigenfunctions *do not correspond* to the same eigenvalue. Equation (2.10) exhibits that as the space coordinate x is changed to $-x$, the wave function $\psi_n(x)$ changes its sign:

$$\psi_n(x) \rightarrow \psi_n(-x) = -\psi_n(x).$$

Such a function is said to be an *odd function* of x and the state represented by this wave function is said to be of *odd* or negative *parity*. If a wave function $\psi(x)$ is such that it remains unchanged as x is changed to $-x$, i.e.

$$\psi_n(x) \rightarrow \psi_n(-x) = \psi_n(x),$$

then it is said to be an *even function* of x and the corresponding state of the system is said to be of *even* or positive *parity*. Thus, the eigenfunctions of a one-dimensional box are of odd parity.

Problem

2.10. Show that if the potential is symmetric, then the wave function representing a particle in an eigenstate must be either even or odd.

The solution of the Schrödinger amplitude equation for a particle moving with constant potential in a one-dimensional box manifests the concept of quantum numbers, shows how, under restraints on a bounded system, discrete energy levels are naturally introduced and exhibits theoretically the existence of zero-point energy in accordance with Heisenberg's principle of uncertainty.

Additional problems

2.11. Show that the boundary conditions for a particle confined within a one-dimensional box cannot be satisfied if $E < 0$.

2.12. Show by explicit calculations that the states ψ_2 and ψ_3 of a particle in a one-dimensional box are orthogonal to each other.

2.13. Show that the eigenfunctions of a particle in a one-dimensional box possess odd parity, the origin of coordinates being taken at the left corner of the box. What will be the situation if the origin is taken at the centre of the box?

2.14. A particle of mass m is confined in a one-dimensional infinite potential well of length a:

$$V(x) = \begin{matrix} 0, & 0 < x < a, \\ \infty, & \text{otherwise.} \end{matrix}$$

If Δx and Δp_x are the uncertainties in the measurement of position and momentum, calculate $\Delta x \Delta p_x$ for a particle in the first excited state.

IOP Publishing

Quantum Mechanics

Mohammad Saleem

Chapter 3

Free particles

3.1 Introduction

The problem of the motion of an electron within a one-dimensional box as analysed in the previous chapter exhibits a characteristic of bounded systems: if classically the motion of a system is confined to a finite region of space, i.e. it is a bounded system, then a physically acceptable solution of the corresponding Schrödinger wave equation yields discrete energy eigenvalues. The set of discrete energy values is also called the discrete spectrum of energy eigenvalues or the discrete energy spectrum. If the classical motion is unbounded, i.e. the particle is not subjected to any external force, it is said to be a free particle and its energy eigenvalues are found to be continuous and are said to form a continuous spectrum. Such a particle has a definite value of momentum, but the position of the particle is completely unknown. The eigenfunctions corresponding to the first category are such that they vanish sufficiently rapidly at large distances so that $\int \Psi^* \Psi \, d\tau$ always converges; the usual normalisation procedure is valid. The eigenfunctions corresponding to the second category, even remaining finite everywhere in space, do not vanish sufficiently rapidly at large distances and therefore the integral $\int \Psi^* \Psi \, d\tau$ taken over an infinite volume diverges. The usual normalisation procedure fails in the case of a free particle or unbounded system. The normalisation difficulty can be overcome in several ways depending upon the nature of the problem. For instance, Ψ may be defined in a finite interval so that it vanishes at the walls or may be assigned a periodic structure at the boundary. This is explained below.

3.2 Free particles

Let us consider a particle which is free to move in space: the force $\mathbf{F}$ acting on it is zero. Then the potential energy V, being related to the force $\mathbf{F}$ by the equation $\mathbf{F} = -\nabla V$, is constant. Taking this constant as zero, the Schrödinger amplitude equation for the motion of a free particle becomes

$$\nabla^2 \psi + \frac{2m}{\hbar^2} E \psi = 0, \tag{3.1}$$

doi:10.1088/978-0-7503-1206-6ch3

or

$$\nabla^2\psi + k^2\psi = 0, \quad k^2 = \frac{2m}{\hbar^2}E. \tag{3.2}$$

If the direction of motion of the particle is taken as the x-axis, this differential equation reduces to

$$\frac{d^2\psi}{dx^2} + k^2\psi = 0. \tag{3.3}$$

Let us find the solutions of this differential equation for $E < 0$ and for $E > 0$.

(i) For $E < 0$, k^2 = a negative number $= -s^2$, where s^2 is positive. Then the corresponding Schrödinger wave equation becomes

$$\frac{d^2\psi}{dx^2} - s^2\psi = 0.$$

Two linearly independent solutions of this differential equation are

$$\psi(x) = A_1 e^{sx}, \ B_1 e^{-sx}$$

which respectively diverge as $x \to +\infty$ and $-\infty$. These solutions, being physically not acceptable, are rejected. This shows that, *in non-relativistic quantum mechanics, a free particle cannot have negative energy.*

(ii) For $E > 0$, the Schrödinger amplitude equation (3.3) gives

$$\frac{d^2\psi}{dx^2} + k^2\psi = 0. \tag{3.3$'$}$$

The general solution of this differential equation is given by

$$\psi(x) = Ae^{ikx} + Be^{-ikx}, \tag{3.4}$$

or

$$\psi(x) = A' \cos kx + B' \sin kx. \tag{3.5}$$

Since the system is unbounded, the energy levels form a continuum. As each positive energy E corresponds to two linearly independent solutions, it is doubly degenerate.

3.3 Normalisation of stationary wave solutions

Equation (3.5) is similar to that of a stationary wave. The function $\psi(x)$ is therefore called a *stationary wave solution* of the Schrödinger amplitude equation. Let us first consider the normalisation of this stationary wave solution

$$\psi(x) = A' \cos kx + B' \sin kx. \tag{3.5$'$}$$

Applying the normalisation condition, we obtain

$$\int_{-\infty}^{\infty} \psi^*(x)\,\psi(x)\mathrm{d}x = \int_{-\infty}^{\infty} (A'^*\cos kx + B'^*\sin kx)(A'\cos kx + B'\,\sin kx)$$

$$= A'^*A' \int_{-\infty}^{\infty} \cos^2 kx\,\mathrm{d}x + B'^*B' \int_{-\infty}^{\infty} \sin^2 kx\,\mathrm{d}x$$

$$+ (A'^*B' + B'^*A') \int_{-\infty}^{\infty} \sin kx \cos kx\,\mathrm{d}x.$$

Owing to the trigonometric functions, $\sin kx$ and $\cos kx$, the value of the integral $\int_{-\infty}^{\infty} \psi^*\psi\,\mathrm{d}x$ is indeterminate so that the values of the constants in the wave function for a free particle cannot be determined in the usual way. This is owing to the fact that the eigenfunctions $\sin kx$ and $\cos kx$ do not tend to zero at large distances; they are oscillatory instead of exponentially decreasing; a typical example of eigenvalue problems with continuous spectra. The difficulty in normalisation can be overcome by assuming that the particle is not completely free to move in the entire space but is confined in a *large* one-dimensional box with rigid walls, large enough not to significantly affect the physical situation under consideration. Since the nodes of a stationary wave are permanently stationary, we can take $\psi(x) = 0$ at the rigid walls of the box. Normalisation is now possible. And as the particle is confined to a finite region of space, the continuum of energy states is replaced by a discrete infinity of energy states.

3.4 Normalisation of progressive wave solutions

The above analysis confines the normalisation procedure to stationary waves. This procedure cannot be applied to a progressive wave as its nodes are in motion and, therefore, rigid walls of the box will not serve our purpose: $\psi(x)$ cannot remain zero at the walls. For this reason, the normalisation technique has to be modified. Let us consider the progressive wave solution

$$\psi(x) = A\mathrm{e}^{\mathrm{i}kx}, \tag{3.6}$$

where linear momentum $\hbar k$ is exactly known. The other independent solution has been taken as zero because the free particle is moving only from left to right. This means that the infinite plane wave of exact linear momentum represented by

$$\Psi(x,\,t) = A\mathrm{e}^{\mathrm{i}kx}\mathrm{e}^{-\mathrm{i}Et/\hbar}$$

is all the time travelling from left to right. The probability density $\Psi^*(x, t)\,\Psi(x, t)$ is independent of not only time but also of the space coordinate x. Consequently, the probability of finding the particle *anywhere* on the x-axis is the same, i.e. the position of the particle on the x-axis is completely unknown. This is in accordance with Heisenberg's principle of uncertainty which requires that if the linear momentum of a particle is known exactly ($\Delta p_x = 0$), then its position should be completely indeterminate ($\Delta x = \infty$). Mathematically, if we try to normalise this wave function, we obtain

$$\int_{-\infty}^{\infty} \psi^*\psi \, dx = A^*A \int_{-\infty}^{\infty} e^{-ikx} e^{ikx} \, dx = A^*A \int_{-\infty}^{\infty} dx.$$

This integral evidently diverges. However, as we know, this is an ideal situation; a particle whose linear momentum is exactly known is moving *somewhere* in a wave of *infinite* length. In actual experiments, the particle is confined to a finite region of space, that is, the laboratory, so both its position and momentum are uncertain in accordance with the principle of uncertainty. The wave functions such as $A \exp(ikx)$ are to be 'normalised' in a sense different from that used in chapter 1. For that purpose, the entire x-axis is divided into equal distances (one-dimensional boxes) each of length a. These distances are large by microscopic standards ($a \gg 10^{-6}$ cm), i.e. compared to all distances of interest to us. Then the influence of the boundary conditions on the motion of the particle in the distance from 0 to a is not important. Let us choose the boundary condition as periodic with period a so that a wave function should have the same value at the walls. Thus, if the extremity of one of the one-dimensional boxes is taken as the origin, then, say,

$$\psi(0) = \psi(a). \tag{3.7}$$

Once the wave functions satisfy this periodic boundary condition, we confine our attention to *one* appropriate one-dimensional box. The normalisation is now achieved by requiring that the integral of $\psi^*\psi$ over *this* one-dimensional box is unity:

$$\int_0^a \psi^*(x)\,\psi(x)\,dx = 1,$$

or

$$A^*A \int_0^a e^{-ikx} e^{ikx} \, dx = 1,$$

or

$$|A|^2 \int_0^a dx = 1,$$

or

$$|A|^2 \, [x]_0^a = 1,$$

or

$$|A|^2 \, a = 1.$$

Since the phase of A does not affect the physical results, we take it as zero. Then the above equation gives

$$A = \frac{1}{\sqrt{a}} \tag{3.8}$$

so that (3.6) takes the form

$$\psi(x) = \frac{1}{\sqrt{a}}\mathrm{e}^{\mathrm{i}kx}. \tag{3.9}$$

This is the normalised wave equation for a free particle. It may be noticed that the probability density is $\psi^*\psi = 1/a$. Therefore, the probability of finding the particle in a box of length a is 1 and is independent of its length. Since $\psi(x)$ is a single-valued function of x and repeats itself after a length a, i.e. $\psi(x)$ is a periodic function of period a, we have

$$\psi(x + a) = \psi(x). \tag{3.10}$$

Now, (3.7) and (3.9) yield

$$\frac{1}{\sqrt{a}}\mathrm{e}^{\mathrm{i}ka} = \frac{1}{\sqrt{a}},$$

or

$$\mathrm{e}^{\mathrm{i}ka} = 1 = \mathrm{e}^{2n\mathrm{i}\pi}, \quad n = 0, \pm1, \pm2, \ldots,$$

or

$$k = \frac{2n\pi}{a}, \quad n = 0, \pm1, \pm2, \ldots,$$

or

$$k^2 = \frac{4n^2\pi^2}{a^2}, \quad n = 0, 1, 2, \ldots$$

The negative values of n have been left out because these do not give any new information about the system. Substituting the expression for k^2 from (3.2) in the above equation, we obtain

$$\frac{2m}{\hbar^2}E = \frac{4n^2\pi^2}{a^2} \quad n = 0, 1, 2, \ldots,$$

or

$$E_n = \frac{n^2h^2}{2ma^2} \quad n = 0, 1, 2, \ldots, \tag{3.11}$$

where to emphasise that energy depends upon n, we have written E_n for E. As a increases, the permitted energy values come closer and closer so that for a macroscopic box, the energy spectrum is *essentially continuous*.

Let us now extend this analysis to three-dimensional space where the Schrödinger wave equation is

$$\nabla^2\psi + k^2\psi = 0, \quad k^2 = \frac{2m}{\hbar^2}E.$$

The solution of this differential equation is:

$$\psi(\mathbf{r}) = A\mathrm{e}^{\mathrm{i}\mathbf{k}\cdot\mathbf{r}}, \tag{3.12}$$

where **k** is in the direction of propagation of the wave and is called the *propagation vector*. Therefore, the normalisation condition, i.e.

$$\int \psi^*(\mathbf{r})\,\psi(\mathbf{r})\,\mathrm{d}\tau = 1$$

in accordance with the procedure adopted above, yields

$$A^2 \int_0^a \mathrm{d}\tau = 1,$$

or

$$A^2(a^3) = 1,$$

or

$$A = \frac{1}{\sqrt{a^3}}.$$

Substituting this value of A in (3.12), we obtain

$$\psi(\mathbf{r}) = \frac{1}{\sqrt{a^3}}\mathrm{e}^{\mathrm{i}\mathbf{k}\cdot\mathbf{r}}. \tag{3.13}$$

By taking $\mathbf{r} = (x, y, z)$ and $\mathbf{k} = (k_x, k_y, k_z)$, it can be easily shown that in this case we must have

$$k_x = \frac{2n_x\pi}{a}, \quad k_y = \frac{2n_y\pi}{a}, \quad k_z = \frac{2n_z\pi}{a}, \quad n_x, n_y, n_z = 0, \pm1, \pm2, \ldots \tag{3.14}$$

It can be seen that the probability density $\psi^*(\mathbf{r})\psi(\mathbf{r})$ is now $1/a^3$. This is independent of **r** as well as t, i.e. the probability is the same at all times and everywhere in the box.

The boundary conditions (3.14) can also be expressed as

$$\mathbf{k} = \frac{2\pi}{a}\mathbf{n}, \quad \mathbf{k} = (k_x, k_y, k_z) \text{ and } \mathbf{n} = (n_x, n_y, n_z).$$

Hence the linear momentum $\mathbf{p} = \hbar\mathbf{k} = (2\pi\hbar/a)\mathbf{n}$ is quantised. The same is valid for the energy $E = \hbar\omega(\mathbf{k})$. As the direction of vector **k** can vary continuously, a positive energy state given by

$$E_n = \frac{\hbar^2 k^2}{4ma^2}$$

is now infinitely degenerate.

For a free particle, when the wave function is not square integrable, i.e.

$$\int_{-\infty}^{\infty} \psi^*(\mathbf{r})\, \psi(\mathbf{r})\, \mathrm{d}\tau$$

diverges, the quantity $\psi^*(\mathbf{r})\psi(\mathbf{r})$ may be interpreted as the *relative probability density*. The ratio of the magnitude of

$$\int \psi^*(\mathbf{r})\, \psi(\mathbf{r})\, \mathrm{d}\tau$$

in two *different* regions of space then determines the relative probability of finding the particle in these regions. Notice that the two wave functions *differing only by a multiplicative constant* will describe the same physical situation.

We will next consider another approach to normalising the eigenfunctions of a free particle, i.e. the eigenfunctions corresponding to a continuous spectrum of eigenvalues. But before that it is essential to introduce Dirac's delta function which is the basis of this approach.

3.5 Dirac's delta function

Dirac formulated relativistic quantum mechanics and shared with Schrödinger, the founder of non-relativistic quantum mechanics, the Nobel Prize for physics in 1933. On the basis of his remarkable contributions, he was elevated in his early thirties to the professorship at Cambridge University which had once been held by Newton. In order to facilitate calculations, Dirac introduced a function which, in his honour, is called *Dirac's delta function* and, in one dimension, is denoted by $\delta(x - x')$. This is defined by

$$\begin{aligned} \delta(x - x') &= 0, \quad \text{for } x \neq x' \\ &= \infty, \quad \text{for } x = x', \end{aligned}$$

in such a way that

$$\int \delta(x - x')\, \mathrm{d}x = 1.$$

The range of integration includes the point x'. In the strict mathematical sense, it is not a function at all: it is an improper function. Nevertheless, it plays a significant role in quantum mechanics. Some of the properties of the delta function are:

$$\delta(x) = \delta(-x),$$

$$\delta'(x) = -\delta'(-x), \quad \text{i.e. } \frac{\mathrm{d}}{\mathrm{d}x}\delta(x) = -\frac{\mathrm{d}}{\mathrm{d}x}\delta(-x),$$

$$x\delta(x) = 0,$$

$$x\delta'(x) = -\delta(x),$$

$$\delta(ax) = \frac{1}{|a|}\delta(x), \quad a \neq 0,$$

$$\delta[F(x)] = \sum_i \frac{\delta(x - x')}{|F'(x_i)|}, \qquad \text{provided } F'(x_i) \neq 0,$$

$$\delta\{(x - a)(x - b)\} = \frac{1}{|a - b|}\{\delta(x - a) + \delta(x - b)\}, \qquad a \neq b,$$

$$f(x)\delta(x - a) = f(a)\delta(x - a),$$

$$f(a) = f(x)\delta(x - a)\,\mathrm{d}x,$$

$$\int \delta(a - x)\delta(x - b)\,\mathrm{d}x = \delta(a - b).$$

It may be noted that the properties of a delta function are realised only when it is used as a factor in an integrand.

We will now establish one of the above relations, i.e.

$$x\delta(x) = -\delta(x).$$

If $f(x)$ is a well-behaved function, then

$$\int_{-\infty}^{\infty} f(x)x\delta'(x)\,\mathrm{d}x = \int_{-\infty}^{\infty} xf(x)\delta'(x)\,\mathrm{d}x = [\delta(x)xf(x)]_{-\infty}^{\infty} - \int_{-\infty}^{\infty} \delta(x)\frac{\mathrm{d}}{\mathrm{d}x}\{xf(x)\}\,\mathrm{d}x.$$

But the first term on the right-hand side vanishes at the boundaries. Therefore,

$$\int_{-\infty}^{\infty} f(x)x\delta'(x)\,\mathrm{d}x = -\int_{-\infty}^{\infty} \delta(x)f(x)\,\mathrm{d}x - \int_{-\infty}^{\infty} \delta(x)x\frac{\mathrm{d}f(x)}{\mathrm{d}x}\,\mathrm{d}x.$$

As the second integral on the right-hand side of this equation is zero, we obtain

$$\int_{-\infty}^{\infty} f(x)x\delta'(x)\,\mathrm{d}x = -\int_{-\infty}^{\infty} f(x)\delta(x)\,\mathrm{d}x.$$

Hence:

$$x\delta(x) = -\delta(x).$$

Problem

3.1. Show that a representation of the Dirac delta function is given by

$$\delta(k - k') = \frac{1}{2\pi}\int_{-\infty}^{\infty} \mathrm{e}^{\mathrm{i}(k-k')x}\,\mathrm{d}x.$$

3.6 Continuous distribution of eigenvalues and Dirac's delta function

When an operator has continuous eigenvalues, then it can be shown that most of the statements valid for a discrete spectrum are also true in this case. For instance, for discrete eigenvalues, the eigenvalue equation is

$$A\psi_i(x) = a_i\psi_i(x)$$

where the subscript i is used to denote a particular eigenvalue as well as the corresponding eigenfunction. For a continuous distribution, this equation is replaced by

$$A\psi_{a'}(x) = a'\psi_{a'}(x)$$

where an eigenvalue in the continuous spectrum is also used as a subscript to designate the corresponding eigenfunction. For a discrete spectrum, the eigenvalues are denoted by a_1, a_2, a_3, ..., while for a continuous spectrum the eigenvalues are designated as a', a'', a''', ...

The orthogonality condition for eigenfunctions corresponding to continuous eigenvalues can be written as

$$\int \psi^*_{a'}(x)\,\psi_a(x)\,\mathrm{d}x = 0, \quad a' \neq a.$$

Since the eigenvalue spectrum is continuous, this integral would be divergent for $a' = a$ because the wave function does not vanish sufficiently rapidly at large distances. Therefore

$$\int \psi^*_{a'}(x)\,\psi_a(x)\,\mathrm{d}x = \infty, \quad a' = a.$$

Hence, we may write

$$\int \psi^*_{a'}(x)\,\psi_a(x)\,\mathrm{d}x = \delta(a - a').$$

Problem

3.2. Discuss the situation when the eigenvalue spectrum of an operator is discrete over a certain range of values and continuous over a separate range of values.

3.7 Eigenfunctions and eigenvalues of the position operator

We have studied some characteristics of continuous eigenvalues including the normalisation procedure. We will now find the eigenfunctions and eigenvalues of the position operator x and show that these eigenvalues are continuous, the Dirac delta function representing the eigenfunctions of x. Let x' be the eigenvalue of x corresponding to the eigenfunction $\psi_{x'}(x)$. Then

$$x\psi_{x'}(x) = x'\psi_{x'}(x),$$

or

$$(x - x')\psi_{x'}(x) = 0.$$

For $x \neq x'$, the function $\psi_{x'}(x)$ must be zero. For $x = x'$, the function $\psi_{x'}(x)$ can have any value subject to the condition that

$$\int_{-\infty}^{\infty} \psi_{x'}^*(x)\psi_{x'}(x)\,\mathrm{d}x = 1.$$

Actually, in order that the integral may exist, $\psi_{x'}(x)$ must approach infinity for $x = x'$. This is because if $\psi_{x'}(x)$ has to represent the wave function immediately after a position measurement giving the value x', it must be extremely large at the point $x = x'$ and vanishingly small everywhere else. Since this analysis is valid for *all finite values* of x', the eigenvalues of x' must form a *continuous spectrum*, i.e. the position coordinate can have any finite value.

The eigenfunction $\psi_{x'}(x)$ of the position operator x has the following properties:

$$\psi_{x'}(x) = 0, \quad \text{for } x \neq x',$$

$$\psi_{x'}(x) = \infty, \quad \text{for } x = x',$$

such that

$$\int \psi_{x'}^*(x)\psi_{x'}(x)\,\mathrm{d}x = 1,$$

the region of integration including the point x'. Hence, it behaves as a delta function:

$$\psi_{x'}(x) = \delta(x - x').$$

We conclude that the eigenfunctions of the position operator are delta functions.

3.8 Eigenfunctions and eigenvalues of the momentum operator

We will now determine the eigenfunctions of the momentum operator $-\mathrm{i}\hbar\nabla$. Let $\mathbf{p}$ represent the eigenvalue of the operator $-\mathrm{i}\hbar\nabla$ corresponding to the eigenfunction u_p. To avoid any ambiguity, we have used the symbol u_p rather than ψ for the momentum eigenfunctions. Then, we have

$$-\mathrm{i}\hbar\nabla u_p = \mathbf{p}u_p.$$

Multiplying scalarly both sides of the above differential equation by $\mathbf{dr}$, we obtain

$$-\mathrm{i}\hbar\mathbf{dr}\cdot\nabla u_p = (\mathbf{dr}\cdot\mathbf{p})u_p.$$

Problem

3.3. Prove that

$$\mathrm{d}\mathbf{r} \cdot \nabla = \mathrm{d}.$$

Making use of the above result and noting that **p** is an eigenvalue, we obtain

$$-\mathrm{i}\hbar\, \mathrm{d}u_p = \mathrm{d}(\mathbf{r} \cdot \mathbf{p})u_p,$$

or

$$-\mathrm{i}h\frac{\mathrm{d}u_p}{u_p} = \mathrm{d}(\mathbf{p} \cdot \mathbf{r}).$$

Integrating the above equation, we obtain

$$-\mathrm{i}\hbar \ln u_p = \mathbf{p} \cdot \mathbf{r} - \mathrm{i}\hbar \ln C,$$

where C is a constant, or

$$-\mathrm{i}\hbar \ln\frac{u_p}{C} = \mathbf{p} \cdot \mathbf{r},$$

or

$$u_p = C\exp[\mathbf{p} \cdot \mathbf{r}/(-\mathrm{i}\hbar)],$$

or

$$u_p = C\mathrm{e}^{\mathrm{i}\mathbf{p}\cdot\mathbf{r}/\hbar} = C\mathrm{e}^{\mathrm{i}\mathbf{k}\cdot\mathbf{r}}, \quad \text{where } \frac{\mathbf{p}}{\hbar} = \mathbf{k}.$$

The eigenfunctions of the momentum operator are formally the same as those of a free particle.

Problem

3.4. Show that the eigenvalues of the momentum operator form a continuous spectrum.

In three–dimensional space, the box normalisation with periodic boundary conditions gives

$$C = \frac{1}{\sqrt{a^3}}.$$

Therefore

$$u_p = \frac{1}{\sqrt{a^3}} e^{i\mathbf{k}\cdot\mathbf{r}}. \tag{3.15}$$

Equations (3.13) and (3.15) show that, for a free particle, the energy and momentum eigenfunctions are formally similar.

It may be emphasised that the eigenfunction $u_p(\mathbf{r}, \mathbf{p})$ of the linear momentum operator does not represent a physical state. This is because it is *not* a vector in the Hilbert space as it does not possess a finite length.

3.9 Normalisation of a free particle eigenfunction using a delta function

Let us next see how, instead of a box normalisation, a delta function can be used for the normalisation of a free particle eigenfunction. We have shown that an eigenfunction of a free particle is given by

$$\psi_k(x) = Ae^{ikx}. \tag{3.6$'$}$$

Then, taking the complex conjugate of this equation, we obtain

$$\psi_k^*(x) = A^* e^{-ikx}.$$

Multiplying the last two equations and integrating with respect to x, we obtain

$$\int \psi_k^*(x)\psi_k(x)\,dx = A^*A \int e^{-ikx}\,e^{ikx}\,dx = A^*A \int dx.$$

Since the limits are from $-\infty$ to ∞, $\int dx$ diverges. Therefore, the function $\psi_k(x)$ cannot be normalised in this manner.

The function $\psi_k(x)$ is said to be k-normalised if

$$\int_{-\infty}^{\infty} \psi_{k'}^*(x)\psi_k(x)\,dx = \delta(k - k'),$$

or

$$A^*A \int_{-\infty}^{\infty} e^{-ik'x}\,e^{ikx}\,dx = \delta(k - k'),$$

or

$$A^*A \int_{-\infty}^{\infty} e^{-i(k'-k)}dx = \delta(k - k'). \tag{3.16}$$

A representation of the delta function $\delta(k - k')$ is given by

$$\delta(k - k') = \frac{1}{2\pi} \int_{-\infty}^{\infty} e^{i(k-k')}\,dx. \tag{3.17}$$

Comparing (3.16) and (3.17), we obtain

$$A^*A = \frac{1}{2\pi},$$

or

$$|A|^2 = \frac{1}{2\pi},$$

or

$$|A| = \frac{1}{\sqrt{2\pi}}.$$

Taking the physically unimportant phase factor as unity, we obtain

$$A = \frac{1}{\sqrt{2\pi}}.$$

The wave function for a free particle is then given by

$$\psi_k(x) = \frac{1}{\sqrt{2\pi}}\mathrm{e}^{ikx}. \tag{3.18}$$

Also, the momentum eigenfunction is given by

$$\psi_p(x) = C\mathrm{e}^{\mathrm{i}px/\hbar}.$$

It is said to be p-normalised if

$$\int \psi_{p'}^*(x)\psi_p(x)\,\mathrm{d}x = \delta(p - p').$$

This yields

$$\begin{aligned}\delta(p - p') &= C^*C\int \mathrm{e}^{-\mathrm{i}p'x/\hbar}\,\mathrm{e}^{\mathrm{i}px/\hbar}\,\mathrm{d}x\\ &= C^*C\int \mathrm{e}^{\mathrm{i}(P'-P)x/\hbar}\,\mathrm{d}x\\ &= C^*C2\pi\hbar\delta(p - p').\end{aligned}$$

This equation shows that

$$C^*C = \frac{1}{2\pi\hbar}$$

$$|C|^2 = \frac{1}{2\pi\hbar} = \frac{1}{h},$$

or

$$C = \frac{1}{\sqrt{h}},$$

the physically unimportant phase factor being taken as unity. The wave function for a free particle is then given by

$$\psi_p(x) = \frac{1}{\sqrt{h}} e^{ipx/\hbar}. \tag{3.19}$$

This completes the discussion on the normalisation of a free particle. We have described the various techniques for the 'normalisation' of a free particle and one of them may be used under appropriate circumstances.

The above analysis shows that when the classical motion of a particle is unbounded, its eigenvalue spectrum is continuous. Owing to the fact that the eigenfunctions do not vanish at large distances, a different normalisation procedure is to be adopted.

Additional problems

3.5. Show that for a free particle the eigenfunctions corresponding to different eigenvalues are orthogonal to each other.

3.6. Show that the eigenfunctions e^{ikx} and e^{-ikx} of a free particle Hamiltonian correspond to the same energy $E = \hbar^2k^2/2m$ and are also the eigenfunctions of the momentum operator but correspond to two different eigenvalues $\hbar k$ and $-\hbar k$.

Chapter 4

Linear harmonic oscillator

The importance of harmonic oscillations dates back to 1581 when, more than four centuries ago, Galileo discovered in a cathedral in Pisa that the time period of a large chandelier was independent of its amplitude. The importance of a harmonic oscillator is still realised today but from a different perspective. It has many applications in different disciplines of physics and in various physical situations. This is because for sufficiently small amplitudes, the vibrations of many physical systems, for instance those of the atoms of crystalline solids, closely approximate simple harmonic vibrations. The solution for linear harmonic oscillators has therefore been applied to understand the behaviour of black body radiation, the structure of nuclei and many other phenomena. Moreover, the discrete energy levels and the uncertainty principle are built into it. No doubt, it is the queen of the quantum mechanical conceptual domain. But let us first discuss its classical theory. We will then analyse its behaviour quantum mechanically and demonstrate its importance. The simple and logical development of the subject with appropriate emphasis on mathematical techniques as well as the concepts introduced, helps the reader not only to follow but also enjoy the subject. The energy spectrum of this classically bound system has discrete eigenvalues.

4.1 Classical theory

According to classical mechanics, a linear harmonic oscillator is a particle of mass m vibrating under the action of a force F such that

$$F = -kx, \tag{4.1}$$

where x is the displacement of the particle from its equilibrium position and k is a constant. Since $F = m\, \mathrm{d}^2x/\mathrm{d}t^2$, the above equation of motion can be written as

$$\frac{\mathrm{d}^2x}{\mathrm{d}t^2} = -\frac{kx}{m}.$$

doi:10.1088/978-0-7503-1206-6ch4

If we impose the condition that $x = 0$ for $\omega t = \pi/2$, where the angular frequency $\omega = 2\pi\nu = \sqrt{k/m}$, ν being the frequency of the oscillations, then the solution of this differential equation is

$$x = a\cos\sqrt{\frac{k}{m}}\,t = a\cos\,\omega t. \tag{4.2}$$

This represents an oscillatory motion of amplitude a with angular frequency ω. If V is the potential energy of the harmonic oscillator, we must have

$$F = -\frac{\mathrm{d}V}{\mathrm{d}x}. \tag{4.3}$$

From (4.1) and (4.3), we obtain

$$\frac{\mathrm{d}V}{\mathrm{d}x} = kx.$$

Integrating with respect to x and taking the constant of integration as zero, we obtain

$$V = \frac{1}{2}kx^2.$$

The kinetic energy of the oscillator is given by

$$T = \frac{1}{2}m\dot{x}^2, \quad \text{where } \dot{x} = \frac{\mathrm{d}x}{\mathrm{d}t}.$$

The total energy of the linear harmonic oscillator is therefore given by

$$E = T + V = \frac{1}{2}m\dot{x}^2 + \frac{1}{2}kx^2 = \frac{1}{2}\left(m\dot{x}^2 + kx^2\right). \tag{4.4}$$

Differentiating equation (4.2) with respect to t and squaring, we find

$$\dot{x}^2 = a^2\omega^2\sin^2\omega t. \tag{4.5}$$

Substituting the expressions for x^2 and $\dot{x}^2$ from (4.2) and (4.5) in (4.4) and simplifying, we obtain

$$E = \frac{1}{2}ka^2.$$

The expression for E shows that it can have continuous values as the amplitude a can take a continuous range of values.

We will now show that quantum mechanically, a solution of the Schrödinger wave equation for a linear harmonic oscillator will be well-behaved only if the oscillator has discrete energy values.

4.2 Quantum theory

The potential energy $V = \frac{1}{2}kx^2$ of a linear harmonic oscillator does not depend upon time explicitly. Therefore, while considering the problem of a harmonic oscillator, it is sufficient to solve the corresponding Schrödinger amplitude equation, obtain its solution $\psi(x)$, multiply it by $\phi(t) = \exp(-\mathrm{i}Et/\hbar)$ and obtain $\Psi(x, t)$ which is the wave function representing the harmonic oscillator.

The Schrödinger amplitude equation in one dimension is

$$\frac{\mathrm{d}^2\psi}{\mathrm{d}x^2} + \frac{2m}{\hbar^2}(E - V)\psi = 0,$$

where for convenience we have written ψ for $\psi(x)$. Substituting the expression $\frac{1}{2}\,kx^2$ for the potential energy V of the harmonic oscillator in the above equation, we obtain

$$\frac{\mathrm{d}^2\psi}{\mathrm{d}x^2} + \frac{2m}{\hbar^2}\left(E - \frac{1}{2}kx^2\right)\psi = 0. \tag{4.6a}$$

This is Schrödinger's amplitude equation for a linear harmonic oscillator. Writing

$$\frac{2m}{\hbar^2}E = \beta \tag{4.6b}$$

and

$$\frac{\sqrt{mk}}{\hbar} = \alpha, \tag{4.6c}$$

this reduces to

$$\frac{\mathrm{d}^2\psi}{\mathrm{d}x^2} + \left(\beta - \alpha^2x^2\right)\psi = 0. \tag{4.7}$$

To simplify it further, let $\sqrt{\alpha}\,x = z$. Then (4.7) becomes:

$$\alpha\frac{\mathrm{d}^2\psi}{\mathrm{d}z^2} + \left(\beta - \alpha z^2\right)\psi = 0,$$

$$\frac{\mathrm{d}^2\psi}{\mathrm{d}z^2} + \left(\frac{\beta}{\alpha} - z^2\right)\psi = 0,$$

or

$$\frac{\mathrm{d}^2\psi}{\mathrm{d}z^2} + (\lambda - z^2)\psi = 0, \quad \lambda = \frac{\beta}{\alpha} = 2\sqrt{\frac{m}{k}}\frac{E}{\hbar}. \tag{4.8}$$

This is known as Weber's equation: this is merely the Schrödinger amplitude equation in a simplified form. It is relatively difficult to solve this differential equation as it stands. We therefore proceed in two steps. *First*, we find a solution of this differential equation for large values of $|z|$. This is called the *asymptotic solution*. *Then* we go ahead by taking the product of the asymptotic solution and some function of z as the general solution of the differential equation.

4.3 The asymptotic solution

To obtain the asymptotic solution, i.e. a solution for large $|z|$, we neglect the constant λ as compared to z^2. Then (4.8) reduces to

$$\frac{\mathrm{d}^2\psi}{\mathrm{d}z^2} - z^2\psi = 0. \tag{4.9}$$

We will now show that

$$\psi = A\exp\left(-\frac{z^2}{2}\right) + B\exp\left(\frac{z^2}{2}\right) \tag{4.10}$$

is the general solution of this differential equation. For this purpose, we differentiate (4.10) with respect to z and obtain

$$\frac{\mathrm{d}\psi}{\mathrm{d}z} = -zA\exp\left(-\frac{z^2}{2}\right) + zB\exp\left(\frac{z^2}{2}\right).$$

Differentiating again with respect to z, we obtain

$$\begin{aligned}\frac{\mathrm{d}^2\psi}{\mathrm{d}z^2} &= z^2A\exp\left(-\frac{z^2}{2}\right) - A\exp\left(-\frac{z^2}{2}\right) + z^2B\exp\left(\frac{z^2}{2}\right) + B\exp\left(\frac{z^2}{2}\right)\\ &= (z^2-1)A\exp\left(-\frac{z^2}{2}\right) + (z^2+1)B\exp\left(\frac{z^2}{2}\right).\end{aligned}$$

For large $|z|$, by neglecting 1 as compared to z^2, this differential equation reduces to

$$\frac{\mathrm{d}^2\psi}{\mathrm{d}z^2} = z^2A\exp\left(-\frac{z^2}{2}\right) + z^2B\exp\left(\frac{z^2}{2}\right) = z^2\psi,$$

where the differential equation has been obtained by using (4.10). This shows that the function ψ as given by (4.10) is a solution of differential equation (4.9) and therefore an asymptotic solution of differential equation (4.8). Is this mathematical solution physically acceptable? In other words, is this wave function ψ well-behaved? To answer this question, we note that the function $\exp(z^2/2)$ tends to infinity as $|z| \to \infty$. Thus if it is retained in the expression for ψ, the latter will not be a well-behaved function and therefore will not be physically acceptable. The other function $\exp(-z^2/2)$, however, tends to zero when $|z| \to \infty$ and will therefore lead to a physically acceptable solution. Thus, the physically acceptable asymptotic solution of differential equation (4.8) is

$$\psi = A\exp\left(-\frac{z^2}{2}\right). \tag{4.11}$$

4.4 The general solution

The asymptotic solution suggests that we may seek the general solution of differential equation (4.8) by writing it in the form

$$\psi = A\exp\left(-\frac{z^2}{2}\right)H(z), \tag{4.12}$$

where $H(z)$ is some function of z. To determine $H(z)$, we proceed as follows. Differentiating (4.12) once and twice with respect to z, we obtain

$$\frac{\mathrm{d}\psi}{\mathrm{d}z} = -zA\exp\left(-\frac{z^2}{2}\right)H(z) + A\exp\left(-\frac{z^2}{2}\right)H'(z),$$

where $H'(z)$ stands for $\mathrm{d}H(z)/\mathrm{d}z$ and

$$\begin{aligned}\frac{\mathrm{d}^2\psi}{\mathrm{d}z^2} = &-A\exp\left(-\frac{z^2}{2}\right)H(z) + z^2A\exp\left(-\frac{z^2}{2}\right)H(z) - zA\exp\left(-\frac{z^2}{2}\right)H'(z)\\ &- zA\exp\left(-\frac{z^2}{2}\right)H'(z) + A\exp\left(-\frac{z^2}{2}\right)H''(z),\end{aligned}$$

where $H''(z)$ stands for $\mathrm{d}^2H/\mathrm{d}z^2$. Substituting the expressions for ψ and $\mathrm{d}^2\psi/\mathrm{d}z^2$ in differential equation (4.8) and simplifying, we obtain

$$H''(z) - 2zH'(z) + (\lambda - 1)H(z) = 0. \tag{4.13}$$

By solving this second-order ordinary differential equation, we obtain the function $H(z)$. By substituting this function in (4.12), we obtain the solution ψ of differential equation (4.8) and hence can write the function $\Psi(x, t)$ representing the linear harmonic oscillator.

In order to solve differential equation (4.13), we suppose that $H(z)$ can be expressed as a power series in z:

$$H(z) = a_0 + a_1z + a_2z^2 + a_3z^3 + a_4z^4 + \cdots.$$

To determine the constants, i.e. a, we obtain the expressions for H' and H'' (in further analysis, we will interchangeably retain or drop the argument z of H and its derivatives) by differentiating H once and twice, respectively. We have

$$H' \equiv \frac{\mathrm{d}H}{\mathrm{d}z} = a_1 + 2a_2z + 3a_3z^2 + 4a_4z^3 + \cdots$$

and

$$H'' \equiv \frac{d^2H}{dz^2} = 2.1a_2 + 3.2a_3z + 4.3a_4z^2 + \cdots.$$

Substituting the expressions for H, H' and H'' in differential equation (4.13), we obtain

$$\left(2.1a_2 + 3.2a_3z + 4.3a_4z^2 + \cdots\right) - 2z\left(a_1 + 2a_2z + 3a_3z^2 + \cdots\right)$$
$$+ (\lambda - 1)\left(a_0 + a_1z + a_2z^2 + \cdots\right) = 0,$$

or

$$\left[2.1a_2 + (\lambda - 1)a_0\right] + \left[3.2a_3 - 2a_1 + (\lambda - 1)a_1\right]z$$
$$+ \left[4.3a_4 - 4a_2 + (\lambda - 1)a_2\right]z^2 + \cdots = 0,$$

or

$$\left[2.1a_2 + (\lambda - 1)a_0\right] + \left[3.2a_3 + (\lambda - 3)a_1\right]z$$
$$+ \left[4.3a_4 + (\lambda - 5)a_2\right]z^2 + \cdots = 0.$$

Since this equation must hold for all values of z, the coefficient of each power of z must vanish separately. Thus, we must have

$$2.1a_2 + (\lambda - 1)a_0 = 0,$$
$$3.2a_3 + (\lambda - 3)a_1 = 0,$$
$$4.3a_4 + (\lambda - 5)a_2 = 0,$$

and so on. Consequently, for the rth power of z, we have

$$(r + 2)(r + 1)a_{r+2} + [\lambda - (2r + 1)]a_r = 0,$$

or

$$a_{r+2} = -\frac{\lambda - (2r + 1)}{(r + 2)(r + 1)}ar,$$
$$r = 0, 1, 2, \ldots \tag{4.14}$$

Such a formula is known as a *recursion formula*. It enables us to calculate $a_2, a_4, a_6, \ldots$ in terms of a_0, and $a_3, a_5, a_7, \ldots$ in terms of a_1. The coefficients a_0 and a_1 are arbitrary. This is what it must be. The general solution of a second-order ordinary differential equation must contain two arbitrary constants. The recursion formula shows that a with even and odd indices are not related. The series $H(z)$ can therefore be written as a sum of two independent series:

$$H(z) = \left(a_0 + a_2z^2 + a_4z^4 + \cdots\right) + \left(a_1z + a_3z^3 + a_5z^5 + \cdots\right)$$
$$= f(z) + g(z),$$

where $f(z)$ is an even while $g(z)$ is an odd function of z:

$$f(z) = a_0 + a_2z^2 + a_4z^4 + \cdots,$$

$$g(z) = a_1z + a_3z^3 + a_5z^5 + \cdots.$$

Equation (4.12) therefore yields

$$\psi = A\exp\left(-\frac{z^2}{2}\right)[f(z) + g(z)]. \tag{4.15}$$

We have thus found a solution of differential equation (4.8). Since the constant A can be absorbed in $f(z)$ as well as $g(z)$, there remain only two arbitrary constants.

4.5 A physically acceptable solution

But is this solution ψ, mathematically representing the harmonic oscillator, physically acceptable? This is possible only if the function ψ is well-behaved, say for instance, it does not diverge for any value of z. To find out the answer to this question, we will examine the behaviour of this function ψ for $|z| \to \infty$. We proceed step by step. We first compare the infinite series $f(z)$ with the series expansion of $\exp(z^2)$ for large $|z|$. We have

$$f(z) = a_0 + a_2z^2 + a_4z^4 + \cdots + a_rz^r + a_{r+2}z^{r+2} + \cdots.$$

By virtue of the recursion formula (4.14), the ratio of successive terms of $f(z)$ is

$$\frac{a_{r+2}z^{r+2}}{a_rz^r} = -\frac{\lambda - (2r+1)}{(r+2)(r+1)}z^2. \tag{4.16}$$

For large values of r, this ratio reduces to $2z^2/r$. The exponential function, $\exp(z^2)$, can also be expanded as a power series:

$$\exp(z^2) = 1 + z^2 + \frac{z^4}{2!} + \frac{z^6}{3!} + \frac{z^8}{4!} + \cdots + \frac{z^r}{\left(\frac{r}{2}\right)!} + \frac{z^{r+2}}{\left(\frac{r}{2}+1\right)!}.$$

The ratio of successive terms in the series expansion of this function is

$$\left[\frac{z^{r+2}}{\left(\frac{r}{2}+1\right)!} \bigg/ \frac{z^r}{\left(\frac{r}{2}\right)!}\right] = \frac{z^2}{\left(\frac{r}{2}+1\right)}.$$

For large values of r, this also reduces to $2z^2/r$. Thus, for *large* r, the ratio of successive terms in the two series for $f(z)$ and $\exp(z^2)$ is the same. Therefore, for *large* $|z|$, for which the role of the terms corresponding to small r is insignificant, the two series converge or diverge together. But $\exp(z^2)$ diverges for large $|z|$.

Hence, $f(z)$ should also diverge for large $|z|$. Similarly, it can be shown that $g(z)$ diverges for large $|z|$. Thus, the solution

$$\psi = A\exp\left(-\frac{z^2}{2}\right)[f(z) + g(z)]$$

turns out to behave like $\exp(z^2/2)$ and is consequently divergent for large $|z|$. Hence the function ψ as given by (4.15) is not well-behaved and is consequently not a physically acceptable solution of the Schrödinger wave equation for a linear harmonic oscillator.

Let us summarise what we have done up to this point. In order to obtain a physically acceptable solution of the Schrödinger amplitude equation for a linear harmonic oscillator, we first obtained a physically acceptable asymptotic solution of this differential equation and then using it as a basis obtained a general solution. However, this general mathematical solution is not physically acceptable; the solution diverges as $|z|$ goes to infinity.

How then can the function ψ be made well-behaved? That is, what steps should be taken so that it remains finite throughout the region of z extending from $-\infty$ to $+\infty$? This can be achieved in the following two ways.

The *first* method is:

(i) Choose $a_1 = 0$, then all a with odd indices will be zero and the series $g(z)$ will vanish so that $\{\exp(-z^2/2)\}g(z)$, one of the two solutions for the harmonic oscillator disappears. The function ψ as given in (4.12) then reduces to

$$\psi = A\left\{\exp\left(-\frac{z^2}{2}\right)\right\}f(z).$$

(ii) Next suppose that for some value of r, say n, where n is zero or an even integer, the quantity $\lambda - (2n + 1)$ involving the energy of the oscillator, and occurring in the numerator of the recursion formula (4.16), becomes zero:

$$\lambda = 2n + 1.$$

The recursion formula tells us that *if this condition is satisfied*, the constant a_{n+2} and consequently all the following constants in the series $f(z)$ will vanish. As a result of this, the infinite series $f(z)$ will reduce to a polynomial of degree n. We denote this polynomial by $f_n(z)$. Equation (4.15) then becomes

$$\psi = A\exp\left(-\frac{z^2}{2}\right)f_n(z).$$

It is evident that for large $|z|$, the exponential function, $\exp(-z^2/2)$, will decrease much more rapidly as compared to the increase in the polynomial $f_n(z)$; consequently, the behaviour of the function ψ will be completely dominated by the function $\exp(-z^2/2)$. Hence the function ψ which has been obtained by imposing the conditions $a_1 = 0$ and $\lambda - (2n + 1) = 0$, $n = 0, 2, 4, \ldots$ will go to zero as $|z| \to \infty$ and is therefore a physically acceptable solution.

The *second* method is exactly similar to the first one. Proceeding on the same lines, it can be shown that

$$\psi = A\exp\left(-\frac{z^2}{2}\right)g_n(z)$$

is also a physically acceptable solution of the Schrödinger wave equation for the harmonic oscillator provided that $\lambda = 2n + 1$, $n = 1, 3, 5, \ldots$.

The above analysis shows that the function

$$\psi = A\exp\left(-\frac{z^2}{2}\right)H_n(z),$$

where

$$H_n(z) = f_n(z), \quad \text{for even } n,$$
$$= g_n(z), \quad \text{for odd } n,$$

is a physically acceptable solution of the Schrödinger wave equation for a harmonic oscillator provided that

$$\lambda = 2n + 1, \quad n = 0, 1, 2, 3, \ldots$$

To emphasise the correspondence between ψ and n, we write ψ_n for ψ:

$$\psi_n = A\exp\left(-\frac{z^2}{2}\right)H_n(z), \quad n = 0, 1, 2, 3, \ldots \tag{4.17a}$$

The eigenfunctions ψ_n described by the above equation tend to zero sufficiently rapidly as z goes to infinity. These are solutions of the Schrödinger wave equation for the linear harmonic oscillator and represent the physical states of the system. The states described by such eigenfunctions are called *bound states.*

4.6 Energy eigenvalues

Let us next determine the energy eigenvalues of a linear harmonic oscillator. It has been shown that the Schrödinger amplitude equation for a harmonic oscillator can have a physically acceptable solution only if

$$2n + 1 = \lambda, \quad n = 0, 1, 2, 3, \ldots$$

$$= \frac{\beta}{\alpha} = \left[\frac{2mE}{\hbar^2}\Big/\frac{2\pi m\nu}{\hbar}\right] = \frac{E}{\pi\hbar\nu} = \frac{2E}{h\nu},$$

or

$$E = (2n + 1)\frac{h\nu}{2}, \quad n = 0, 1, 2, 3, \ldots \tag{4.17b}$$

To emphasise the correspondence between E and n, we will use E_n for E. Hence, we write

$$E_n = (2n + 1)\frac{h\nu}{2}, \quad n = 0, 1, 2, 3, \ldots \tag{4.17c}$$

This shows that, quantum mechanically, the existence of *physically acceptable solutions* for a harmonic oscillator requires it to possess a *discrete energy spectrum.* The energy values of a harmonic oscillator in the nth and $(n + 1)$th states are given by

$$E_n = (2n + 1)\frac{h\nu}{2},$$

$$E_{n+1} = (2n + 3)\frac{h\nu}{2}.$$

Therefore the difference in adjacent energy levels is

$$E_{n+1} - E_n = h\nu.$$

This shows that all adjacent levels of the discrete energy spectrum of a linear harmonic oscillator are equally separated. This is exhibited in figure 4.1. The discrete energy spectrum is observed at the microscopic level but not at the macroscopic level where the energy spectrum is continuous. Why is this so? Is quantum mechanics not valid at the macroscopic level? This is not the case. In fact, for large values of n when the macroscopic level is approached, the energies of the states become so high that the tiny constant energy difference $h\nu$ between the adjacent states becomes negligible and hence the energy spectrum appears continuous. This is in accordance with the

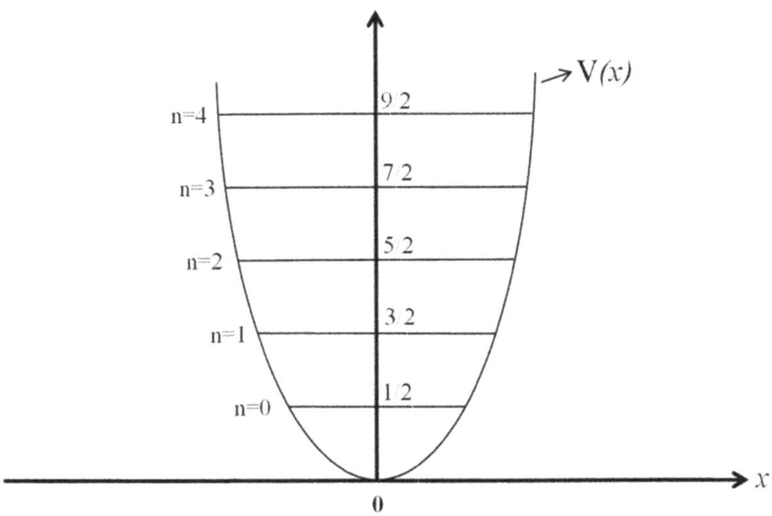

Figure 4.1. Energy levels for harmonic oscillator.

correspondence principle which states that for $\hbar \to 0$ quantum mechanics reduces to classical mechanics.

It can be mentioned that the absence of a continuous energy spectrum and equal spacing of discrete levels are special features of a harmonic oscillator.

As the angular frequency ω is equal to $2\pi\nu$, (4.17*c*) can also be written as

$$E_n = (2n + 1)\frac{\hbar\omega}{2}. \tag{4.18}$$

For the lowest energy of the linear harmonic oscillator, called the ground state, $n = 0$ and therefore (4.17*c*) yields

$$E_0 = \frac{h\nu}{2}.$$

This is called the *zero-point energy* of the harmonic oscillator and is a consequence of the uncertainty principle. This means that even in the ground state, a harmonic oscillator is not at rest; it is always in oscillatory motion. Actually, a harmonic oscillator in thermodynamic equilibrium with its environment at a temperature approaching absolute zero would be in the ground state. In classical mechanics, the lowest energy state of a harmonic oscillator occurs when it is at rest in its equilibrium position. That is, when its position and momentum are simultaneously zero. In quantum mechanics, the Heisenberg uncertainty principle does not allow this configuration, as position and momentum cannot be measured precisely at the same time. Therefore, a quantum mechanically harmonic oscillator vibrates, with zero-point energy, even in its ground state. This result plays an important role in the theory of metals.

Equations (4.17*b*) and (4.17*a*) show that both the eigenvalues and the eigenfunctions are associated with only one quantum number n. Therefore, for each eigenvalue, there is only one corresponding eigenfunction, i.e. two or more linearly independent eigenfunctions do not correspond to the same eigenvalue; the eigenvalues are *not* degenerate. This property seems to be a common characteristic of bound states for a one-dimensional potential that remains finite for all finite values of the space coordinate x.

4.7 Hermite polynomials

The polynomials $H_n(z)$ which occur in the expression for the wave function $\psi_n(x)$ representing the harmonic oscillator and are given in (4.17*a*) are called the *Hermite polynomials* of nth degree. They were known to mathematicians long before the advent of quantum mechanics and are also defined by the equation

$$\sum_{n=0}^{\infty} \frac{H_n(z)}{n!} s^n = \exp\left[z^2 - (s - z)^2\right] = \exp\left(2zs - s^2\right) \tag{4.19}$$

where s and z are independent variables. The function $\exp[z^2 - (s - z)^2]$ is said to be the *generating function* for the Hermite polynomials.

Problem

4.1. Show that the Hermite polynomials defined by their generating function satisfy differential equation (4.13).

The first few Hermite polynomials of degrees 0, 1, 2, 3, 4 and 5 are

$$H_0(z) = 1$$
$$H_1(z) = 2z,$$
$$H_2(z) = 4z^2 - 2,$$
$$H_3(z) = 8z^3 - 12z,$$
$$H_4(z) = 16z^4 - 48z^2 + 12,$$
$$H_5(z) = 32z^5 - 160z^3 + 120z.$$

4.8 The normalisation process

We have seen that in the case of a linear harmonic oscillator, the solution of the Schrödinger amplitude equation is given by

$$\psi_n(x) = A\exp\left(-\frac{z^2}{2}\right)H_n(z), \quad z = \sqrt{\alpha}x, \tag{4.17a}$$

or

$$\psi_n(x) = A\exp\left(-\frac{\alpha x^2}{2}\right)H_n(\sqrt{\alpha}x). \tag{4.20}$$

To find the expression for A, we make use of the normalisation condition. This is

$$\int_{-\infty}^{\infty} \psi_n^*(x)\psi_n(x)\,\mathrm{d}x = 1.$$

Substituting the expression for ψ_n from (4.20) in the above equation, we obtain

$$\int_{-\infty}^{\infty} A^*\exp\left(-\frac{\alpha x^2}{2}\right)H_n(\sqrt{\alpha}x)A\exp\left(-\frac{\alpha x^2}{2}\right)H_n(\sqrt{\alpha}x)\,\mathrm{d}x = 1,$$

or

$$|A|^2\int_{-\infty}^{\infty} \exp(-z^2)H_n^2(z)\frac{1}{\sqrt{\alpha}}\,\mathrm{d}z = 1,$$

or

$$\int_{-\infty}^{\infty} \exp(-z^2) H_n^2(z) \mathrm{d}z = \frac{\sqrt{\alpha}}{A^2}, \tag{4.21}$$

where the physically uninteresting phase of A has been taken as zero. We will evaluate this integral with the help of generating functions $S(s, z)$ and $T(t, z)$ defined by the equations

$$S(s, z) = \sum_{n=0}^{\infty} \frac{H_n(z)}{n!} s^n = \exp[z^2 - (s - z)^2], \tag{4.22}$$

$$T(s, z) = \sum_{m=0}^{\infty} \frac{H_m(z)}{m!} t^m = \exp[z^2 - (t - z)^2]. \tag{4.23}$$

For this purpose, let us consider the integral

$$I = \int_{-\infty}^{\infty} S(s, z) T(t, z) \exp(-z^2) \, \mathrm{d}z. \tag{4.24}$$

Substituting the expressions for $S(s, z)$ and $T(t, z)$ from (4.22) and (4.23) in relation (4.24), we obtain

$$\begin{aligned} I &= \int_{-\infty}^{\infty} \sum_{n=0}^{\infty} \frac{H_n(z)}{n!} s^n \sum_{m=0}^{\infty} \frac{H_m(z)}{m!} t^m \exp(-z^2) \, \mathrm{d}z \\ &= \sum_{n=0}^{\infty} \sum_{m=0}^{\infty} \frac{s^n t^m}{n! m!} \int_{-\infty}^{\infty} H_n(z) H_m(z) \exp(-z^2) \, \mathrm{d}z. \end{aligned} \tag{4.25}$$

By making use of the generating functions, we can write the integral I as

$$\begin{aligned} I &= \int_{-\infty}^{\infty} \exp\left[z^2 - (s - z)^2\right] \exp\left[z^2 - (t - z)^2\right] \exp(-z^2) \, \mathrm{d}z \\ &= \exp(2st) \int_{-\infty}^{\infty} \exp\left[-(z - s - t)^2\right] \mathrm{d}z \\ &= \exp(2st) \int_{-\infty}^{\infty} \exp\left[-(z - s - t)^2\right] \mathrm{d}(z - s - t). \end{aligned} \tag{4.26}$$

The integral on the right-hand side is of the form

$$\int_{-\infty}^{\infty} [\exp(-u^2)] \, \mathrm{d}u$$

and is therefore equal to $\sqrt{\pi}$. Equation (4.26) then yields

$$I = [\exp(2st)] \sqrt{\pi}.$$

By expanding $\exp(2st)$, the above equation gives

$$I = \sqrt{\pi}\left[1 + \frac{2st}{1!} + \frac{(2st)^2}{2!} + \cdots + \frac{(2st)^n}{n!} + \cdots\right]. \tag{4.27}$$

From (4.25) and (4.27), we find that as the left-hand sides are the same, the right-hand sides must be equal. Thus, we have

$$\sum_{n=0}^{\infty}\sum_{m=0}^{\infty}\frac{s^n t^m}{n!m!}\int_{-\infty}^{\infty} H_n(z)H_m(z)\exp(-z^2)\,\mathrm{d}z$$
$$= \sqrt{\pi}\left[1 + \frac{2st}{1!} + \frac{(2st)^2}{2!} + \cdots + \frac{(2st)^n}{n!} + \cdots\right]. \tag{4.28}$$

Comparing the coefficients of $s^n t^n$ on the two sides of this equation, we obtain

$$\int_{-\infty}^{\infty} \exp(-z^2)H_n^2(z)\,\mathrm{d}z = \sqrt{\pi}2^n n!.$$

Substituting this value of the integral in (4.21) and simplifying, we obtain

$$A_n = \sqrt{\frac{\sqrt{\alpha}}{\sqrt{\pi}\;2^n\;n!}}. \tag{4.29}$$

where in order to emphasise the dependence of A on n, we have written A_n for A. The normalised solution $\psi_n(x)$ of the Schrödinger amplitude equation for the harmonic oscillator can therefore be written as

$$\psi_n(x) = \sqrt{\frac{\sqrt{\alpha}}{\sqrt{\pi}\;2^n\;n!}}\exp\left(-\frac{\alpha x^2}{2}\right)H_n\left(\sqrt{\alpha}\,x\right). \tag{4.30}$$

The wave function for the ground state of the harmonic oscillator is obtained by putting $n = 0$. This yields

$$\Psi_0(x) = \left(\frac{\alpha}{\pi}\right)^{1/4}\exp\left(-\frac{\alpha x^2}{2}\right)H_0\left(\sqrt{\alpha}\,x\right).$$

Since the Hermite polynomial of zero degree is unity, the above equation reduces to

$$\Psi_0(x) = \left(\frac{\alpha}{\pi}\right)^{1/4}\exp\left(-\frac{\alpha x^2}{2}\right).$$

This is the normalised eigenfunction of a linear harmonic oscillator in the ground state. Similarly, it can be shown that the eigenfunction for the first excited state is given by

$$\Psi_1(x) = \left(\frac{\alpha}{\pi}\right)^{1/4}\frac{1}{\sqrt{2}}\exp\left(-\frac{\alpha x^2}{2}\right)H_1\left(\sqrt{\alpha}\,x\right).$$

Writing $\sqrt{\alpha}x = z$, we may write equation (4.30) as

$$\psi_n(z) = \sqrt{\frac{\sqrt{\alpha}}{\sqrt{\pi}\,2^n\,n!}}\exp\left(-\frac{z^2}{2}\right)H_n(z).$$

Problem

4.2. The wave function $\psi_0(x, t)$ corresponding to the ground state of a linear harmonic oscillator is given by

$$\Psi_0(x, t) = \left(\frac{\alpha}{\pi}\right)^{1/4}\exp\left(-\frac{\alpha x^2}{2}\right)\exp(-\mathrm{i}E_0 t/\hbar), \quad \alpha = \frac{\sqrt{mk}}{\hbar}.$$

Evaluate $\langle x\rangle$, $\langle p_x\rangle$, $\langle x^2\rangle$, $\langle p_x^2\rangle$ and hence show that

$$\Delta x\,\Delta p_x = \frac{\hbar}{2}.$$

(It may be assumed that

$$\int_{-\infty}^{\infty} \mathrm{e}^{-\alpha x^2}\,\mathrm{d}x = \sqrt{\frac{\pi}{\alpha}},$$

$$\int_{-\infty}^{\infty} x^2\mathrm{e}^{-\alpha x^2}\,\mathrm{d}x = \sqrt{\frac{\pi}{4\alpha^3}},$$

$$\int_{-\infty}^{\infty} f(x)\,\mathrm{d}x = 0$$

where $f(x)$ is an odd function of x.)

4.9 Probability distributions

Figures 4.2(*a*)–(*c*) exhibit the probability distributions plotted against the displacement. The solid and broken curves in each figure give the probability distribution for quantum mechanical and classical oscillators with the same energy. The vertical lines indicate the limits of the classical motion in each case. We know that classically the velocity of a harmonic oscillator is maximum at the position of equilibrium and reduces to zero as turning points are approached. If the location of the particle is noted *at random*, it will have a greater chance of being observed in intervals where the velocity is relatively smaller. It is less likely to be observed at places where it is moving with greater speeds. The harmonic oscillator will therefore have the minimum chance of being observed in the vicinity of its two extremities.

A comparison of quantum mechanical and classical probability curves shows that for small values of n there is little correspondence between them. For instance, for $n = 0$, the probability density in the ground state, $\psi_0^*\,\psi_0$, has a maximum at $x = 0$ while the classical probability distribution has a minimum at that point. However, as the quantum number n increases in value, the average quantum mechanical probability approaches the classical value. This is an example of the *correspondence principle* which states that for large quantum numbers, quantum mechanics reduces to classical mechanics.

The following points can be noted.

1. The probability density distribution as exhibited in figure 4.2 shows that even at high values of n, there are many values of x where the oscillating particle

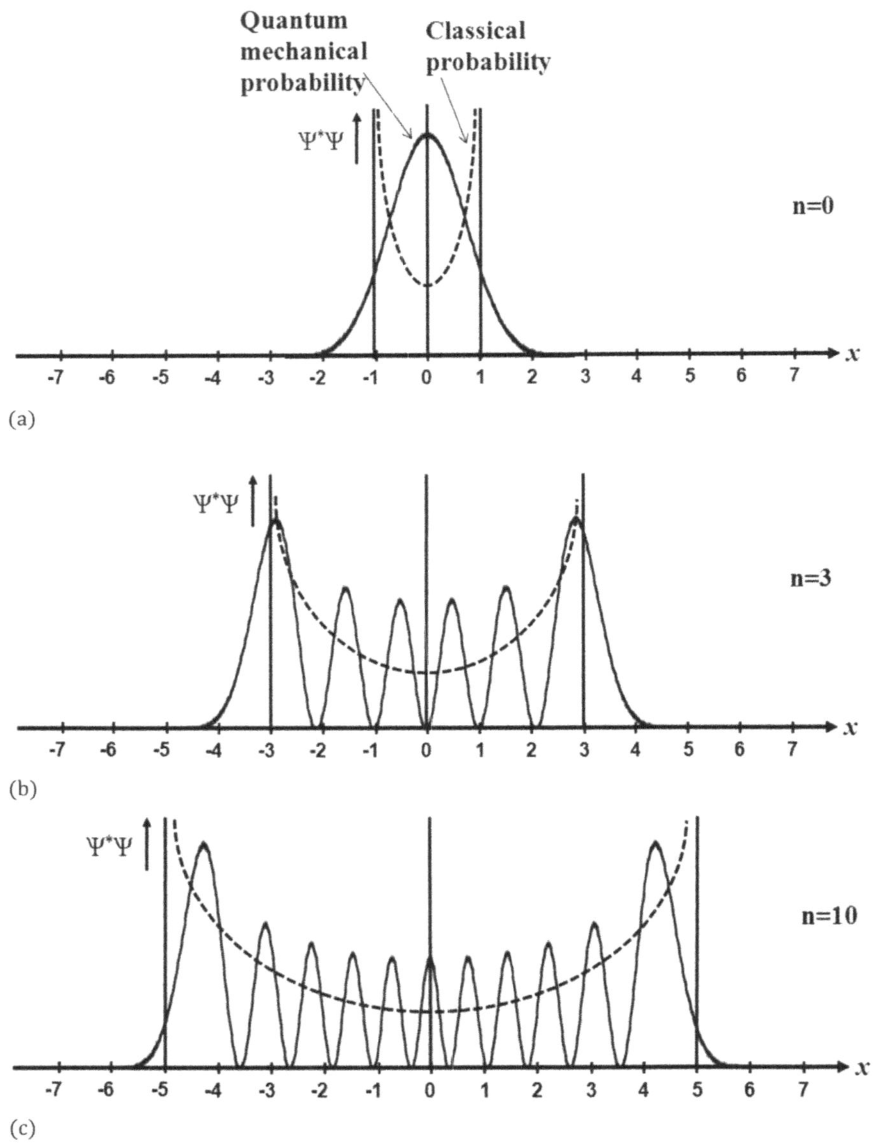

Figure 4.2. Probability distribution functions for a harmonic oscillator plotted against x.

will *never be found*. This means that quantum mechanically, in such a situation the linear harmonic oscillator will be jumping from one location to another, without ever passing through the intermediate location where $\Psi^*\Psi = 0$. It is this quantum jumping with which Schrödinger and Einstein could never reconcile themselves.

2. According to classical theory, a harmonic oscillator *cannot* reach a distance greater than its amplitude which is its maximum range. Quantum mechanically, there is a finite, although small, probability of finding the oscillator even beyond the classical limits.

4.10 The importance of the harmonic oscillator

The harmonic oscillator problem is of great importance for the following reasons:

1. A quantised radiation field behaves as if it were a collection of an infinite number of harmonic oscillators.
2. The motion of an arbitrary physical system executing small oscillations about a position of stable equilibrium approximates to that of a harmonic oscillator.

To prove the second characteristic, let $V(x)$ be the potential of a system executing small oscillations about a position of stable equilibrium, $x = a$. Then the potential energy is minimum at this point. Let us expand $V(x)$ in a Taylor series about the point $x = a$:

$$V(x) = V(a) + (x - a)V(a) + \frac{(x - a)^2}{2!}V'(a) + \cdots, \tag{4.58}$$

where

$$V(a) = (\mathrm{d}V/\mathrm{d}x)x = a, \text{ etc.}$$

The potential energy of the system at $x = a$, i.e. $V(a)$, may be taken as zero of the energy scale. Then $V(a) = 0$ and (4.58) reduces to

$$V(x) = (x - a)V(a) + \frac{(x - a)^2}{2!}V'(a) + \cdots. \tag{4.59}$$

Since the potential energy of the system is minimum at $x = a$, we must have $V'(a) = 0$ and $V'(a) \geqslant 0$, so that (4.59) becomes

$$V(x) = \frac{(x - a)^2}{2!}V'(a) + \cdots. \tag{4.60}$$

Moreover, as the oscillations of the system about the point $x = a$ are small, higher order terms can be neglected provided $V''(a) > 0$. Then, we can write

$$V(x) = \frac{(x - a)^2}{2!}V'(a). \tag{4.61}$$

If point a is chosen as the origin of the coordinate system, we have

$$V(x) = \frac{1}{2}x^2V'(0),$$

or

$$V(x) = \frac{1}{2}kx^2, \quad k = V''(0). \tag{4.62}$$

Hence, except for $k = 0$, the motion is that of a harmonic oscillator.

The vibrational spectra manifested for instance by molecules and nuclei can be interpreted by assuming equal spacing of energy levels, a characteristic of harmonic oscillations.

4.11 Parity

A function $f(x)$ is said to be an even or odd function of x if it is unaltered or changes sign by changing x to $-x$. Thus, for an even function $f(-x) = f(x)$ while for an odd function $f(-x) = -f(x)$. The parity of any state of a system is said to be even or odd according to whether the eigenfunction corresponding to that state is even or odd. We have shown (see (4.30)) that the eigenfunction of a harmonic oscillator corresponding to its nth state is given by

$$\psi_n(x) = \sqrt{\frac{\sqrt{\alpha}}{\sqrt{\pi}\ 2^n\ n!}}\ \exp\left(-\alpha x^2/2\right)H_n\left(\sqrt{\alpha}\,x\right). \tag{4.30$'$}$$

Now $\exp(-\alpha x^2/2)$ is an even function of x. Therefore, $\psi_n(x)$ will be an even or odd function of x according to whether $H_n(\sqrt{\alpha}\,x)$ is an even or odd function of x, as the product of an even function with an even (odd) function is an even (odd) function. But we have shown that $H_n(\sqrt{\alpha}\,x)$ is an even or odd function of x according to whether n is zero/even or odd. Thus, $\psi_n(x)$ will be an even or odd function of x according to whether n is zero/even or odd. Hence, an eigenstate of a harmonic oscillator has a *definite* parity; it is even for zero/even n and odd for odd n. This fact is also expressed by stating that the states labelled with zero or even values of n have positive parity while the states labelled with odd values of n have negative parity.

It may be noted that the potential energy $V(x)$ of a linear harmonic oscillator is invariant under parity: $V(x) = 1/2kx^2 = V(-x)$, the potential energy remains unchanged as x is changed to $-x$. Under such a symmetric potential, if $\psi(x)$ is a solution of the Schrödinger amplitude equation, then $\psi(-x)$ is also a solution of the amplitude equation corresponding to the same energy.

Example 4.1.

Problem

Find the probability that the motion of a linear harmonic oscillator in the ground state would be within the classical limits.

Solution

When a harmonic oscillator is in the ground state, its energy is $h\nu/2$. In this state, the wave function for the harmonic oscillator is given by

$$\psi_0(x) = \left(\frac{\alpha}{\pi}\right)^{1/4} \exp\left(-\alpha x^2/2\right), \quad \alpha = \frac{\sqrt{mk}}{\hbar}.$$

Consequently, the probability density is

$$\psi_0^*(x)\psi_0(x) = \left(\frac{\alpha}{\pi}\right)^{1/2} \exp\left(-\alpha x^2\right). \tag{4.63}$$

If a is the classical amplitude of the harmonic oscillator, then the probability of finding the particle within the classical limits $-a$ and a is

$$\begin{aligned}\int_{-a}^{a} \psi_0^*(x)\psi_0(x)\, \mathrm{d}x &= \left(\frac{\alpha}{\pi}\right)^{1/2} \int_{-a}^{a} \exp(-\alpha x^2)\, \mathrm{d}x \\ &= \left(\frac{\alpha}{\pi}\right)^{1/2} \int_{-a}^{a} \exp(-\alpha x^2)\, \mathrm{d}x \\ &= \left(\frac{\alpha}{\pi}\right)^{1/2} I,\end{aligned} \tag{4.64}$$

where

$$I = \int_{-a}^{a} \exp(-\alpha x^2)\, \mathrm{d}x. \tag{4.65}$$

Now, classically, the velocity and consequently the kinetic energy of the harmonic oscillator is zero at the classical limits a and $-a$. Therefore, for $x = a$, the total energy E is given by

$$E = V = \frac{1}{2}ka^2,$$

or

$$a^2 = \frac{2E}{k}. \tag{4.66}$$

For the ground state, the energy is given by

$$E = \frac{h\nu}{2}. \tag{4.67}$$

Substituting the value of E from (4.67) in (4.66), we obtain

$$a^2 = \frac{h\nu}{k} = \frac{1}{\alpha},$$

or

$$a = \frac{1}{\sqrt{\alpha}}. \tag{4.68}$$

But $\sqrt{\alpha}\,x = z$. This gives $\sqrt{\alpha}\,\mathrm{d}x = \mathrm{d}z$. Consequently, $z \to 1$ as $x \to a = 1/\sqrt{\alpha}$. The integral I can then be written as

$$I = \int_{-1}^{1} \mathrm{e}^{-z^2} \frac{1}{\sqrt{\alpha}} \mathrm{d}z = \frac{1}{\sqrt{\alpha}} \int_{-1}^{1} \mathrm{e}^{-z^2} \mathrm{d}z.$$

Since e^{-z^2} is an even function of z, the above integral can be written as

$$\begin{aligned}
I &= \frac{2}{\sqrt{\alpha}} \int_0^1 \mathrm{e}^{-z^2}\,\mathrm{d}z \\
&= \frac{2}{\sqrt{\alpha}} \int_0^1 \left[1 - z^2 + \frac{z^4}{2!} - \frac{z^6}{3!} + \cdots\right] \mathrm{d}z \\
&= \frac{2}{\sqrt{\alpha}} \left[z - \frac{z^3}{3} + \frac{z^5}{5}\frac{1}{2!} - \frac{z^7}{7}\frac{1}{3!} + \frac{z^9}{9}\frac{1}{4!} - \cdots\right]_0^1 \\
&= \frac{2}{\sqrt{\alpha}} \left[1 - \frac{1}{3} + \frac{1}{5}\frac{1}{2!} - \frac{1}{7}\frac{1}{3!} + \frac{1}{9}\frac{1}{4!} - \cdots\right] \\
&= \frac{2}{\sqrt{\alpha}} \left[1 - \frac{1}{3} + \frac{1}{10} - \frac{1}{42} + \frac{1}{216} - \cdots\right] \\
&= \frac{2}{\sqrt{\alpha}} [1 - 0.333 + 0.100 - 0.024 + 0.005 - \cdots] \\
&= \frac{2}{\sqrt{\alpha}} (1.105 - 0.357) \\
&= \frac{2}{\sqrt{\alpha}} \times 0.748 \\
&= \frac{1.496}{\sqrt{\alpha}}.
\end{aligned}$$

Substituting this value of I in (4.64), we obtain

$$\begin{aligned}
\int_{-a}^{a} \psi_0^*(x)\psi_0(x)\,\mathrm{d}x &= \sqrt{\frac{\alpha}{\pi}}\,I = \sqrt{\frac{\alpha}{\pi}} \cdot \frac{1.496}{\sqrt{\alpha}} \\
&= \frac{1.496}{1.772} = 0.844.
\end{aligned} \tag{4.69}$$

This gives us the probability of finding the harmonic oscillator within the classical limits.

This completes the analysis of a linear harmonic oscillator that has rightly been called the queen of quantum mechanical problems. The bounded system exhibits equally spaced energy levels in the eigenvalue spectrum. There is a finite probability of finding the oscillator outside the classical limits. A quantised radiation field behaves as if it were a collection of an infinite number of harmonic oscillators. And the most significant aspect is that any system executing small oscillations about a position of stable equilibrium is in its behaviour similar to a harmonic oscillator.

Additional problems

4.3. Calculate the value of $\Delta x \Delta p_x$ for the harmonic oscillator in the nth state.

4.4. Comment on the statement that for a linear harmonic oscillator, there is no continuous energy spectrum, even for large values of n.

4.5. Show by explicit calculations that the solutions of a linear harmonic oscillator corresponding to the ground state and the first excited state are orthogonal to each other.

4.6. Show that the eigenvalues corresponding to the bound states of a one-dimensional problem are always non-degenerate.

4.7. Calculate the quantum mechanical probability that a linear harmonic oscillator in its first excited state will be found outside the limits of its classical motion.

4.8. The expectation value of x^2 of a linear harmonic oscillator in the nth state is

$$\left\langle x^2 \right\rangle = \frac{2n+1}{2\alpha}, \qquad \alpha = \frac{\sqrt{mk}}{\hbar}.$$

Calculate its potential and kinetic energies.

4.9. Show that a *non-degenerate* eigenfunction of a linear harmonic oscillator should be either an even or odd function of n.

4.10. Show that for a symmetric potential $V(-x) = V(x)$, the symmetric or antisymmetric solutions of the amplitude equation can always be constructed so that each one of them has a definite parity.

IOP Publishing

Quantum Mechanics

Mohammad Saleem

Chapter 5

The role of Hermitian operators

In quantum mechanics, operators play a unique role as the observables, i.e. the physical quantities which can be measured, are represented by them. As has already been mentioned in chapter 1, it is *assumed* that the measurement of a physical quantity must yield one of the eigenvalues of the operator representing that physical quantity. Since the observables are always real, the operators representing them should be such that their eigenvalues are invariably real. The *Hermitian operators*, named after the nineteenth century French mathematician Charles Hermite (who was, it is stated, so much involved in research even at an early age that he had difficulties passing his obligatory examinations!), possess this characteristic. First, we will define a linear operator and then define a Hermitian operator and discuss in detail its various characteristics. Finally, we will prove certain theorems involving such operators and analyse the results wherever necessary.

5.1 Linear operators

An operator A is said to be a *linear operator* if it satisfies the equations

$$A(\psi_i + \psi_j) = A\psi_i + A\psi_j,$$
$$A(c\psi_k) = cA\psi_k$$

where ψ are arbitrary functions and c is a constant. For instance, the differential operator $\mathrm{d}/\mathrm{d}x$ is a linear operator. In this book, we will use the word 'operator' for 'linear operator'.

5.2 Hermitian operators

Let us next define a Hermitian operator. We proceed step by step and first define the Hermitian conjugate of an operator. Let ψ and ϕ be two arbitrary functions with

doi:10.1088/978-0-7503-1206-6ch5

integrable squares. Let A be a linear operator. Suppose that there exists another linear operator B such that

$$\int \psi A^* \phi \, \mathrm{d}\tau = \int \phi B \psi \, \mathrm{d}\tau$$

where the asterisk (*) denotes the complex conjugate and $\mathrm{d}\tau = \mathrm{d}x\mathrm{d}y\mathrm{d}z$. Then the operator B is called the *Hermitian conjugate* or *adjoint* of A and is written as $A^\dagger$. Thus, if $A^\dagger$ denotes the Hermitian conjugate of A, the above equation takes the form

$$\int \psi A^* \phi \, \mathrm{d}\tau = \int \phi A^\dagger \psi \, \mathrm{d}\tau. \tag{5.1}$$

Now we can define a Hermitian operator. If the Hermitian conjugate of an operator is equal to the operator itself, i.e. if $A^\dagger = A$, the operator A is called a *Hermitian* or *self-adjoint* operator. Equation (5.1) then reduces to

$$\int \psi A^* \phi \, \mathrm{d}\tau = \int \phi A \psi \, \mathrm{d}\tau. \tag{5.2}$$

This equation may be taken as the definition of a Hermitian operator. Nevertheless, it should always be kept in mind that this definition is deeply rooted in the square integrable characteristic of the functions ϕ and ψ.

Example 5.1.

Problem

Show that the position operator x is Hermitian.

Solution

Let us consider the position operator x. By the definition of the Hermitian conjugate of an operator as applied to x, we have

$$\int \psi x^* \phi \, \mathrm{d}x = \int \phi x^\dagger \psi \, \mathrm{d}x.$$

Interchanging the integrals on the two sides of the above equation, we obtain

$$\int \phi x^\dagger \psi \, \mathrm{d}x = \int \psi x^* \phi \, \mathrm{d}x = \int \psi x \phi \, \mathrm{d}x = \int \phi x \psi \, \mathrm{d}x$$

where we have made use of the fact that the multiplicative operator x is real. Since the functions ϕ and ψ are arbitrary, the above equation will hold good only if

$$x^\dagger = x.$$

This shows that x is a Hermitian operator.

Problem

5.1. Find out whether any real operator $f(x)$ is Hermitian or not.

Example 5.2.

Problem

Show that the linear momentum operator $p_x \equiv -i\hbar\partial/\partial x$ is Hermitian.

Solution

Let us consider the operator $p_x \equiv -i\hbar\partial/\partial x$. By the very definition of the Hermitian conjugate of an operator, we have

$$\begin{aligned}\int \phi p_x^\dagger \psi \, dx &= \int \psi p_x^* \phi \, dx = \int \psi \left(-i\hbar \frac{\partial}{\partial x} \right)^* \phi \, dx \\ &= \int \psi \left(i\hbar \frac{\partial}{\partial x} \right) \phi \, dx = i\hbar \int \psi \left(\frac{\partial \phi}{\partial x} \right) dx \\ &= i\hbar [\psi \, \phi]_{-\infty}^{\infty} - i\hbar \int \frac{\partial \psi}{\partial x} \phi \, dx\end{aligned}$$

where the limits are from $-\infty$ to $+\infty$. The functions ψ and ϕ having integrable squares must both vanish at these limits because otherwise the normalisation integrals will diverge. Therefore, the above equation yields

$$\begin{aligned}\int \phi p_x^\dagger \psi \, dx &= -i\hbar \int \phi \frac{\partial \psi}{\partial x} dx \\ &= \int \phi \left(-i\hbar \frac{\partial}{\partial x} \right) \psi \, dx \\ &= \int \phi p_x \psi \, dx.\end{aligned}$$

Since ϕ and ψ are arbitrary, the above relation shows that $p_x^\dagger = p_x$, i.e. the linear momentum operator p_x is Hermitian. It is significant to note that it has been possible to prove that $p_x \equiv -i\hbar\partial/\partial x$ is a Hermitian operator *only because* we have assumed that the functions ϕ and ψ have integrable squares and consequently vanish at infinity.

Problem

5.2. Show that the operator p_x^2 is Hermitian.

We will now derive some important relations.

Theorem 5.1.

Prove that $(AB)^\dagger = B^\dagger A^\dagger$.

Proof

Consider the operator AB which is the product of two operators A and B. By the definition of the Hermitian conjugate of an operator, we have

$$\int \psi(AB)^* \phi \, d\tau = \int \phi(AB)^\dagger \psi \, d\tau. \tag{5.3}$$

But

$$\int \psi(AB)^* \phi \, d\tau = \int \psi A^* B^* \phi \, d\tau = \int \psi A^* \left(B^* \phi\right) d\tau$$
$$= \int \left(B^* \phi\right) A^\dagger \psi \, d\tau$$

where in the last step we have made use of the definition of Hermitian conjugate of an operator. Now the above equation can be written as

$$\int \psi(AB)^* \phi \, d\tau = \int \left(A^\dagger \psi\right) B^* \phi \, d\tau = \int \phi B^\dagger A^\dagger \psi \, d\tau, \tag{5.4}$$

where in the last step we have again made use of the definition of the Hermitian conjugate of an operator. Comparing (5.3) and (5.4), we obtain

$$\int \phi(AB)^\dagger \psi \, d\tau = \int \phi B^\dagger A^\dagger \psi \, d\tau.$$

Since the functions ψ and ϕ are arbitrary, the above equation will hold only if

$$(AB)^\dagger = B^\dagger A^\dagger. \tag{5.5}$$

This gives us the desired relationship.

If A and B are two Hermitian operators, we have $A^\dagger = A$ and $B^\dagger = B$. Equation (5.5) then becomes

$$(AB)^\dagger = BA.$$

If the two operators *commute*, i.e. $BA = AB$, then the above equation reduces to

$$(AB)^\dagger = AB.$$

The above constraints show that a product of two *Hermitian operators* is Hermitian only if they *mutually commute*.

The operator $AB - BA$ is called the *commutator* of A and B and is denoted by $[A, B]$. If A and B commute, then $[A, B] = 0$. If $AB = -BA$, then the two operators are said to anticommute and the expression $AB + BA$ is called the anticommutator of A and B and is written as $\{A, B\}$ or $[A, B]_+$.

Problem

5.3. Prove that

$$\left(A^{\dagger}\right)^{\dagger} = A$$

and

$$(A + B)^{\dagger} = A^{\dagger} + B^{\dagger}.$$

Theorem 5.2.

Prove that $(A^{*})^{\dagger} = (A^{\dagger})^{*}$.

Proof

Consider an operator A. By the definition of the Hermitian conjugate of an operator, we have

$$\int \psi A^{*} \phi \, \mathrm{d}\tau = \int \phi A^{\dagger} \psi \, \mathrm{d}\tau.$$

Taking the complex conjugate of both sides of this equation, we obtain

$$\int \psi^{*} (A^{*})^{*} \phi^{*} \, \mathrm{d}\tau = \int \phi^{*} \left(A^{\dagger}\right)^{*} \psi^{*} \, \mathrm{d}\tau. \tag{5.6}$$

By using the definition of the Hermitian conjugate of an operator as applied to the operator A^{*}, the integral on the left-hand side of (5.6) is given by

$$\int \psi^{*} (A^{*})^{*} \phi^{*} \, \mathrm{d}\tau = \int \phi^{*} (A^{*})^{\dagger} \psi^{*} \, \mathrm{d}\tau. \tag{5.7}$$

Comparing (5.7) and (5.6), we obtain

$$\int \phi^{*} (A^{*})^{\dagger} \psi^{*} \, \mathrm{d}\tau = \int \phi^{*} \left(A^{\dagger}\right)^{*} \psi^{*} \, \mathrm{d}\tau.$$

Since the functions ψ and ϕ are arbitrary, this equation will hold good only if

$$(A^{*})^{\dagger} = \left(A^{\dagger}\right)^{*}.$$

This gives the desired result.

Example 5.3.

Problem

Show that the adjoint $\lambda^\dagger$ of a complex number λ is its complex conjugate λ^*.

Solution

Let A be an arbitrary operator. Then, by definition, we have

$$\int \psi A^* \phi \, \mathrm{d}\tau = \int \phi A^\dagger \psi \, \mathrm{d}\tau$$

where $A^\dagger$ is the Hermitian conjugate of A. Interchanging the two sides of this equation and multiplying both sides by a complex number λ^*, we obtain

$$\int \phi \lambda^* A^\dagger \psi \, \mathrm{d}\tau = \int \psi \lambda^* A^* \phi \, \mathrm{d}\tau = \int \psi (\lambda A)^* \phi \, \mathrm{d}\tau = \int \phi (\lambda A)^\dagger \psi \, \mathrm{d}\tau$$

where in obtaining the last integral we have made use of the definition of the Hermitian conjugate of an operator. Since ϕ and ψ are arbitrary, the equation will hold only if

$$(\lambda A)^\dagger = \lambda^* A^\dagger$$

or

$$A^\dagger \lambda^\dagger = \lambda^\dagger A^\dagger = \lambda^* A^\dagger$$

where we have made use of the relation $(AB)^\dagger = B^\dagger A^\dagger$. Since the operator A is arbitrary, the above equation will hold only if

$$\lambda^\dagger = \lambda^*.$$

This shows that the Hermitian conjugate of a complex number is equal to its complex conjugate.

If λ is real, then $\lambda^* = \lambda$ and the above equation yields $\lambda^\dagger = \lambda$. That is, the Hermitian conjugate of a real number is equal to the number itself.

Problems

5.4. Prove that
 (i) $(\mathrm{i}A)^\dagger = -\mathrm{i}A^\dagger$;
 (ii) $(A + \mathrm{i}B)^\dagger = A^\dagger - \mathrm{i}B^\dagger$.

5.5. If A and B are Hermitian operators so that $A^\dagger = A$ and $B^\dagger = B$, then prove that

$$(A + \mathrm{i}B)^\dagger = A - \mathrm{i}B.$$

5.6. Prove that any operator A can be written as a linear combination of two Hermitian operators B and C:

$$A = B + \mathrm{i}C,$$

where

$$B = (A + A^{\dagger})/2 \qquad \text{and} \qquad C = (A - A^{\dagger})/2\mathrm{i}.$$

Example 5.4.

Problem

Show that $\overline{A^{\dagger}A} \geqslant 0$.

Solution

By the definition of the expectation value of any operator, we have

$$\begin{aligned}\overline{A^{\dagger}A} &= \int \psi^*(A^{\dagger}A)\Psi \, \mathrm{d}\tau = \int \Psi^* A^{\dagger}(A\Psi) \, \mathrm{d}t \\ &= \int (A\Psi)A^*\Psi^* \, \mathrm{d}\tau = \int (A\Psi)(A\Psi)^* \, \mathrm{d}t \\ &= \int |A\Psi|^2 \, \mathrm{d}\tau \geqslant 0\end{aligned}$$

where in deriving the result we have made use of the definition of the Hermitian conjugate of an operator.

Corollary

If A is Hermitian, then $A^{\dagger} = A$, and the above inequality yields

$$\overline{A^2} \geqslant 0.$$

Next we prove some important theorems.

Theorem 5.3.

The eigenvalues of a Hermitian operator are always real.

Proof

Let H be a Hermitian operator. Let ψ be an eigenfunction of H corresponding to the eigenvalue λ. Then

$$H\psi = \lambda\psi. \tag{5.8}$$

Taking the complex conjugate of the above equation, we obtain

$$H^*\psi^* = \lambda^*\psi^*. \tag{5.9}$$

Multiplying (5.9) from the left by ψ and integrating with respect to τ, we obtain

$$\int \psi H^*\psi^* \, d\tau = \lambda^* \int \psi^*\psi \, d\tau. \tag{5.10}$$

Multiplying (5.8) from the left by ψ^* and integrating with respect to τ, we obtain

$$\lambda \int \psi^*\psi \, d\tau = \int \psi^* H\psi \, d\tau = \int \psi^* H^\dagger \psi \, d\tau = \int \psi H^* \psi^* \, d\tau \tag{5.11}$$

because H is Hermitian. Comparing (5.10) and (5.11), we obtain

$$\lambda^* \int \psi^*\psi \, d\tau = \lambda \int \psi^*\psi \, d\tau.$$

Since the integrated probability $\int \psi^*\psi \, d\tau$ cannot be zero, it can be cancelled on both sides. The above equation then yields

$$\lambda^* = \lambda.$$

This proves the theorem: the eigenvalues of a Hermitian operator are always real. It is for this reason that Hermitian operators are used in quantum mechanics to represent physical quantities. The outcome of a physical measurement must be a real quantity. Since, in quantum mechanics, the measurement of a physical quantity must yield one of the eigenvalues of the operator representing that quantity, the eigenvalues of the operator must be real. This is ensured if the operator is Hermitian. It may be remarked that the converse of this theorem is not true. That is, it is *not essential* that if the eigenvalue of an operator is real, then the operator should be Hermitian. A non-Hermitian operator can have both real and complex eigenvalues. Therefore, it is convenient to impose the condition of being Hermitian on quantum mechanical operators. Actually, the Hermitian property is a stronger condition on the operator than the reality of eigenvalues.

Theorem 5.4.

The linearly independent eigenfunctions of a Hermitian operator corresponding to different eigenvalues are orthogonal to each other.

Proof

Let H be a Hermitian operator. Let ψ_m and ψ_n be two linearly independent eigenfunctions of this operator with different eigenvalues λ_m and λ_n, respectively. Then

$$H\psi_m = \lambda_m \psi_m \tag{5.12}$$

and

$$H\psi_n = \lambda_n \psi_n. \tag{5.13}$$

Multiplying (5.13) from the left by ψ_m^* and then integrating with respect to τ, we obtain

$$\int \psi_m^* H \psi_n \, \mathrm{d}\tau = \lambda_n \int \psi_m^* \psi_n \, \mathrm{d}\tau. \tag{5.14}$$

Taking the complex conjugate of (5.12), we obtain

$$H^* \psi_m^* = \lambda_m^* \psi_m^*. \tag{5.15}$$

Since the eigenvalues of a Hermitian operator are always real, we must have $\lambda_m^* = \lambda_m$. Equation (5.15) therefore gives

$$H^* \psi_m^* = \lambda_m \psi_m^*.$$

Interchanging sides and multiplying the above equation from the right by ψ_n and integrating with respect to τ, we obtain

$$\begin{aligned} \lambda_m \int \psi_m^* \psi_n \, \mathrm{d}\tau &= \int \left(H^* \psi_m^* \right) \psi_n \, \mathrm{d}\tau = \int \psi_n H^* \psi_m^* \, \mathrm{d}\tau \\ &= \int \psi_m^* H^\dagger \psi_n \, \mathrm{d}\tau = \int \psi_m^* H \psi_n \, \mathrm{d}\tau \end{aligned} \tag{5.16}$$

because H is Hermitian. Subtracting (5.16) from (5.14), we obtain

$$\left(\lambda_n - \lambda_m \right) \int \psi_m^* \psi_n \, \mathrm{d}\tau = 0. \tag{5.17}$$

Since the eigenvalues λ_m and λ_n are distinct, $\lambda_n \neq \lambda_m$. Therefore, we must have

$$\int \psi_m^* \psi_n \, \mathrm{d}\tau = 0.$$

This shows that the linearly independent eigenfunctions ψ_m and ψ_n of a Hermitian operator corresponding to different eigenvalues are orthogonal to each other.

Theorem 5.5.

If λ is a degenerate eigenvalue of an operator A corresponding to linearly independent eigenfunctions ψ_1, ψ_2, …, then every linear combination of this set of functions is also an eigenfunction of A and corresponds to the same eigenvalue λ.

Proof

Consider an operator A. Let λ be a degenerate eigenvalue of the operator A corresponding to linearly independent eigenfunctions ψ_1, ψ_2, ... Then, we must have a set of distinct equations

$$A\psi_1 = \lambda\psi_1,$$
$$A\psi_2 = \lambda\psi_2$$
$$\ldots.$$

Multiplying these equations by a_1, a_2, ..., respectively, and adding, we obtain

$$A(a_1\psi_1 + a_2\psi_2 + \cdots) = (a_1\lambda\psi_1 + a_2\lambda\psi_2 + \cdots)$$
$$= \lambda(a_1\psi_1 + a_2\psi_2 + \cdots). \tag{5.18}$$

This shows that $(a_1\psi_1 + a_2\psi_2 + \cdots)$, a linear combination of linearly independent eigenfunctions of an operator A corresponding to a degenerate eigenvalue λ, is also an eigenfunction of the operator A and corresponds to the same eigenvalue λ.

Theorem 5.6.

The eigenfunctions of a Hermitian operator with a degenerate eigenvalue can be made orthogonal.

Proof

We have already shown that

$$(\lambda_n - \lambda_m)\int \psi_m^*\psi_n \,\mathrm{d}\tau = 0. \tag{5.17'}$$

If $\lambda_n \neq \lambda_m$, then $\int \psi_m^*\psi_n \,\mathrm{d}\tau = 0$, so that the eigenfunctions are mutually orthogonal.

Let us now examine the situation when $\lambda_m = \lambda_n = \lambda$ (say), i.e. when the eigenvalue λ of the Hermitian operator H corresponding to linearly independent eigenfunctions ψ_m and ψ_n is doubly degenerate. Then, as is evident from (5.17′), the integral

$$\int \psi_m\psi_n^* \,\mathrm{d}\tau$$

may or may not be zero. If it is zero, ψ_m and ψ_n are orthogonal to each other. If it is *not zero*, let us form two linear combinations of the functions ψ_m and ψ_n:

$$a\psi_m + b\psi_n\ (= f,\ \text{say}), \tag{5.19}$$

$$c\psi_m + d\psi_n\ (= g,\ \text{say}). \tag{5.20}$$

Then, as shown in the last theorem, both f and g, being different linear combinations of the eigenfunctions ψ_m and ψ_n belonging to the same eigenvalue $\lambda\,(=\lambda_m=\lambda_n)$, must also be eigenfunctions of the operator H corresponding to the same degenerate eigenvalue λ. If we wish f and g to be orthogonal, we must have

$$\int f^* g \, \mathrm{d}\tau = 0. \tag{5.21}$$

Moreover, if f and g are to be individually normalised, we must have

$$\int f^* f \, \mathrm{d}\tau = 1, \tag{5.22}$$

$$\int g^* g \, \mathrm{d}\tau = 1. \tag{5.23}$$

Even if we take the physically unimportant phase factor as unity, the last three equations are not sufficient to determine the values of the four constants a, b, c, d; they can fix their ratios only. The value of any one of the constants can be chosen arbitrarily; the remaining three constants will then be determined by these equations. Consequently, these three constants can be selected in an infinite number of ways so that we may form an infinite number of linear combinations of ψ_m and ψ_n with the desired property of orthogonality and corresponding to the degenerate eigenvalue λ. Hence *all* eigenfunctions of a Hermitian operator are either orthogonal or can be made orthogonal; of course we have only to replace ψ_m and ψ_n by f and g in the above example.

Theorem 5.7.

The Hamiltonian operator H of a system is Hermitian provided that the integrated probability is conserved.

Proof

We know that if Ψ is the wave function representing a system, then the integrated probability P is given by

$$P = \int \psi^* \psi \, \mathrm{d}\tau. \tag{5.24}$$

In order that this probability may be conserved, we must have

$$\frac{\mathrm{d}P}{\mathrm{d}t} = 0,$$

or

$$\frac{\mathrm{d}}{\mathrm{d}t}\int \Psi^*\Psi\,\mathrm{d}\tau = 0,$$

or

$$\int\left(\frac{\partial\Psi^*}{\partial t}\Psi + \Psi^*\frac{\partial\Psi}{\partial t}\right)\mathrm{d}\tau = 0. \tag{5.25}$$

Now according to the time-dependent Schrödinger wave equation, we have

$$H\Psi = \mathrm{i}\hbar\frac{\partial\Psi}{\partial t}. \tag{5.26}$$

Taking the complex conjugate of both sides of this equation, we obtain

$$H^*\Psi^* = -\mathrm{i}\hbar\frac{\partial\Psi^*}{\partial t}. \tag{5.27}$$

Substituting the expressions for $\partial\Psi/\partial t$ and $\partial\Psi^*/\partial t$ from (5.26) and (5.27) in (5.25), we obtain

$$\int\left\{(H^*\Psi^*)\Psi - \Psi^*(H\Psi)\right\}\mathrm{d}\tau = 0,$$

or

$$\int(\Psi H^*\Psi^* - \Psi^* H\Psi)\mathrm{d}\tau = 0. \tag{5.28}$$

By making use of the definition of the Hermitian conjugate of an operator, the first integral in (5.28) takes the form

$$\int \Psi^* H^\dagger\Psi\,\mathrm{d}\tau$$

and therefore (5.28) becomes

$$\int(\Psi^* H^\dagger\Psi - \Psi^* H\Psi)\mathrm{d}\tau = 0,$$

or

$$\int \Psi^*(H^\dagger - H)\Psi\,\mathrm{d}\tau = 0.$$

This would hold only if $H^\dagger - H = 0$, i.e. if $H^\dagger = H$. This shows that H is Hermitian. Evidently, the Hamiltonian operator H, being Hermitian, possesses all the properties of a Hermitian operator. For example, the energy E, the eigenvalue of the operator H, is real and eigenfunctions of H are or can be made orthogonal.

5.3 The closure relation

Consider a Hermitian operator A representing an observable. The eigenfunctions of this operator form a complete set so that every function can be expanded in terms of them. Thus, if $\psi_i(x)$ form a complete set, then an arbitrary function $\phi(x)$ can be expressed as a linear combination of them:

$$\phi(x) = \sum_i a_i \psi_i(x).$$

Multiplying this equation by $\psi_j^*(x)$ and integrating with respect to x, we obtain

$$\begin{aligned}\int \psi_j^*(x)\phi(x)\,\mathrm{d}x &= \int \psi_i^*(x)\sum_i a_i\psi_i(x)\,\mathrm{d}x \\ &= \sum_i a_i \int \psi_j^*(x)\psi_i(x)\,\mathrm{d}x \\ &= \sum_i a_i\delta_{ij} \\ &= a_j.\end{aligned}$$

Changing j to i, this yields

$$a_i = \int \psi_i^*(x)\phi(x)\,\mathrm{d}x.$$

Substituting this expression for a_i in the expansion equation for $\phi(x)$, we obtain

$$\begin{aligned}\phi(x) &= \sum_i \left[\int \psi_i^*(x')\phi(x')\,\mathrm{d}x'\right]\psi_i(x) \\ &= \int\left[\sum_i \psi_i^*(x')\psi_i(x)\right]\phi(x')\,\mathrm{d}x'.\end{aligned}$$

This shows that

$$\sum_i \psi_i^*(x')\psi_i(x) = \delta(x - x')$$

because it also reduces the right-hand side to $\phi(x)$. This relation is called the *closure relation*. The expansion theorem in the case of a continuous distribution of eigenvalues may be written as

$$\psi = \int c_a\psi_a\,\mathrm{d}a$$

or, using a slightly different notation, as

$$\psi = \int c(a)\psi(a)\,\mathrm{d}a.$$

This replaces

$$\psi = \sum_i c_i \psi_i.$$

The closure relation for a discrete spectrum,

$$\sum_i \psi_i^*(x')\psi_i(x) = \delta(x - x'),$$

is then replaced by

$$\int_{-\infty}^{\infty} \psi_a^*(x')\psi_a(x)\,\mathrm{d}x = \delta(x - x').$$

Theorem 5.8.

If two operators have a complete set of simultaneous eigenfunctions, they must commute.

Proof

Let A and B be two given operators and ϕ_i a complete set of their simultaneous eigenfunctions corresponding to the eigenvalues a_i and b_i, respectively. Then

$$A\phi_i = a_i\phi_i, \tag{5.29}$$

$$B\phi_i = b_i\phi_i. \tag{5.30}$$

Operating (5.29), from the left, by B, we obtain

$$BA\phi_i = a_i B\phi_i = a_i b_i \phi_i, \tag{5.31}$$

where we have made use of (5.30). Operating (5.30), from the left, by A, we obtain

$$AB\phi_i = b_i A\phi_i = b_i a_i \phi_i = a_i b_i \phi_i, \tag{5.32}$$

where we have made use of (5.29). From (5.31) and (5.32), we have

$$AB\phi_i = BA\phi_i,$$

or

$$(AB - BA)\phi_i = 0. \tag{5.33}$$

Now an arbitrary function ψ can be expressed as a linear combination of the complete set of functions ϕ_i. Therefore, we may write

$$\psi = \sum_i a_i \phi_i.$$

Multiplying (5.33) by a_i and summing over i, we obtain

$$\sum_i a_i(AB - BA)\phi_i = 0,$$

or

$$(AB - BA)\sum_i a_i\phi_i = 0,$$

or

$$(AB - BA)\psi = 0.$$

Since ψ is an arbitrary function, the above equation will hold if

$$AB = BA.$$

That is, in this case the operators A and B must commute. This proves the theorem.

Theorem 5.9.

If two Hermitian operators commute, then a common complete set of eigenfunctions can always be found for them.

Proof

Let A and B be two commuting Hermitian operators. Then

$$A^\dagger = A, \qquad B^\dagger = B$$

and

$$AB = BA.$$

Let ϕ_i be an eigenfunction of A corresponding to the eigenvalue a_i. Then, we have

$$A\phi_i = a_i\phi_i. \tag{5.34}$$

We will first prove that ϕ_i is also an eigenfunction of B provided that the eigenvalue a_i of the operator A is *non-degenerate*. We proceed as follows. Operating (5.34) on the left by B, we obtain

$$BA\phi_i = a_i B\phi_i.$$

Since the operators A and B commute, the above equation may be written as

$$AB\phi_i = a_i B\phi_i,$$

or

$$A(B\phi_i) = a_i(B\phi_i).$$

This equation shows that $B\phi_i$ is also an eigenfunction of the operator A and corresponds to the same eigenvalue a_i. If the eigenvalue a_i is degenerate, then the corresponding eigenfunctions ϕ_i and $B\phi_i$ will be linearly independent. But it is given that the eigenvalue a_i is non-degenerate. Therefore, ϕ_i and $B\phi_i$ are linearly dependent. Hence one must be a constant multiple of the other,

$$B\phi_i = b_i\phi_i,$$

where b_i is a constant. The above equation shows that ϕ_i is also an eigenfunction of B. Since ϕ_i is arbitrary, every eigenfunction of A is also an eigenfunction of B. Therefore, the complete set of eigenfunctions of the Hermitian operator A is also a set of eigenfunctions of B. Similarly, the complete set of eigenfunctions of the Hermitian operator B is also a set of eigenfunctions of A. This is possible only if the two Hermitian operators have a common complete set of eigenfunctions. This proves the theorem for non-degenerate eigenvalues.

Let us next prove the theorem when the eigenvalue a_i of the operator A is *degenerate*.

To make the calculations simple, suppose that the eigenvalue a_i is two-fold degenerate, i.e. by definition, two eigenfunctions of A corresponding to the same eigenvalue a_i are linearly independent. This means that ϕ_i and $B\phi_i$ must be linearly independent: $B\phi_i$ cannot be a constant multiple of ϕ_i. However, a linear combination of the eigenfunctions ϕ_i and $B\phi_i$ of the operator A belonging to its degenerate eigenvalue a_i is itself an eigenfunction of A and belongs to the same eigenvalue. That is:

$$A(c_1\phi_i + c_2 B\phi_i) = a_i(c_1\phi_i + c_2 B\phi_i).$$

Can this linear combination be an eigenfunction of the operator B? If this is so, then we must have

$$B(c_1\phi_i + c_2 B\phi_i) = b_i(c_1\phi_i + c_2 B\phi_i). \tag{5.35}$$

This would mean that $(c_1\phi_i + c_2 B\phi_i)$ is also an eigenfunction of B and consequently a simultaneous eigenfunction of A and B. To determine the values of c_1 and c_2 for which such a situation may exist, we proceed as follows.

Suppose for convenience that ϕ_i and $B\phi_i$ are individually normalised and mutually orthogonal. Multiplying (5.35) from the left by ϕ_i^* and integrating with respect to τ, we obtain

$$c_1\int \phi_i^* B\phi_i \, d\tau + c_2\int \phi_i^* B(B\phi_i)\, d\tau = b_i c_1\int \phi_i^*\phi_i \, d\tau + b_i c_2\int \phi_i^*(B\phi_i)\, d\tau = b_i c_1.$$

Writing B_{11} for $\int \phi_i^* B\phi_i \, d\tau$ and B_{12} for $\int \phi_i^* B(B\phi_i) \, d\tau$ in the above equation, we obtain

$$c_1 B_{11} + c_2 B_{12} = b_i c_1,$$

or

$$(B_{11} - b_i)c_1 + B_{12}c_2 = 0. \tag{5.36}$$

Similarly, multiplying (5.35) on the left by $(B\phi_i^*)$ and integrating with respect to τ, we obtain

$$B_{21}c_1 + (B_{22} - b_i)c_2 = 0. \tag{5.37}$$

In order that a non-trivial solution of (5.36) and (5.37) may exist, we must have

$$\begin{vmatrix} B_{11} - b_1 & B_{12} \\ B_{21} & B_{22} - b_1 \end{vmatrix} = 0. \tag{5.38}$$

This quadratic equation determines the two values of b_i. For each value of b_i, the values of c_1 and c_2 are computed with the help of (5.36) and (5.37). Hence, we obtain two different linear combinations of ϕ_i and $B\phi_i$ as simultaneous eigenfunctions of A and B. This proves the theorem.

It is significant to note that the above analysis does not imply that if two operators A and B commute, then every eigenfunction of A is also necessarily an eigenfunction of B. Our analysis only shows that it is always possible to find a complete set of simultaneous eigenfunctions of A and B.

The above analysis can be easily extended to the case when more than one eigenvalue of A has different degrees of degeneracy. Suppose that the choice of an operator B which commutes with A does not remove the degeneracy completely. We can then choose another operator C which commutes with A as well as B. We may find a complete set of simultaneous eigenfunctions of these three operators A, B, C whose eigenvalues are not degenerate. This process may be continued till there is no more degeneracy. The set of mutually commuting operators A, B, C, …, M which have a complete set of eigenfunctions corresponding to non-degenerate eigenvalues is called a *complete set of commuting operators* or a *complete set of commuting observables.* For a given system, except for the functions of these operators, this is indeed the largest set of commuting operators. At any time, this would give us the maximum possible information about the system.

5.4 Constants of motion

In classical physics, if a physical quantity does not change with time, it is said to be a constant of motion. In fact, classically, geometrical symmetries in various mechanical systems are associated with the constants of motion, i.e. with the quantities which are conserved in time. For instance, if the motion of a particle is symmetric under translations in the x-direction, then p_x, the x-component of its linear momentum, is conserved. In quantum mechanics, if the *expectation value* of a physical quantity represented by an operator does not change with time, then the physical quantity is said to be a constant of the motion. Let us now determine the conditions which must be satisfied so that in quantum mechanics a physical quantity may be a constant of motion. By definition, the expectation value $\langle A \rangle$ of an observable represented by an operator A is given by

$$\langle A \rangle = \int \Psi^*(\mathbf{r}, t) A \Psi(\mathbf{r}, t)\, \mathrm{d}\tau \tag{5.39}$$

where Ψ is the normalised wave function representing the system on which measurements are made. Differentiating with respect to time t, we obtain

$$\frac{\mathrm{d}}{\mathrm{d}t}\langle A \rangle = \int \frac{\partial \Psi^*}{\partial t} A \Psi\, \mathrm{d}\tau + \int \Psi^* \frac{\partial A}{\partial t} \Psi\, \mathrm{d}\tau + \int \Psi^* A \frac{\partial \Psi}{\partial t}\, \mathrm{d}\tau. \tag{5.40}$$

Now the time-dependent Schrödinger wave equation is

$$\mathrm{i}\hbar \frac{\partial \Psi}{\partial t} = H\Psi.$$

Taking the complex conjugate of both sides of the above equation, we obtain

$$-\mathrm{i}\hbar \frac{\partial \Psi^*}{\partial t} = H^*\Psi^*.$$

Substituting the expressions for $\partial\Psi/\partial t$ and $\partial\Psi^*/\partial t$ from the above two equations in (5.40), we obtain

$$\frac{\mathrm{d}}{\mathrm{d}t}\langle A \rangle = \frac{\mathrm{i}}{\hbar}\int (H^*\Psi^*) A \Psi\, \mathrm{d}\tau + \int \Psi^* \frac{\partial A}{\partial t} \Psi\, \mathrm{d}\tau - \frac{\mathrm{i}}{\hbar}\int \Psi^* A (H\Psi)\, \mathrm{d}\tau. \tag{5.41}$$

Now the first integral on the right-hand side of the above equation can be written as

$$\int (H^*\Psi^*) A \Psi\, \mathrm{d}\tau = \int (A\Psi) H^*\Psi^*\, \mathrm{d}\tau = \int \Psi^* H A \Psi\, \mathrm{d}\tau$$

where we have made use of the fact that H is Hermitian. Then (5.41) takes the form

$$\frac{\mathrm{d}}{\mathrm{d}t}\langle A\rangle = \frac{\mathrm{i}}{\hbar}\int \Psi^* HA\Psi\,\mathrm{d}\tau - \frac{\mathrm{i}}{\hbar}\int \Psi^* AH\Psi\,\mathrm{d}\tau + \int \Psi^* \frac{\partial A}{\partial t}\,\mathrm{d}\tau$$

$$= -\frac{\mathrm{i}}{\hbar}\int \Psi^*(AH - HA)\Psi\,\mathrm{d}\tau + \int \Psi^* \frac{\partial A}{\partial t}\Psi\,\mathrm{d}\tau$$

$$= -\frac{\mathrm{i}}{\hbar}\int \Psi^*[A, H]\Psi\,\mathrm{d}\tau + \int \Psi^* \frac{\partial A}{\partial t}\Psi\,\mathrm{d}\tau,$$

or

$$\frac{\mathrm{d}}{\mathrm{d}t}\langle A\rangle = -\frac{\mathrm{i}}{\hbar}\langle [A, H]\rangle + \left\langle \frac{\partial A}{\partial t}\right\rangle. \tag{5.42}$$

]This is known as the *equation of motion* for any operator A. If A does not depend upon time explicitly, then $\partial A/\partial t = 0$. Moreover, if it commutes with H, then $[A, H] = 0$. Consequently, (5.42) yields

$$\frac{\mathrm{d}}{\mathrm{d}t}\langle A\rangle = 0. \tag{5.43}$$

Integrating with respect to t, we obtain $\langle A\rangle$ = constant in time. This result shows that a physical quantity represented by an operator which does not depend explicitly upon time and commutes with the Hamiltonian of the system is a constant of motion, i.e. it does not change with time. For instance, if the energy E of a system is represented by a Hamiltonian operator H that does not depend upon time explicitly, then as $\partial H/\partial t = 0$, and $[H, H]$ is always zero, the expectation value $\langle H\rangle$ of H should not change with time. Hence, in this case the energy of the system is a constant of the motion.

An interesting question arises: when will an operator A commute with H ($\equiv p^2/2m + V(\mathbf{r})$)? The answer depends upon the nature of the operator A. Suppose that it is p_x, the x-component of linear momentum. Now, if $V(\mathbf{r}) \equiv V(y, z)$, i.e. the potential energy does *not* depend upon the variable x, then the operator p_x which commutes with p^2 will also commute with $V(y, z)$ and hence with the Hamiltonian operator H. The commutator of p_x and H is therefore zero. Moreover, as p_x does not depend upon time explicitly, by virtue of relation (5.42), it will be a *constant of motion*. This fact may also be expressed by stating that the system is symmetric under translation in the x-direction. This demonstrates the relationship between symmetry and the law of conservation even in quantum mechanics.

5.5 The classical limit of quantum mechanics: the Ehrenfest theorem

It is probably the destiny of all theories that with the passage of time, which may extend over decades and even centuries, they cannot confront the new experimental data successfully. This paves the way for new concepts and novel ideas. But, after all, these old theories have been confronting experimental data for a long time. Therefore, in classical limits, a new theory must reduce to the old one so that it may also explain the old results. Thus, for macroscopic objects, quantum mechanics must reduce to classical mechanics which has been successful for such objects for centuries. However, in quantum mechanics, the prediction of events is probabilistic in nature while classical mechanics predicts the exact behaviour and precise result of each individual measurement. This means that in the classical limit the expectation values in quantum mechanics should show the same behaviour as the individual events in classical mechanics. Hence, in the classical limit the expectations values in quantum mechanics must obey Newton's second law of motion which governs the behaviour of physical systems in classical mechanics. We will prove this. We first show that

$$m\frac{\mathrm{d}}{\mathrm{d}t}\langle \mathbf{r}\rangle = \langle \mathbf{p}\rangle. \tag{5.44}$$

Equation (5.42) shows that, as the position vector **r** does not depend upon time explicitly, the rate of change of the expectation value of position vector **r** is given by

$$\frac{\mathrm{d}}{\mathrm{d}t}\langle \mathbf{r}\rangle = -\frac{\mathrm{i}}{\hbar}[\mathbf{r}, H]. \tag{5.45}$$

Let us first consider the x-component of **r**:

$$\frac{\mathrm{d}}{\mathrm{d}t}\langle x\rangle = -\frac{\mathrm{i}}{\hbar}\langle [x, H]\rangle. \tag{5.46}$$

Let us evaluate $[x, H]$. We have

$$(xH - Hx)\phi = x\left(-\frac{\hbar^2}{2m}\nabla^2 + V\right)\phi - \left(-\frac{\hbar^2}{2m}\nabla^2 + V\right)(x\phi), \tag{5.47}$$

but

$$x\nabla^2\phi = x\left(\frac{\partial^2\phi}{\partial x^2} + \frac{\partial^2\phi}{\partial y^2} + \frac{\partial^2\phi}{\partial z^2}\right)$$

and

$$\begin{aligned}
\nabla^2(x\phi) &= \left(\frac{\partial^2}{\partial x^2} + \frac{\partial^2}{\partial y^2} + \frac{\partial^2}{\partial z^2}\right)(x\phi) \\
&= \frac{\partial}{\partial x}\frac{\partial}{\partial x}(x\phi) + \frac{\partial}{\partial y}\frac{\partial}{\partial y}(x\phi) + \frac{\partial}{\partial z}\frac{\partial}{\partial z}(x\phi) \\
&= \frac{\partial}{\partial x}\left(\phi + x\frac{\partial\phi}{\partial x}\right) + \frac{\partial}{\partial y}\left(x\frac{\partial\phi}{\partial y}\right) + \frac{\partial}{\partial z}\left(x\frac{\partial\phi}{\partial z}\right) \\
&= \frac{\partial\phi}{\partial x} + \frac{\partial\phi}{\partial x} + x\frac{\partial^2\phi}{\partial x^2} + x\frac{\partial^2\phi}{\partial y^2} + x\frac{\partial^2\phi}{\partial z^2}.
\end{aligned}$$

Substituting these expressions in (5.47) and simplifying, we obtain

$$(xH - Hx)\phi = \frac{\hbar^2}{m}\frac{\partial \phi}{\partial x},$$

or

$$(xH - Hx)\ \frac{\hbar^2}{m}\frac{\partial}{\partial x}.$$

By virtue of this relation, (5.46) becomes

$$\frac{\mathrm{d}}{\mathrm{d}t}\langle x\rangle = \frac{1}{m}\left(-\mathrm{i}\hbar\frac{\partial}{\partial x}\right),$$

or

$$\frac{\mathrm{d}}{\mathrm{d}t}\langle x\rangle = \frac{1}{m}\left\langle p_x\right\rangle.$$

Similarly, we obtain

$$\frac{\mathrm{d}}{\mathrm{d}t}\langle y\rangle = \frac{1}{m}\left\langle p_y\right\rangle,$$

$$\frac{\mathrm{d}}{\mathrm{d}t}\langle z\rangle = \frac{1}{m}\left\langle p_z\right\rangle.$$

Combining the last three equations, we obtain

$$\frac{\mathrm{d}}{\mathrm{d}t}\langle \mathbf{r}\rangle = \frac{1}{m}\langle \mathbf{p}\rangle. \tag{5.44'}$$

This shows that in quantum mechanics, in terms of expectation values, the linear momentum and rate of change of position vector are related in the same manner as in classical physics.

We now come to the proof of the second law of motion for expectation values in quantum mechanics.

Let us consider the linear momentum operator $\mathbf{p}$. Then by virtue of (5.42), we have

$$\frac{\mathrm{d}}{\mathrm{d}t}\langle \mathbf{p}\rangle = -\frac{\mathrm{i}}{\hbar}\langle[\mathbf{p}, H]\rangle + \left\langle \frac{\partial \mathbf{p}}{\partial t}\right\rangle. \tag{5.48}$$

Since $\mathbf{p}$ does not depend upon time explicitly, we have

$$\frac{\partial \mathbf{p}}{\partial t} = 0.$$

Equation (5.48) therefore reduces to

$$\frac{\mathrm{d}}{\mathrm{d}t}\langle \mathbf{p}\rangle = -\frac{\mathrm{i}}{\hbar}\langle[\mathbf{p}, H]\rangle = -\frac{\mathrm{i}}{\hbar}\langle[-\mathrm{i}\hbar\nabla, H]\rangle = -\langle[\nabla, H]\rangle. \tag{5.49}$$

Let us now calculate the commutator of ∇ and H. We have

$$[\nabla, H]\phi = \left[\nabla, \left(-\frac{\hbar^2}{2m}\nabla^2 + V\right)\right]\phi$$
$$= \nabla\left(-\frac{\hbar^2}{2m}\nabla^2 + V\right)\phi - \left(-\frac{\hbar^2}{2m}\nabla^2 + V\right)(\nabla\phi)$$
$$= \nabla\left(-\frac{\hbar^2}{2m}\nabla^2\phi + (V\phi)\right) - \left(-\frac{\hbar^2}{2m}\nabla^2 + V\right)(\nabla\phi)$$
$$= \left(-\frac{\hbar^2}{2m}\nabla^3\phi\right) + (\nabla V)\phi + V\nabla\phi) - \left(-\frac{\hbar^2}{2m}\nabla^3\phi + V\nabla\phi\right)$$
$$= (\nabla V(\mathbf{r}))\phi.$$

This yields

$$[\nabla, H] = (\nabla V(\mathbf{r})).$$

Substituting this expression for $[\nabla, H]$ in (5.49), we obtain

$$\frac{d}{dt}\langle\mathbf{p}\rangle = \langle(-\nabla V(\mathbf{r}))\rangle, \tag{5.50}$$

or

$$\frac{d}{dt}\langle\mathbf{p}\rangle = \langle\mathbf{F}(\mathbf{r})\rangle. \tag{5.51}$$

This is called the *Ehrenfest theorem*. Paul Ehrenfest was a close friend of Einstein who appreciated Ehrenfest's 'unusually well-developed faculty to grasp the essence of a theoretical notion, to strip a theory of its mathematical accoutrements until the simple basic idea emerged with clarity'. 'This capacity' according to Einstein 'made him…the best teacher in our profession whom I have ever known'. In one dimension, the theorem takes the form

$$\frac{d}{dt}\langle p_x\rangle = \langle F(x)\rangle. \tag{5.52}$$

We should be very careful in its interptretation. The right-hand side of this equation is the average value of force. However, as we are considering the expectation values of x and p_x, we actually have to consider the value of force at the expectation value of x: $F(\langle x\rangle)$. Therefore, before giving any interpretation to this equation, we have to find a relation between $\langle F(x)\rangle$ and $F(\langle x\rangle)$. The force operator $F(x)$ can be written as

$$F(x) = F(x - \langle x\rangle + \langle x\rangle) = F(\langle x\rangle + \Delta x),$$

where $\Delta x = x - \langle x \rangle$. Expanding it in a Taylor series about the point $x = \langle x \rangle$, we obtain

$$F(x) = F\langle x \rangle + (\Delta x)F'(\langle x \rangle) + \frac{(\Delta x)^2}{2}F''(\langle x \rangle) + \cdots.$$

Taking the average of both sides of this equation, we obtain

$$\langle F(x) \rangle = \left\langle F\langle x \rangle + \langle (\Delta x) \rangle F'(\langle x \rangle) + \frac{\langle (\Delta x)^2 \rangle}{2}F''(\langle x \rangle) + \cdots \right\rangle.$$

Since $\langle (\Delta x) \rangle$ is always zero, the above equation reduces to

$$\langle F(x) \rangle = \left\langle F\langle x \rangle + \frac{\langle (\Delta x)^2 \rangle}{2}F''(\langle x \rangle) + \cdots \right\rangle.$$

Neglecting $(\Delta x)^3$ and higher order terms, we obtain

$$\langle F(x) \rangle = \left\langle F\langle x \rangle + \frac{\langle (\Delta x)^2 \rangle}{2}F''(\langle x \rangle) \right\rangle.$$

The quantum mechanical equation of motion (5.52) therefore takes the form

$$\frac{\mathrm{d}}{\mathrm{d}t}\langle p_x \rangle = \langle F(x) \rangle = F(\langle x \rangle) + \frac{\langle (\Delta x)^2 \rangle}{2}F''(\langle x \rangle). \tag{5.53}$$

The second term on the right-hand side of (5.53), i.e. the expression

$$\frac{\langle (\Delta x)^2 \rangle}{2}F''(\langle x \rangle),$$

is the quantum mechanical correction to Newton's classical equation of motion. When it can be ignored compared to the first term on the right-hand side of the equation, we obtain

$$\frac{\mathrm{d}}{\mathrm{d}t}\langle p_x \rangle = \langle F(x) \rangle = F(\langle x \rangle). \tag{5.54}$$

When can the second term on the right-hand side of (5.53) be ignored compared to the first? This is allowed when Δx is microscopic in scale and the force varies appreciably only over macroscopic distances. It is only in that situation that the second law of motion is valid for expectation values in quantum mechanics. We take this opportunity to emphasise that one should be careful when turning classical equations to the corresponding expectation value equations in quantum mechanics.

5.6 The virial theorem

The virial theorem gives a general relation between the expectation value of the kinetic energy T of a particle and the expectation values of a quantity involving its potential energy V when the potential energy does not depend upon time explicitly. This is consequently helpful in simplifying certain computations. Let us derive this theorem. We have shown that if A is any operator, then

$$\frac{\mathrm{d}}{\mathrm{d}t}\langle A\rangle = -\frac{\mathrm{i}}{\hbar}\langle[A, H]\rangle + \left\langle\frac{\partial A}{\partial t}\right\rangle. \tag{5.42'}$$

Consider the case when A is the operator $\mathbf{r}\cdot\mathbf{p}$. Since $\mathbf{r}\cdot\mathbf{p}$ does not depend upon time explicitly, its partial derivative with respect to time is zero and therefore the last term in the above equation disappears. We then have

$$\begin{aligned}\frac{\mathrm{d}}{\mathrm{d}t}\langle\mathbf{r}\cdot\mathbf{p}\rangle &= -\frac{\mathrm{i}}{\hbar}\langle[\mathbf{r}\cdot\mathbf{p}, H]\rangle \\ &= -\frac{\mathrm{i}}{\hbar}\left\langle\left[xp_x + yp_y + zp_z, H\right]\right\rangle\end{aligned} \tag{5.55a}$$

$$= -\frac{\mathrm{i}}{\hbar}\left\{\left\langle\left[xp_x, H\right]\right\rangle + \left\langle\left[yp_y, H\right]\right\rangle + \left\langle\left[zp_z, H\right]\right\rangle\right\}. \tag{5.55b}$$

Let us first calculate the commutator of xp_x and H. We have

$$\begin{aligned}\left[xp_x, H\right]\phi &= \left(xp_x H - Hxp_x\right)\phi \\ &= \left[x\left\{(-\mathrm{i}\hbar)\frac{\partial}{\partial x}\right\}\left(-\frac{\hbar^2}{2m}\nabla^2 + V\right)\right]\phi \\ &\quad -\left(-\frac{\hbar^2}{2m}\nabla^2 + V\right)\left\{x\left(-\mathrm{i}\hbar\frac{\partial}{\partial x}\right)\right\}\phi\end{aligned}$$

$$= -\mathrm{i}\hbar\left[x\frac{\partial}{\partial x}\left(-\frac{\hbar^2}{2m}\nabla^2\phi + V\phi\right) - \left(-\frac{\hbar^2}{2m}\nabla^2 + V\right)\left(x\frac{\partial\phi}{\partial x}\right)\right], \tag{5.56}$$

but

$$\begin{aligned}\nabla^2\left(x\frac{\partial\phi}{\partial x}\right) &= \left(\frac{\partial^2}{\partial x^2} + \frac{\partial^2}{\partial y^2} + \frac{\partial^2}{\partial z^2}\right)\left(x\frac{\partial\phi}{\partial x}\right) \\ &= \frac{\partial}{\partial x}\frac{\partial}{\partial x}\left(x\frac{\partial\phi}{\partial x}\right) + \frac{\partial}{\partial y}\frac{\partial}{\partial y}\left(x\frac{\partial\phi}{\partial x}\right) + \frac{\partial}{\partial z}\frac{\partial}{\partial z}\left(x\frac{\partial\phi}{\partial x}\right) \\ &= \frac{\partial}{\partial x}\left(\frac{\partial\phi}{\partial x} + x\frac{\partial^2\phi}{\partial x^2}\right) + \frac{\partial}{\partial y}\left(x\frac{\partial^2\phi}{\partial x\partial y}\right) + \frac{\partial}{\partial z}\left(x\frac{\partial^2\phi}{\partial z\partial x}\right) \\ &= \frac{\partial^2\phi}{\partial x^2} + \frac{\partial^2\phi}{\partial x^2} + x\frac{\partial^3\phi}{\partial x^3} + x\frac{\partial^3\phi}{\partial x\partial y^2} + x\frac{\partial^3\phi}{\partial x\partial z^2}.\end{aligned}$$

Combining the first two and the last three terms separately, we obtain

$$\nabla^2\left(x\frac{\partial\phi}{\partial x}\right) = 2\frac{\partial^2\phi}{\partial x^2} + x\frac{\partial}{\partial x}\nabla^2\phi.$$

Substituting the expression for $\nabla^2(x\frac{\partial\phi}{\partial x})$ in (5.56), we obtain

$$\begin{aligned}\left[xp_x, H\right]\phi &= -\mathrm{i}\hbar\left[\left\{-\frac{\hbar^2}{2m}x\frac{\partial}{\partial x}\nabla^2\phi + x\frac{\partial V}{\partial x}\phi + Vx\frac{\partial\phi}{\partial x}\right\}\right.\\ &\quad\left.+\left\{\frac{\hbar^2}{2m}x\frac{\partial}{\partial x}\nabla^2\phi + \frac{\hbar^2}{m}\frac{\partial^2\phi}{\partial x^2} - Vx\frac{\partial\phi}{\partial x}\right\}\right]\\ &= -\mathrm{i}\hbar\left[\frac{\hbar^2}{m}\frac{\partial^2\phi}{\partial x^2} + x\frac{\partial V}{\partial x}\phi\right],\end{aligned}$$

therefore,

$$\left[xp_x, H\right] = -\mathrm{i}\hbar\left(\frac{\hbar^2}{m}\frac{\partial^2}{\partial x^2} + x\frac{\partial V}{\partial x}\right). \tag{5.57a}$$

Similarly,

$$\left[yp_y, H\right] = -\mathrm{i}\hbar\left(\frac{\hbar^2}{m}\frac{\partial^2}{\partial y^2} + y\frac{\partial V}{\partial y}\right) \tag{5.57b}$$

$$\left[zp_z, H\right] = -\mathrm{i}\hbar\left(\frac{\hbar^2}{m}\frac{\partial^2}{\partial z^2} + z\frac{\partial V}{\partial z}\right). \tag{5.57c}$$

Combining the last three equations, we obtain

$$\left(xp_x + yp_y + zp_z, H\right) = -\mathrm{i}\hbar\left(\frac{\hbar^2}{m}\nabla^2 + \mathbf{r}\cdot\text{grad } V\right),$$

or

$$[\mathbf{r}\cdot\mathbf{p}, H] = -\mathrm{i}\hbar\left\{2\left(\frac{\hbar^2}{m}\nabla^2 + \mathbf{r}\cdot\text{grad } V\right) = \mathrm{i}\hbar(2T - \mathbf{r}\cdot\nabla V),\right.$$

or

$$-\frac{\mathrm{i}}{\hbar}[\mathbf{r}\cdot\mathbf{p}, H] = 2T - \mathbf{r}\cdot\nabla V, \tag{5.58}$$

but

$$\frac{\mathrm{d}}{\mathrm{d}t}\langle\mathbf{r}\cdot\mathbf{p}\rangle = -\frac{\mathrm{i}}{\hbar}\langle[\mathbf{r}\cdot\mathbf{p}, H]\rangle. \tag{5.55a$'$}$$

Therefore (5.55a') and (5.58) yield:

$$\frac{\mathrm{d}}{\mathrm{d}t}\langle \mathbf{r}\cdot\mathbf{p}\rangle = 2\langle T\rangle - \langle \mathbf{r}\cdot \text{grad } V\rangle. \tag{5.59}$$

For a stationary state:

$$\frac{\mathrm{d}}{\mathrm{d}t}\langle \mathbf{r}\cdot\mathbf{p}\rangle = 0. \tag{5.60}$$

Therefore (5.59) yields

$$2\langle T\rangle = \langle \mathbf{r}\cdot \nabla V\rangle. \tag{5.61}$$

This equation expresses what is known as the *virial theorem*. As mentioned at the beginning of this chapter, it gives a relationship between the expectation values of quantities involving T and V of a particle in a stationary state, and is helpful in simplifying certain computations.

Problem

5.7. Show that

$$[x, p_x] = \mathrm{i}\hbar.$$

Example 5.5.

Problem

Show that the difference between $\mathbf{r}\cdot\mathbf{p}$ and $\mathbf{p}\cdot\mathbf{r}$ is a constant.

Solution

We have

$$\begin{aligned}(\mathbf{r}\cdot\mathbf{p} - \mathbf{p}\cdot\mathbf{r})\Psi &= \left(xp_x + yp_y + zp_z\right)\Psi - \left(p_x x + p_y y + p_z z\right)\Psi \\ &= \left(xp_x - p_x x\right)\Psi + \left(yp_y - p_y y\right)\Psi + (zp_z - p_z z)\Psi \\ &= \mathrm{i}\hbar\Psi + \mathrm{i}\hbar\Psi + \mathrm{i}\hbar\Psi = 3\mathrm{i}\hbar\Psi,\end{aligned}$$

where the commutation relations $[x, p_x] = \mathrm{i}\hbar$, … have been used. Hence, we obtain

$$(\mathbf{r}\cdot\mathbf{p} - \mathbf{p}\cdot\mathbf{r})\Psi = 3\mathrm{i}\hbar\Psi,$$

therefore,

$$\mathbf{r} \cdot \mathbf{p} - \mathbf{p} \cdot \mathbf{r} = 3i\hbar.$$

This shows that the difference between $\mathbf{r} \cdot \mathbf{p}$ and $\mathbf{p} \cdot \mathbf{r}$ is constant. Therefore, in the proof of the virial theorem, it is immaterial whether we start with $\mathbf{r} \cdot \mathbf{p}$ or $\mathbf{p} \cdot \mathbf{r}$ because the difference between them, being constant, commutes with H.

Problems

5.8. For a spherically symmetric potential $V = r^n$, find out the relation between the expectation values of T and V.

5.9. Prove that

$$m\frac{\mathrm{d}}{\mathrm{d}t}\langle x^2\rangle = \langle p_x x\rangle + \langle x p_x\rangle.$$

Example 5.6.

Problem

The normalised energy eigenfunction corresponding to the first excited state of a linear harmonic oscillator is given by

$$\psi_1(x) = \left(\frac{4\alpha^3}{\pi}\right)^{1/4} x \exp\frac{\alpha x^2}{2}, \quad \alpha = \frac{\sqrt{mk}}{2}.$$

By using the virial theorem, or otherwise, show that $\Delta x \Delta p_x = \frac{3\hbar}{2}$. Also calculate the expectation values of the kinetic and potential energies. (It may be asumed that

$$\int \mathrm{e}^{-\alpha x^2}\,\mathrm{d}x = \sqrt{\frac{\pi}{\alpha}},$$

$$\int x^2 \mathrm{e}^{-\alpha x^2}\,\mathrm{d}x = \sqrt{\frac{\pi}{4\alpha^3}},$$

$$\int x^4 \mathrm{e}^{-\alpha x^2}\,\mathrm{d}x = \sqrt{\frac{9\pi}{16\alpha^5}},$$

$$\int f(x)\,\mathrm{d}x = 0,$$

where $f(x)$ is an odd function of x. The integration is from $-\infty$ to $+\infty$.)

Solution

The eigenfunction ψ_1 of a harmonic oscillator in its first excited state is given by

$$\psi_1(x) = \left(\frac{4\alpha^3}{\pi}\right)^{1/4} x \exp\left(-\frac{\alpha x^2}{2}\right), \quad \alpha = \frac{\sqrt{mk}}{2}.$$

Then the expectation value of x is given by

$$\langle x \rangle = \int \psi_1^*(x) x \psi_1(x)\,\mathrm{d}x.$$

Since all three factors in the integrand in the above integral are odd functions of x, their product is therefore also an odd function of x. The integral of an odd function between $-\infty$ and ∞ is zero. Hence $\langle x \rangle = 0$.

The expectation value of x^2 is given by

$$\begin{aligned}\langle x \rangle^2 &= \int \psi_1^*(x) x^2 \psi_1(x)\,\mathrm{d}x \\ &= \int \left(\frac{4\alpha^3}{\pi}\right)^{1/2} x^4 \exp\left(-\alpha x^2\right)\mathrm{d}x \\ &= \frac{3}{2\alpha}.\end{aligned}$$

The expectation value of p_x is given by

$$\langle p_x \rangle = \int \psi_1^*(x) p_x \psi_1(x)\,\mathrm{d}x.$$

Since $\psi_1(x)$ is an odd function of x, its derivative p_x [$\equiv -\mathrm{i}\hbar \mathrm{d}/\mathrm{d}x$]$\psi_1(x)$ must be an even function of x. Therefore, its product with an odd function of x, i.e. $\psi_1^*(x)$, will yield an odd function of x. Hence the integral and consequently $\langle p_x \rangle = 0$.

Now by virtue of the virial theorem, for a harmonic oscillator, the expectation value of the kinetic energy is given by

$$\langle T \rangle = \frac{1}{2}\left\langle kx^2 \right\rangle = \frac{k}{2}\left\langle x^2 \right\rangle. \tag{5.62}$$

We have already proved that

$$\left\langle x^2 \right\rangle = \frac{3}{2\alpha}.$$

Substituting this value of $\langle x^2 \rangle$ in (5.62), we obtain

$$\langle T \rangle = \frac{3k}{4\alpha}.$$

As $T = p^2/2m$, we have

$$\langle p_x^2 \rangle = 2m\langle T \rangle = \frac{3mk}{2\alpha}, \tag{5.63}$$

but

$$\alpha = \frac{\sqrt{mk}}{\hbar}.$$

Substituting this value in (5.63), we obtain

$$\langle p_x^2 \rangle = \frac{3\alpha\hbar^2}{2}.$$

We can now obtain the values of Δx and Δp. We have

$$\Delta x = +\sqrt{\langle x^2 \rangle - \langle x \rangle^2} = \sqrt{\frac{3}{2\alpha}}.$$

Similarly,

$$\Delta p_x = +\sqrt{\langle p_x^2 \rangle - \langle p_x \rangle^2} = \sqrt{\frac{3\alpha\hbar^2}{2}}.$$

Multiplying the equations for Δx and Δp_x, we obtain

$$\Delta x \Delta p_x = \frac{3}{2}\hbar.$$

Since $V = \frac{1}{2}kx^2$, the expectation value of V is the same as that of T:

$$\langle T \rangle = \langle V \rangle = \frac{3k}{4\alpha}.$$

Problem

5.10. Show that the value of $\Delta x \Delta p_x$ for the harmonic oscillator in the nth state is given by

$$\Delta x \Delta p_x = \left(n + \frac{1}{2}\right)\hbar.$$

5.7 Heisenberg's uncertainty principle

Let us now derive mathematically Heisenberg's principle of uncertainty. Consider an operator $A + i\lambda B$, where A and B are Hermitian operators and λ is a real number. Then

$$\overline{(A + i\lambda B)(A + i\lambda B)^\dagger} \geqslant 0$$

where $\langle A \rangle \equiv \overline{A}$ denotes the expectation value. (We will be using the two notations interchangeably.) Or

$$\overline{(A + i\lambda B)\left(A^{\dagger} - i\lambda B^{\dagger}\right)} \geqslant 0,$$

because λ being a real number, $\lambda^{\dagger} = \lambda$, and i being a complex number $i^{\dagger} = -i$. Since the operators A and B are Hermitian, we have $A^{\dagger} = A$ and $B^{\dagger} = B$. Then the above inequality gives

$$\overline{(A + i\lambda B)(A - i\lambda B)} \geqslant 0,$$

or

$$\overline{A^2 - i\lambda AB + i\lambda BA) + \lambda^2 B^2} \geqslant 0,$$

or

$$\overline{A^2} + \overline{\lambda^2 B^2} - i\lambda\overline{(AB - BA)} \geqslant 0, \tag{5.64}$$

or

$$f(\lambda) \geqslant 0,$$

where

$$f(\lambda) = \overline{A^2} + \overline{\lambda^2 B^2} - i\lambda\overline{(AB - BA)}. \tag{5.65}$$

Differentiating with respect to λ, we obtain

$$f'(\lambda) = 2\lambda\overline{B^2} - i\overline{(AB - BA)}. \tag{5.66}$$

Differentiating again, we obtain

$$f''(\lambda) = 2\overline{B^2}.$$

But as B is Hermitian, $\overline{B^2} \geqslant 0$. Therefore, $f''(\lambda) \geqslant 0$. Since $f''(\lambda)$ cannot be negative, $f(\lambda)$ has no maximum. For the minimum value of $f(\lambda)$, its differential coefficient with respect to λ must be zero:

$$f'(\lambda) = 0,$$

or

$$2\lambda\overline{B^2} - i\overline{(AB - BA)} = 0,$$

or

$$\lambda = \frac{i}{2}\frac{\overline{AB - BA}}{\overline{B^2}}.$$

Substituting this value of λ in (5.64), we obtain

$$\overline{A^2} - \frac{1}{4}\frac{\overline{(AB - BA)}^2}{\overline{B^2}^2}\overline{B^2} + \frac{1}{2}\frac{\left(\overline{AB - BA}\right)^2}{\overline{B^2}} \geqslant 0,$$

or

$$\overline{A^2}\,\overline{B^2} \geqslant -\frac{1}{4}\overline{(AB - BA)}^2. \tag{5.67}$$

This relation is valid for all Hermitian operators.

Let us next define operators P and Q and their expectation values $\langle P\rangle$ and $\langle Q\rangle$ such that

$$A = P - \langle P\rangle$$

and

$$B = Q - \langle Q\rangle.$$

Then for the operator A, we have

$$A^2 = P^2 + \langle P\rangle^2 - 2\langle P\rangle P$$

so that

$$\begin{aligned}\langle A^2\rangle &= \langle P^2\rangle + \langle P\rangle^2 - 2\langle P\rangle\langle P\rangle \\ &= \langle P^2\rangle + \langle P\rangle^2 - 2\langle P\rangle^2 \\ &= \langle P^2\rangle - \langle P\rangle^2 \\ &= (\Delta P)^2,\end{aligned}$$

where ΔP is the uncertainty in the measurement of the physical quantity represented by the operator P. Similarly, if ΔQ is the uncertainty in the measurement of the physical quantity represented by Q, then

$$\langle B^2\rangle = \langle \Delta Q\rangle^2.$$

Moreover:

$$\begin{aligned}AB - BA &= (P - \langle P\rangle)(Q - \langle Q\rangle) - (Q - \langle Q\rangle)(P - \langle P\rangle) \\ &= PQ - P\langle Q\rangle - \langle P\rangle Q + \langle P\rangle\langle Q\rangle \\ &\quad - QP + P\langle Q\rangle + \langle P\rangle Q - \langle P\rangle\langle Q\rangle \\ &= PQ - QP = [P, Q].\end{aligned}$$

Substituting the expressions involving operators A and B and the expectation values of A^2 and B^2 in relation (5.67), we obtain

$$(\Delta P)^2(\Delta Q)^2 \geqslant -\frac{1}{4}(\langle [P, Q]\rangle)^2. \tag{5.68}$$

Let us now consider the position and momentum operators x and p_x. The commutator of x and p_x is

$$[x, p_x] = i\hbar.$$

Then, by virtue of (5.68), writing $P = x$ and $Q = p_x$, we obtain

$$(\Delta x)^2 \left(\Delta p_x\right)^2 \geqslant -\frac{1}{4}\left(\left\langle [x, p_x] \right\rangle\right)^2,$$

or

$$(\Delta x)^2 \left(\Delta p_x\right)^2 \geqslant -\frac{1}{4}(\langle i\hbar \rangle)^2,$$

or

$$(\Delta x)^2 \left(\Delta p_x\right)^2 \geqslant \frac{\hbar^2}{4},$$

or

$$\Delta x \Delta p_x \geqslant \frac{\hbar}{2}. \tag{5.69}$$

If the operators P and Q commute, then $[P, Q] = 0$. According to (5.68), this yields

$$\Delta P \Delta Q \geqslant 0.$$

This shows that the product $\Delta P \Delta Q$ of uncertainties in the measurement of the observables represented by the Hermitian operators P and Q *can* be zero, i.e. both the observables, represented by these operators, can be measured precisely at the same time. Such observables are said to be *compatible*. Thus, the *compatible observables* must be represented by *commuting operators*. The measurement of any one of these observables does not affect the values assigned to others by preceding measurements. For instance, the three components of linear momentum are compatible as the operators representing them mutually commute. If the commutator is different from zero, a simultaneous precise measurement of observables represented by these operators is impossible. In such a case, the observables, such as position and momentum, are said to be *incompatible*. Extending the result to the compatibility of a number of observables, we see that a simultaneous precise measurement can be made only for those observables of a system whose operators mutually commute. But what is the maximum number of attributes which can be determined precisely at the same time? *Nobody knows*. However, we *assume* that there is a maximum number of attributes which can be determined precisely at the same time. These must be represented by mutually commuting Hermitian operators. The appearance of $\hbar$ in the inequality (5.69) demonstrates the incompatibality of observables as essentially a quantum mechanical phenomenon. It does not occur in classical physics where, for instance, the state of a particle is determined by a simultaneous precise measurement of its position and momentum. As $\hbar \to 0$, then as required by Bohr's correspondence principle, the quantum mechanical phenomenon disappears. If the Hamiltonian of a physical system is one of the complete set of commuting observables, then all the others are constants of the motion.

5.8 The parity operator

We have defined the parity of any state of a system as even or odd according to whether the wave function representing that state is an even or odd function of space variables. Let us next define a parity operator. Consider an operator P which is defined by the equation

$$P\psi(x) = \psi(-x),$$

i.e. it changes the sign of the space coordinate of the function on which it acts. Operating on both sides of the above equation from the left by P, we obtain

$$P[P\psi(x)] = P\psi(-x),$$

or

$$P^2\psi(x) = \psi(x).$$

The function $\psi(x)$ is, therefore, an eigenfunction of P^2 with eigenvalue 1. We write it as

$$P^2 = 1,$$

or

$$P = \pm 1.$$

Thus the operator P can have two eigenvalues, +1 and −1. For the eigenvalue +1, we have

$$P\psi(x) = \psi(x),$$

but, by definition:

$$P\psi(x) = \psi(-x).$$

A comparison of the last two equations shows that

$$\psi(-x) = \psi(x).$$

Thus, an even function of x is always an eigenfunction of P with eigenvalue +1. That is, the state of even parity is associated with the +1 eigenvalue of the operator P. Similarly we can show that the state of odd parity corresponds to the eigenvalue −1 of the operator P. Hence, we can state that a system has an even or odd parity according to whether the eigenvalue of the operator P is +1 or −1. The operator P is called the *parity operator*. Can this operator represent a physical quantity? For this, the operator should be linear and Hermitian. We will first show that P is a linear operator. Let $u(x)$ and $v(x)$ be two arbitrary functions of x. Then

$$\begin{aligned} P\{u(x) + v(x)\} &= u(-x) + v(-x) \\ &= Pu(x) + Pv(x) \end{aligned}$$

and

$$P\{cu(x)\} = cu(-x) = cPu(x).$$

We next show that the linear operator P is Hermitian: $P^\dagger = P$. By the definition of the Hermitian conjugate of an operator, we have

$$\int \psi(x) P^* \phi^*(x)\,\mathrm{d}x = \int \phi^*(x) P^\dagger \psi(x)\,\mathrm{d}x. \tag{5.70}$$

The left-hand side of the above equation can be written as

$$\int \psi(x)\{P\phi(x)\}^*\,\mathrm{d}x = \int \psi(x)\phi^*(-x)\,\mathrm{d}x.$$

Changing the variable from x to $u = -x$ in the right-hand side of the above equation, we have

$$\begin{aligned}\int \psi(x)\{P\phi(x)\}^*\,\mathrm{d}x &= \int \psi(-u)\phi^*(u)\,\mathrm{d}(-u)\\ &= \int \psi(-u)\phi^*(u)\,\mathrm{d}u\\ &= \int \phi^*(u)\psi(-u)\,\mathrm{d}u\\ &= \int \phi^*(x)\psi(-x)\,\mathrm{d}x,\end{aligned}$$

because renaming the variable of integration does not change the value of the integral. Thus, the above equation yields

$$\int \psi(x) P^* \phi^*(x)\,\mathrm{d}x = \int \phi^*(x) P\psi(x)\,\mathrm{d}x. \tag{5.71}$$

Comparing (5.70) and (5.71), we obtain

$$P^\dagger = P.$$

This shows that the linear operator P is Hermitian and therefore can represent a physical quantity. We *assume* that P does represent a physical quantity. This quantity has already been labelled as parity. The parity of a system can have only two values: ± 1.

Is this parity a constant of motion? We will show that this is the case provided the Hamiltonian of the system is an even function of x, i.e. if $H(-x) = H(x)$. For then

$$P[H(x)\psi(x)] = H(-x)\psi(-x) = H(x)\psi(-x) = H(x)P\psi(x)$$

so that

$$PH = HP,$$

or

$$[P, H] = 0.$$

This shows that P commutes with H. Moreover, P does not depend explicitly upon time. It is, therefore, a constant of the motion.

Now, by definition, the parity operator P transforms a wave function into its mirror image: $P\psi(x) = \psi(-x)$. Therefore, if the mirror image of a physical system can be realised in nature with equal probability, we say that parity is conserved, and the system is said to be invariant under parity transformation. It had been assumed up to 1956 that parity was conserved under all circumstances. In 1956, Lee and Yang showed that this was not true. Parity is not conserved in weak interactions. To test the conservation of parity in weak interactions, we must set up an experimental arrangement which contains weak interactions and is *not* identical to its mirror image. If the mirror image is realised in nature with equal probability, parity is conserved; otherwise it is violated.

It can be metioned that an arbitrary function $\psi(x)$ can always be written as

$$\begin{aligned}\psi(x) &= \frac{1}{2}\psi(x) + \frac{1}{2}\psi(x) - \frac{1}{2}\psi(-x) + \frac{1}{2}\psi(-x)\\ &= \frac{1}{2}[\psi(x) + \psi(-x)] + \frac{1}{2}[\psi(x) - \psi(-x)]\\ &= \text{an even function} + \text{an odd function.}\end{aligned}$$

That is, it can be expressed as a sum of an even and an odd function.

Example 5.7.

Problem

If $u_+ = p + \mathrm{i}q$ and $u_- = p - \mathrm{i}q$, where p and q are Hermitan operators such that $[p, q] = -\mathrm{i}k$ and k is a positive number, prove that

$$u_-u_+ - u_+u_- = 2k.$$

Also show that if λ is an eigenvalue of the operator u_+u_- corresponding to the eigenfunction ϕ, then $(u_+\phi)$ is also an eigenfunction of u_+u_- and corresponds to the eigenvalue $\lambda + 2k$. Show that the possible eigenvalues of u_+u_- are λ, $\lambda + 2k$, $\lambda + 4k$, $\lambda + 6k$, ...

Solution

We have

$$\begin{aligned}u_-u_+ - u_+u_- &= (p - \mathrm{i}q)(p + \mathrm{i}q) - (p + \mathrm{i}q)(p - \mathrm{i}q)\\ &= p^2 + \mathrm{i}pq - \mathrm{i}qp + q^2 - p^2 + \mathrm{i}pq - \mathrm{i}qp - q^2\\ &= 2\mathrm{i}pq - 2\mathrm{i}qp = 2\mathrm{i}(pq - qp)\\ &= 2\mathrm{i}[p, q] = 2\mathrm{i}(-\mathrm{i}k) = 2k.\end{aligned}$$

This is the required solution of the first part of the problem.

To prove the second part, we note that if λ is the eigenvalue of the operator u_+u_- corresponding to the eigenfunction ϕ, then we may write

$$u_+u_-\phi = \lambda\phi.$$

By making use of the relation

$$u_-u_+ - u_+u_- = 2k$$

we have

$$u_-u_+ = u_+u_- + 2k.$$

Operating upon ϕ, we obtain

$$\begin{aligned} u_-u_+\phi &= u_+u_-\phi + 2k\phi \\ &= \lambda\phi + 2k\phi. \end{aligned}$$

Operating from the left by u_+, we obtain

$$u_+u_-(u_+\phi) = (\lambda + 2k)(u_+\phi).$$

By repeating the operation by u_+u_-, it can be easily deduced that the possible eigenvalues of u_+u_- are λ, $\lambda + 2k$, $\lambda + 4k$, $\lambda + 6k$, ….

This proves the second part.

Example 5.8.

Problem

Show that the eigenvalues of the operator $A^\dagger A$ are non-negative.

Solution

Let λ be the eigenvalue of the given operator $A^\dagger A$ corresponding to the eigenfunction ψ. Then, the eigenvalue equation is

$$\left(A^\dagger A\right)\psi = \lambda\psi.$$

Then, the expectation value of this operator is given by

$$\left\langle A^\dagger A\right\rangle = \int \psi^* A^\dagger A\psi \, \mathrm{d}\tau.$$

By using the eigenvalue equation, the above relation yields

$$\left\langle A^\dagger A\right\rangle = \int \psi^* \lambda\psi \, \mathrm{d}\tau = \lambda.$$

This shows that the expectation value of the operator $A^\dagger A$ is also λ. But the expectation value of the operator $A^\dagger A$ is either equal to or greater than zero:

$$\left\langle A^\dagger A \right\rangle \geqslant 0.$$

Hence, by virtue of the last two relations, we obtain $\lambda \geqslant 0$. This is the desired result.

Example 5.9.

Problem

Show that a unitary operaor U can be expressed as

$$U = \frac{I + \mathrm{i}K}{I - \mathrm{i}K}$$

where K is a Hermitian operator and the operator $1/(I - \mathrm{i}K)$ is the inverse of $(I - \mathrm{i}K)$ and stands either to the left or right of $(I + \mathrm{i}K)$.

Solution

It has been given that the inverse of $(I - \mathrm{i}K)$ exists. Then, writing $1/(I - \mathrm{i}K)$ to the right of $(I + \mathrm{i}K)$, we obtain

$$U = (I + \mathrm{i}K)\frac{1}{1 - \mathrm{i}K} = (I + \mathrm{i}K)(I - \mathrm{i}K)^{-1}.$$

Taking the Hermitian conjugate of both sides, we obtain

$$\begin{aligned} U^\dagger &= \left[(I - \mathrm{i}K)^{-1}\right]^\dagger (I + \mathrm{i}K)^\dagger \\ &= \left[(I - \mathrm{i}K)^\dagger\right]^{-1} (I + \mathrm{i}K)^\dagger \\ &= \left(I + \mathrm{i}K^\dagger\right)^{-1}\left(I - \mathrm{i}K^\dagger\right) \\ &= (I + \mathrm{i}K)^{-1}(I - \mathrm{i}K), \end{aligned}$$

because K is Hermitian: $K^\dagger = K$. Now, the product of $U^\dagger$ and U gives

$$U^\dagger U = (I + \mathrm{i}K)^{-1}(I - \mathrm{i}K)(I + \mathrm{i}K)(I - \mathrm{i}K)^{-1}.$$

Since $(I - \mathrm{i}K)$ and $(I + \mathrm{i}K)$ commute, the above equation can be written as

$$U^\dagger U = (I + \mathrm{i}K)^{-1}(I + \mathrm{i}K)(I - \mathrm{i}K)(I - \mathrm{i}K)^{-1} = I \cdot I = I$$

Similarly, it can be shown that

$$UU^\dagger = I.$$

Proceeding on the same lines, we can show that the relation $UU^\dagger = I = U^\dagger U$ holds good even when $1/(I - \mathrm{i}K)$ is on the left side of $(I + \mathrm{i}K)$.

Remarks

The exponential operator $\exp(-\mathrm{i}Ht/\hbar)$ is an abbreviation for

$$1 - \frac{\mathrm{i}Ht}{\hbar} + \frac{\left(\frac{\mathrm{i}Ht}{\hbar}\right)^2}{2!\hbar} - \frac{\left(\frac{\mathrm{i}Ht}{\hbar}\right)^3}{3!\hbar} + \cdots.$$

5.9 Antilinear operators

An operator A is said to be *antilinear* if it satisfies the equation

$$A\left(c_1\psi_1 + c_2\psi_2\right) = c_1^* A\psi_1 + c_2^*\psi_2,$$

where ψ_1 and ψ_2 are two arbitrary functions and c_1 and c_2 are in general complex numbers.

Problem

5.11. Show that if A and B are two antilinear operators, then the product AB is a linear operator.

Antilinear transformations are important in describing the behaviour of a system under time-reversal.

5.10 Antiunitary operators

An *antilinear operator A* is said to be *antiunitary* if

$$A^\dagger A = I = AA^\dagger.$$

Problem

5.12. Show that if a transformation preserves the absolute value of the scalar product, then it is either unitary or antiunitary.

Additional problems

5.13. Show that every unitary operator U can be written in the form

$$U = A + iB$$

where A and B are commuting Hermitian operators. Should A and B satisfy some equation?

5.14. Show that a unitary operator U can always be expressed in the form

$$U = e^{iK},$$

where K is a Hermitian operator.

5.15. Find the eigenvalues of an operator A which is such that $A^3 = I$. What will be its allowed eigenvalue if it is Hermitian?

5.16. Prove that the position and linear momentum operators, x_i and p_j satisfy the following commutation relations:

$$[x_i, x_j] = [p_i, p_j] = 0,$$

$$[p_i, x_j] = -[x_j, p_i] = -i\hbar\delta_{ij}.$$

5.17. Find out whether the matrix

$$A = \begin{bmatrix} 3 & i \\ -i & 7 \end{bmatrix}$$

can represent an observable in quantum mechanics or not.

5.18. Show that the integrated probability is conserved in time provided that the total energy operator is Hermitian. Also prove its converse.

5.19. State and prove the virial theorem.

5.20. Define parity and show that it is a constant of motion.

5.21. Prove the Ehrenfest theorem. Does it differ in terms of expectation values from Newton's second law of motion?

5.22. Show that if ΔE and Δt are uncertainties in the measurement of energy and time, then

$$\Delta E \Delta t \geqslant \hbar.$$

5.23. Show that starting from a set of n linearly independent vectors in an n-dimensional space, an orthogonal basis can always be found by the Schmidt orthogonalisation method.

5.24. Let 'Q' be a quantum mechanical operator corresponding to a general physical quantity. Prove that the time rate of change of the expectation value of 'Q' is given by

$$\frac{d}{dx}\langle Q\rangle = \frac{i}{\hbar}\langle [H, Q]\rangle + \frac{\partial Q}{\partial t}$$

where $[H, Q] = HQ - QH$.

Chapter 6

Potentials with finite discontinuities

We will now consider the motion of a *microscopic entity*, say an electron, moving in a one-dimensional space which is divided into two or more regions, in each of which the potential is finite and constant and different from adjacent regions. It means that in going from one region to the other, the potential changes suddenly through *finite* values. We shall consider three such cases and see how the results obtained quantum mechanically differ from or are similar to those obtained from classical theory. Such sudden changes in potential do not occur in physical systems but are approximated by them and therefore are used to explain such physical phenomena which cannot be interpreted by classical physics.

We will consider a uniform beam of non-interacting electrons so that whenever desirable, we may concentrate on a single electron for the analysis of a problem. If the motion of a microscopic entity, say an electron, is confined to a finite region, the microscopic entity is said to be in a bound state.

6.1 Potential steps

We will first consider a microscopic entity, say an electron of mass m, moving from left to right with energy E in a one-dimensional space taken as the x-axis. The space is divided into two infinite regions with different constant potentials, the potential changing discontinuously through a finite jump, say, from $V = 0$ to $V = V_0$, at the point where the two regions meet. This point, called the *turning point*, is taken as the origin ($x = 0$) so that, as shown in figure 6.1, the potential changes in going from one region to the other according to

$$\begin{aligned} V(x) &= 0, \quad \text{for } x < 0, \\ &= V_0, \quad \text{for } x > 0, \end{aligned} \tag{6.1}$$

where V_0 *is a positive number*. This is called a *potential step* problem. It may be mentioned that a potential step does not actually exist but it is approximated by

doi:10.1088/978-0-7503-1206-6ch6

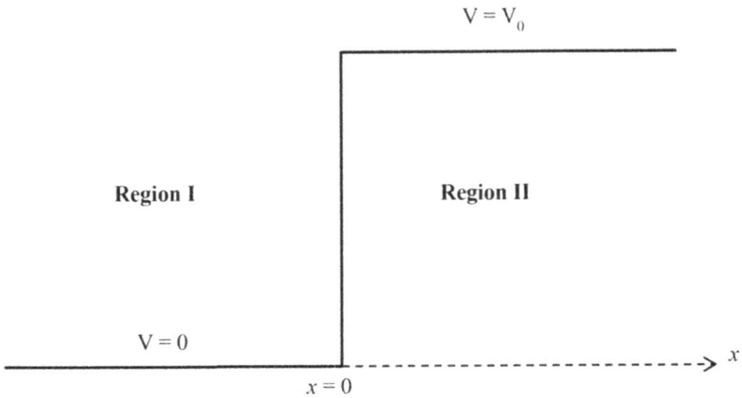

Figure 6.1. Potential step.

physical problems such as the surface potential step of a metal. Let us see what classical mechanics and quantum mechanics predict about the behaviour of an electron in such a problem. Since the potential energy, being a constant, does not depend upon time explicitly, in quantum mechanics it is sufficient to consider the amplitude equation. The solution $\psi(x)$ when multiplied by the time factor $e^{-iEt/\hbar}$, where E is the total energy of the electron, gives the wave function $\Psi(x, t)$ representing the electron. In the two regions, I and II, the amplitude equation can be written as

$$\text{Region I:} \quad \frac{d^2\psi_1}{dx^2} + \frac{2m}{\hbar^2}E\psi_1 = 0, \quad \text{for } x < 0, \tag{6.2}$$

$$\text{Region II:} \quad \frac{d^2\psi_2}{dx^2} + \frac{2m}{\hbar^2}(E - V_0)\psi_2 = 0, \quad \text{for } x > 0. \tag{6.3}$$

The Schrödinger amplitude equation in each region is treated as independent of the other.

Case 1. $E > V_0 > 0$.
Let us first consider the case when E, the total energy of the electron, is greater than the potential energy V_0, i.e. $E > V_0$, where V_0 is positive. This is shown in figure 6.2. Classically, all the electrons coming from region I will be transmitted and none will be reflected at $x = 0$ as each one of them has enough energy to enter region II for which $x > 0$. Let us see what quantum mechanics tells us about this process. Writing

$$\alpha = +\frac{\sqrt{2mE}}{\hbar} \quad \text{and} \quad \beta = +\frac{\sqrt{2m(E - V_0)}}{\hbar}, \tag{6.4}$$

where α^2 and β^2 are both positive, the Schrödinger amplitude equation in the two regions takes the form given below:

$$\text{Region I:} \quad \frac{d^2\psi_1}{dx^2} + \alpha^2\psi_1 = 0, \quad \text{for } x < 0, \tag{6.5}$$

$$\text{Region II:} \quad \frac{d^2\psi_2}{dx^2} + \beta^2\psi_2 = 0, \quad \text{for } x > 0. \tag{6.6}$$

Then the solutions of the Schrödinger amplitude equation in the two regions are

$$\text{Region I:} \quad \psi_1 = Ae^{i\alpha x} + Be^{-i\alpha x}, \tag{6.7}$$

$$\text{Region II:} \quad \psi_2 = Ce^{i\beta x} + De^{-i\beta x}, \tag{6.8}$$

where A, B, C, D are constants of integration. The functions $e^{i\alpha x}$ and $e^{-i\alpha x}$ when multiplied by the time factor $e^{-iEt/\hbar}$ represent the electrons moving in region I towards the right and towards the left, respectively. However, as is customary in the literature, we will be talking about the moving particles or waves without mentioning the time factor. In region I, the electrons will be moving not only from left to right but also from right to left, the latter after reflection, if any, from $V = V_0$ at $x = 0$. Similarly, $e^{i\beta x}$ and $e^{-i\beta x}$ represent waves travelling, respectively, towards the right and towards the left in region II. Since there is no potential step in region II, no electrons will be moving from right to left in this region. We would therefore take $D = 0$. Equations (6.7) and (6.8) then become

$$\text{Region I:} \quad \psi_1 = Ae^{i\alpha x} + Be^{-i\alpha x}, \tag{6.9}$$

$$\text{Region II:} \quad \psi_2 = Ce^{i\beta x}. \tag{6.10}$$

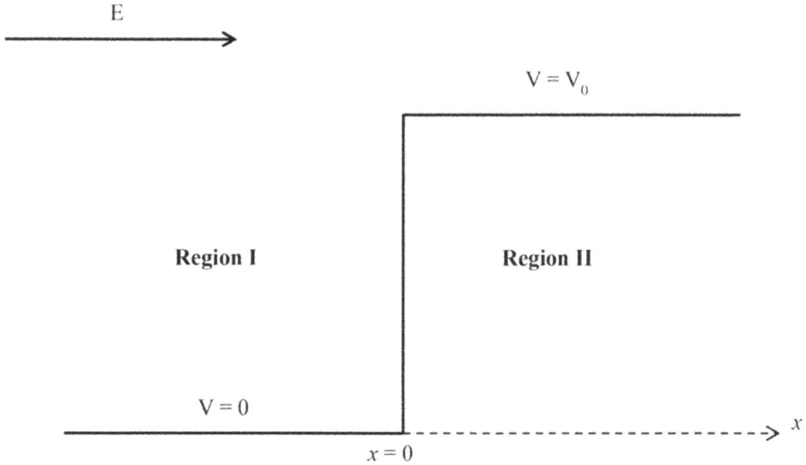

Figure 6.2. Potential step with $E > V_0$

Now the derivatives of ψ_1 and ψ_2 with respect to x are:

$$\text{Region I:} \quad \frac{d\psi_1}{dx} i\alpha A e^{i\alpha x} - i\alpha B e^{-i\alpha x}, \tag{6.11}$$

$$\text{Region II:} \quad \frac{d\psi_2}{dx} = i\beta C e^{i\beta x}. \tag{6.12}$$

The potential has a finite discontinuity at the step $x = 0$ but the amplitude and its first derivative should be continuous even at the step. This requires that

$$\psi_1 = \psi_2, \quad \text{at } x = 0,$$

and

$$\frac{d\psi_1}{dx} = \frac{d\psi_2}{dx}, \quad \text{at } x = 0.$$

These conditions as applied to (6.9) and (6.10), and to (6.11) and (6.12) yield

$$A + B = C \tag{6.13}$$

and

$$i\alpha(A - B) = i\beta C. \tag{6.14}$$

Substituting the expression for C from (6.13) in (6.14), we obtain

$$i\alpha(A - B) = i\beta(A + B),$$

or

$$(\alpha - \beta)A = (\alpha + \beta)B,$$

or

$$B = \frac{\alpha - \beta}{\alpha + \beta} A. \tag{6.15}$$

Substituting the expression for B from (6.13) in (6.14), we obtain

$$i\alpha(A + A - C) = i\beta C,$$

or

$$\alpha(2A - C) = \beta C,$$

or

$$C = \frac{2\alpha}{\alpha + \beta} A. \tag{6.16}$$

The wave functions ψ_1 and ψ_2 are then given by

$$\psi_1 = Ae^{i\alpha x} + Be^{-i\alpha x} = Ae^{i\alpha x} + \frac{\alpha - \beta}{\alpha + \beta}e^{-i\alpha x}, \quad x < 0,$$

$$\psi_2 = Ce^{i\beta x} = \frac{2\alpha}{\alpha + \beta}Ae^{i\beta x}. \quad x > 0.$$

We know that the probability flux **j**, i.e. the probability per unit time that a microscopic entity represented by the wave function $\Psi(x, t)$ will cross the point x, is given by

$$\begin{aligned}\mathbf{j} &= \frac{\hbar}{m}\,\mathrm{Im}\left(\Psi^*\frac{\partial\Psi}{\partial x}\right)\\ &= \frac{\hbar}{m}\,\mathrm{Im}\left(\psi^* e^{iEt/\hbar}e^{-iEt/h}\right)\frac{d\psi}{dx}, \quad \text{as } \Psi(x, t) = \psi(x)e^{-iEt/\hbar},\\ &= \frac{\hbar}{m}\,\mathrm{Im}\left(\psi^*\frac{d\psi}{dx}\right). \end{aligned} \tag{6.17}$$

Let us now calculate the incident probability flux, S_{incident}, i.e. the probability per unit time that the incident electron represented by the wave function Ψ will cross the point $x = 0$. We have

$$\begin{aligned}S_{\text{incident}} &= \frac{\hbar}{m}\mathrm{Im}\left(\psi^*_{\text{incident}}\frac{d\psi_{\text{incident}}}{dx}\right)\\ &= \frac{\hbar}{m}\,\mathrm{Im}\left(A^* e^{-i\alpha x} i\alpha A e^{i\alpha x}\right)\\ &= \frac{\hbar}{m}\,\mathrm{Im}\left(i\alpha |A|^2\right)\\ &= \frac{\hbar}{m}\alpha |A|^2 . \end{aligned} \tag{6.18}$$

The reflected probability flux, $S_{\text{reflected}}$, is given by

$$\begin{aligned}S_{\text{reflected}} &= \frac{\hbar}{m}\mathrm{Im}\left(\psi^*_{\text{reflected}}\frac{d\psi_{\text{reflected}}}{dx}\right)\\ &= \frac{\hbar}{m}\,\mathrm{Im}\left\{B^* e^{i\alpha x}(-i\alpha)Be^{-i\alpha x}\right\}\\ &= \frac{\hbar}{m}\,\mathrm{Im}\left(-i\alpha |B|^2\right)\\ &= -\frac{\hbar}{m}\alpha |B|^2, \end{aligned} \tag{6.19}$$

the negative sign merely indicating that after reflection the electron moves along the negative direction of the x-axis.

Similarly, the transmitted probability flux, $S_{\text{transmitted}}$, is given by

$$S_{\text{transmitted}} = \frac{\hbar}{m}\,\text{Im}\left(\psi^*_{\text{transmitted}}\frac{d\psi_{\text{transmitted}}}{dx}\right)$$

$$= \frac{\hbar}{m}\,\text{Im}\left\{C^*e^{-i\beta x}(i\beta)Ce^{i\beta x}\right\}$$

$$= \frac{\hbar}{m}\,\text{Im}\left(-i\beta|C|^2\right)$$

$$= \frac{\hbar}{m}\beta|C|^2. \tag{6.20}$$

The ratio of the reflected probability flux to the incident probability flux is called the *coefficient of reflection* or *reflectivity* and is denoted by R. It represents the fraction of electrons which are reflected. Thus, we have

$$R = \frac{S_{\text{reflected}}}{S_{\text{incident}}} = \frac{|B|^2}{|A|^2}$$

the negative sign being ignored as in $S_{\text{reflected}}$, it merely indicates that after reflection the electron moves along the negative direction of the x-axis. By virtue of (6.15), it reduces to

$$R = \left(\frac{\alpha - \beta}{\alpha + \beta}\right)^2. \tag{6.21}$$

This gives the probability that an electron coming from the left will be reflected at the interface $x = 0$.

Similarly, the ratio of the transmitted probability flux to the incident probability flux is called the *coefficient of transmission* or *transmissivity* and is denoted by T. Hence, we have

$$T = \frac{S_{\text{transmitted}}}{S_{\text{incident}}} = \frac{\beta\,|C|^2}{\alpha\,|A|^2}.$$

Substituting the expression for C from (6.16) in the above equation, we obtain

$$T = \frac{\beta}{\alpha}\frac{4\alpha^2}{(\alpha + \beta)^2} = \frac{4\alpha\beta}{(\alpha + \beta)^2}. \tag{6.22}$$

This gives the probability of transmission.

Problems

6.1. What is the physical interpretation of the symmetry of expressions for R and T as given in (6.21) and (6.22) with respect to interchange of α and β?

6.2. What will happen when $\beta = \alpha$?

Since the electron is either reflected or transmitted, the total probability must be one. It can be easily verified that

$$R + T = 1$$

as it must be. Thus, although the energy E of the incident electron is greater than the potential energy V_0 at the step, i.e. $E > V_0$, quantum mechanically there is a probability of an electron being reflected as well as transmitted. This reflection of electrons with $E > V_0 > 0$ from a sharp change in potential is a quantum mechanical phenomenon. In fact, the above quantum mechanical result shows that under these circumstances, the electron is exhibiting those characteristics which in everyday life are attributed to waves; partial reflection of light from an interface between two media is a familiar example of this wave characteristic. If the potential does *not* change appreciably within a wavelength of the electron wave, there will be practically no reflection.

Let us now consider two special cases:

(i) If E, the energy of the incident electron, although *greater than* V_0, is nearly equal to it, i.e. E goes to V_0 from above, then $E \approx V_0$ and we may write

$$\alpha = \frac{\sqrt{2mE}}{\hbar} \approx \frac{\sqrt{2mV_0}}{\hbar}$$

and

$$\beta = \frac{\sqrt{2m(E - V_0)}}{\hbar} \approx 0.$$

Substituting this expression for β in (6.21) and (6.22), we obtain

$$R = \left(\frac{\alpha - \beta}{\alpha + \beta}\right)^2 \approx 1,$$

$$T = \frac{4\alpha\beta}{(\alpha + \beta)^2} \approx 0.$$

That is, when the energy of the incident electron is close to the potential energy V_0 at the step, the probability of its reflection approaches unity; there is practically no transmission. It is important to note that classically all such electrons are transmitted and there is no reflection. According to terminology used in the literature, the quantum mechanical behaviour, different from classical results, is exhibited by the microscopic entity when the potential changes appreciably within the wavelength $\lambda = h/p$ of this entity. Then the electron exhibits those properties which in everyday life are attributed to waves.

(ii) If $E \gg V_0$, i.e. the energy of the incident electron is much greater than the potential V_0 at and beyond the step, then V_0 can be neglected compared to E and we have

$$\beta = \frac{\sqrt{2m(E - V_0)}}{\hbar} \approx \frac{\sqrt{2mE}}{\hbar} = \alpha.$$

Substituting this value of β in (6.21) and (6.22), we obtain

$$R = \left(\frac{\alpha - \beta}{\alpha + \beta}\right)^2 \approx 0,$$

$$T = \frac{4\alpha\beta}{(\alpha + \beta)^2} \approx 1.$$

Thus, for $EE \gg V_0 V_0$, the probability of reflection of the electron is negligible. This is close to the classical result where, in such a situation, no electrons are reflected.

Thus, for $E > V_0$, classically all the electrons are transmitted; no electron is reflected. However, the above analysis shows that quantum mechanically, the electrons are reflected as well as transmitted; the values of reflectivity R and transmissivity T depending upon E and V_0.

Case 2. $0 < E < V_0$.
Let us next consider the case for which $0 < E < V_0$. This is shown in figure 6.3. Classically, the particle would proceed towards the right with constant velocity until it reached the origin where the potential rises suddenly. This potential will act like a barrier of height V_0 which a particle with energy $E < V_0$ cannot penetrate; it will be

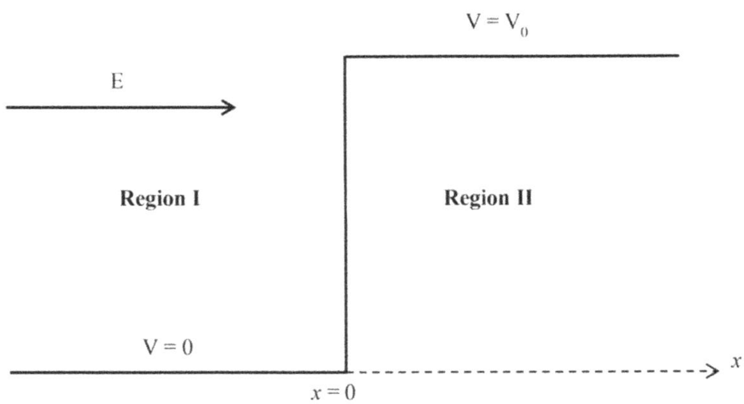

Figure 6.3. Potential step with $0 < E < V_0$.

reflected. Let us see what quantum mechanics predicts about this phenomenon. The Schrödinger amplitude equation in the two regions is given by

$$\text{Region I:} \quad \frac{d^2\psi_1}{dx^2} + \frac{2m}{\hbar^2}E\psi_1 = 0, \quad \text{for } x < 0,$$

$$\text{Region II:} \quad \frac{d^2\psi_2}{dx^2} + \frac{2m}{\hbar^2}(E - V_0)\psi_2 = 0, \quad \text{for } x > 0.$$

Writing

$$\alpha = \frac{\sqrt{2mE}}{\hbar} \qquad \text{and} \qquad \gamma = \frac{\sqrt{2m(V_0 - E)}}{\hbar}$$

where α^2 and γ^2 are both positive, we obtain

$$\text{Region I:} \quad \frac{d^2\psi_1}{dx^2} + \alpha^2\psi_1 = 0, \quad \text{for } x < 0,$$

$$\text{Region II:} \quad \frac{d^2\psi_2}{dx^2} - \gamma^2\psi_2 = 0, \quad \text{for } x > 0.$$

Then the solutions of the Schrödinger amplitude equation in the two regions are

$$\text{Region I:} \quad \psi_1 = Ae^{i\alpha x} + Be^{-i\alpha x},$$

$$\text{Region II:} \quad \psi_2 = Ce^{\gamma x} + De^{-\gamma x}.$$

where A, B, C, D are constants of integration. Note that in region II, the solutions of the Schrödinger amplitude equation are *real exponentials* rather than complex exponentials. Since for $x \to \infty$, the term $Ce^{\gamma x} \to \infty$, it is rejected and the above equations become

$$\text{Region I:} \quad \psi_1 = Ae^{i\alpha x} + Be^{-i\alpha x},$$

$$\text{Region II:} \quad \psi_2 = De^{-\gamma x}.$$

Now the derivatives of ψ_1 and ψ_2 with respect to x are:

$$\text{Region I:} \quad \frac{d\psi_1}{dx} = i\alpha Ae^{i\alpha x} - i\alpha Be^{-i\alpha x},$$

$$\text{Region II:} \quad \frac{d\psi_2}{dx} = -\gamma De^{-\gamma x}.$$

The continuity of the wavefunction and its derivative at the step $x = 0$ requires that

$$\psi_1 = \psi_2, \quad \text{at } x = 0,$$

$$\frac{d\psi_1}{dx} = \frac{d\psi_2}{dx}, \quad \text{at } x = 0.$$

These conditions yield

$$A + B = D \tag{6.23}$$

and

$$i\alpha(A - B) = -\gamma D. \tag{6.24}$$

Substituting the expression for D from (6.23) in (6.24), we obtain

$$i\alpha(A - B) = -\gamma(A + B),$$

or

$$(i\alpha + \gamma)A = (i\alpha - \gamma)B,$$

or

$$B = \frac{i\alpha + \gamma}{i\alpha - \gamma}A. \tag{6.25}$$

Substituting the expression for B from (6.23) in (6.24), we obtain

$$i\alpha(A + A - D) = -\gamma D,$$

or

$$i\alpha 2A = (i\alpha - \gamma)D,$$

or

$$D = \frac{2i\alpha}{i\alpha - \gamma}A. \tag{6.26}$$

The wave functions ψ_1 and ψ_2 are then given by

$$\begin{aligned}\psi_1 &= Ae^{i\alpha x} + Be^{-i\alpha x}\\ &= Ae^{i\alpha x} + \frac{i\alpha + \gamma}{i\alpha - \gamma}Ae^{-i\alpha x}, \quad x < 0,\end{aligned}$$

and

$$\psi_2 = De^{-\gamma x} = \frac{2i\alpha}{i\alpha - \gamma}Ae^{-\gamma x}, \quad x > 0.$$

The incident probability flux S_{incident} is given by

$$\begin{aligned} S_{\text{incident}} &= \frac{\hbar}{m}\ \text{Im}\left(\psi^*_{\text{incident}}\frac{\mathrm{d}\psi_{\text{incident}}}{\mathrm{d}x}\right) \\ &= \frac{\hbar}{m}\ \text{Im}\left(A^*\mathrm{e}^{-\mathrm{i}\alpha x}\mathrm{i}\alpha A\mathrm{e}^{\mathrm{i}\alpha x}\right) \\ &= \frac{\hbar}{m}\ \text{Im}\left(\mathrm{i}\alpha|A|^2\right) \\ &= \frac{\hbar}{m}\alpha|A|^2. \end{aligned}$$

The reflected probability flux, $S_{\text{reflected}}$, is given by

$$\begin{aligned} S_{\text{reflected}} &= \frac{\hbar}{m}\ \text{Im}\left(\psi^*_{\text{reflected}}\frac{\mathrm{d}\psi_{\text{reflected}}}{\mathrm{d}x}\right) \\ &= \frac{\hbar}{m}\ \text{Im}\left\{B^*e^{\mathrm{i}\alpha x}(-\mathrm{i}\alpha)B\mathrm{e}^{-\mathrm{i}\alpha x}\right\} \\ &= \frac{\hbar}{m}\ \text{Im}\left(-\mathrm{i}\alpha|B|^2\right) \\ &= -\frac{\hbar}{m}\alpha|B|^2 \end{aligned}$$

the negative sign merely indicating that after reflection the electron moves along the negative direction of the x-axis.

The transmitted probability flux, $S_{\text{transmitted}}$, is given by

$$\begin{aligned} S_{\text{tramsmitted}} &= \frac{\hbar}{m}\ \text{Im}\left(\psi^*_{\text{transmitted}}\frac{\mathrm{d}\psi_{\text{transmitted}}}{\mathrm{d}x}\right) \\ &= \frac{\hbar}{m}\ \text{Im}\left\{D^*\mathrm{e}^{-\gamma x}(-\gamma)D\mathrm{e}^{-\gamma x}\right\} \\ &= \frac{\hbar}{m}\ \text{Im}\left(\mathrm{e}^{-2\gamma x}|D|^2\right) \\ &= 0. \end{aligned}$$

The *coefficient of reflection* R is given by

$$R = \frac{S_{\text{reflected}}}{S_{\text{incident}}} = \frac{|B|^2}{|A|^2},$$

ignoring the negative sign in $S_{\text{reflected}}$, because it only indicates the direction of motion of the electron. Now, from (6.25), we have

$$B = \frac{\mathrm{i}\alpha + \gamma}{\mathrm{i}\alpha - \gamma}A, \tag{6.25$'$}$$

therefore,

$$B^* = \frac{-i\alpha + \gamma}{-i\alpha - \gamma} A^*.$$

The last two equations show that

$$\begin{aligned} R &= \frac{|B|^2}{|A|^2} \\ &= \frac{-i\alpha + \gamma}{-i\alpha - \gamma} \frac{i\alpha + \gamma}{i\alpha - \gamma} \\ &= 1. \end{aligned}$$

And the ratio T of the transmitted probability flux to the incident probability flux, i.e. the *coefficient of transmission*, is given by

$$T = \frac{S_{\text{transmitted}}}{S_{\text{incident}}} = 0$$

because $S_{\text{transmitted}}$ has been shown to be equal to zero. Since the electron can either be reflected or transmitted, the total probability must be one. This is as it must be.

This behaviour is the same as expected classically. All the electrons, each with energy $E < V_0$, should be reflected. However, the *important point* is this: although the current flowing through the region $x > 0$ is zero, i.e. the electrons are completely reflected at the step, *the probability of finding an electron in region II is not zero.* This can be seen as follows. We have

$$\psi_2 = De^{-\gamma x} = \frac{2i\alpha}{i\alpha - \gamma} Ae^{-\gamma x}, \quad x > 0,$$

therefore,

$$\psi_2^* \psi_2 = D^* e^{-\gamma x} = |D|^2 e^{-2\gamma x}. \tag{6.27}$$

This shows that the probability density in region II is different from zero. Thus, there is a finite probability of finding the electron in a region which is classically inaccessible. *Again, the electron is manifesting the properties of waves.* The transmission current is zero but, as shown in figure 6.4, part of the wave penetrates into

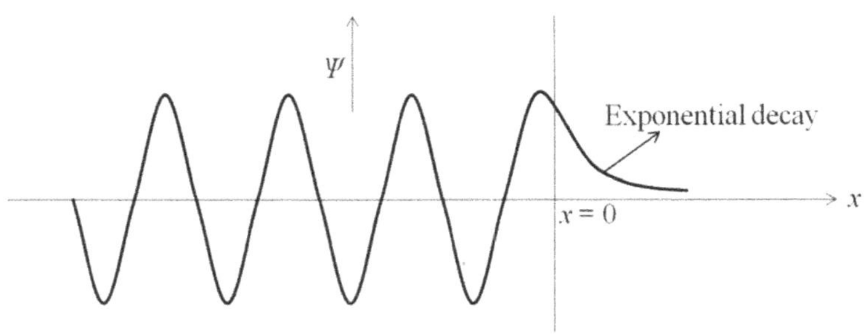

Figure 6.4. Penetration of electrons in the classically forbidden region.

the classically forbidden region although the wave entering the forbidden region is exponentially damped. In the classically inaccessible region, the total energy E is less than the potential energy V_0. Therefore, the kinetic energy of the electron is negative. How can it be detected then? Will it not violate the law of conservation of energy? Actually, in order to observe the electron, we will have to shine so much light on it that its kinetic energy becomes positive. Hence no contradiction would arise from the observation of the electron in that region.

It is significant to note that the expressions for R and T are independent of $\hbar$. What does this mean? We know that as $\hbar \to 0$, the quantum mechanical results should reduce to the classical values. But as the expressions for R and T do not contain $\hbar$, it might be concluded that the results are also valid classically. No! In fact, for a classical situation, the de Broglie wavelength λ ($= h/p$) should be small compared to the relevant dimension of the system. In this case, the relevant size of the region is that in which the potential suddenly changes from zero to V_0. This size is zero in the limiting case. That is, for this sharp rise in potential, the de Broglie wavelength is *greater* than the size of the region in which this sharp change in potential occurs. There is no classical region and hence *no paradox*.

Problems

6.3. What will be the situation when $E < 0$?

6.4. For $0 < E < V_0$, to avoid any penetration of electrons in the classically forbidden region, can the constant D be taken equal to zero?

6.5. Why do we not obtain discrete values of energy here?

6.2 The potential barrier

Let us next consider the electrons coming from the left towards a **potential barrier**, i.e. a one-dimensional space divided into three parts each having a constant potential, the values of potential $V(x)$ changing suddenly at their points of contact. As shown in figure 6.5, the point of contact between regions I and II is taken as the origin $x = 0$. The region in the middle is finite in length, say equal to a, while the

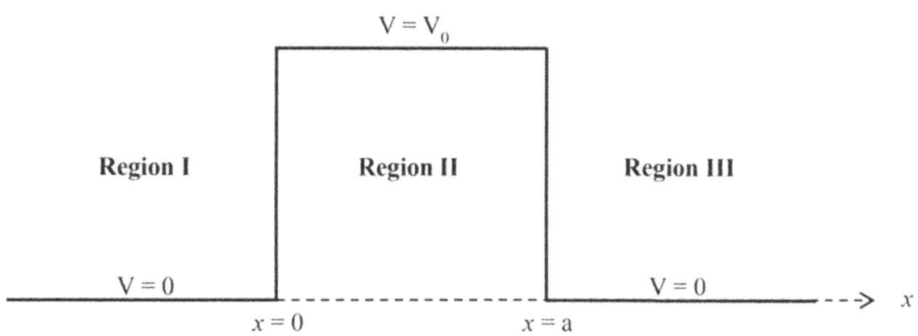

Figure 6.5. The potential barrier.

remaining two regions extend to infinity, on one side each. The potential in the case of the barrier is defined as

$$\begin{aligned} V(x) &= 0, \quad \text{for } x < 0, \\ &= V_0, \quad \text{for } 0 < x < a, \\ &= 0, \quad \text{for } x > a, \end{aligned} \tag{6.28}$$

where V_0 is a positive number. The Schrödinger amplitude equation in the three regions can be written as

$$\text{Region I:} \quad \frac{d^2\psi_1}{dx^2} + \frac{2m}{\hbar^2}E\psi_1 = 0, \quad \text{for } x < 0,$$

$$\text{Region II:} \quad \frac{d^2\psi_2}{dx^2} + \frac{2m}{\hbar^2}(E - V_0)\psi_2 = 0, \quad \text{for } 0 < x < a,$$

$$\text{Region III:} \quad \frac{d^2\psi_3}{dx^2} + \frac{2m}{\hbar^2}E\psi_3 = 0, \quad \text{for } x > a.$$

Case 3: $E > V_0$.
Let us first consider the case when $E > V_0$. This is shown in figure 6.6. Classically, all the electrons will be transmitted from region I to region II. To find the quantum mechanical behaviour, we proceed as follows. Writing

$$\alpha = \frac{\sqrt{2mE}}{\hbar} \quad \text{and} \quad \beta = \frac{\sqrt{2m_0(E - V_0)}}{\hbar},$$

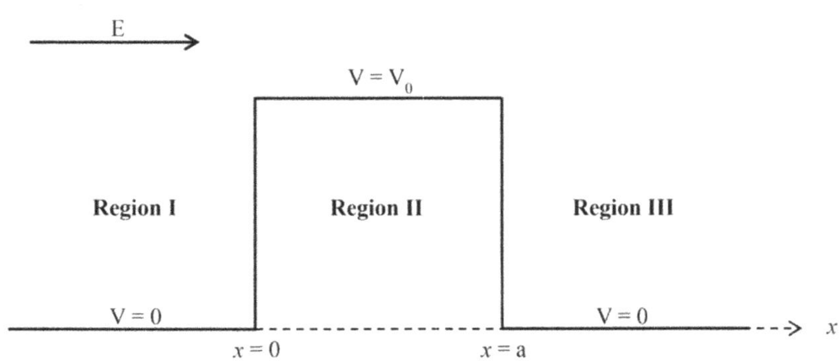

Figure 6.6. Potential barrier with $E > V_0$.

where α^2 and β^2 are both positive, we obtain

$$\text{Region I:} \quad \frac{d^2\psi_1}{dx^2} + \alpha^2\psi_1 = 0, \quad \text{for } x < 0,$$

$$\text{Region II:} \quad \frac{d^2\psi_2}{dx^2} + \beta^2\psi_2 = 0, \quad \text{for } 0 < x < a,$$

$$\text{Region III:} \quad \frac{d^2\psi_3}{dx^2} + \alpha^2\psi_3 = 0, \quad \text{for } x > a.$$

Then the solutions of the Schrödinger amplitude equation in the three regions are

$$\text{Region I:} \quad \psi_1 = Ae^{i\alpha x} + Be^{-i\alpha x}, \tag{6.29}$$

$$\text{Region II:} \quad \psi_2 = Ce^{i\beta x} + De^{-i\beta x}, \tag{6.30}$$

$$\text{Region III:} \quad \psi_3 = Fe^{i\alpha x} + Ge^{-i\alpha x}, \tag{6.31}$$

where A, B, C, D, F, G are constants of integration. Since in region III, no electrons are moving from right to left, the last term in (6.31) must vanish. This is achieved by taking $G = 0$. The last equation therefore reduces to

$$\text{Region III:} \quad \psi_3 = Fe^{i\alpha x}. \tag{6.32}$$

The values of the various constants in the solutions ψ_1, ψ_2, ψ_3 of the Schrödinger wave equation in the three regions can be calculated by making use of the boundary conditions. The functions and their derivatives must be continuous at the relevant boundaries. Differentiating (6.29), (6.30) and (6.32) with respect to x, we obtain

$$\text{Region I:} \quad \frac{d\psi_1}{dx} = i\alpha Ae^{i\alpha x} - i\alpha Be^{-i\alpha x}, \tag{6.33}$$

$$\text{Region II:} \quad \frac{d\psi_2}{dx} = i\beta Ce^{i\beta x} - i\beta De^{-i\beta x}, \tag{6.34}$$

$$\text{Region III:} \quad \frac{d\psi_3}{dx} = i\alpha Fe^{i\alpha x}. \tag{6.35}$$

Since the potential $V(x)$ has finite discontinuities at the points $x = 0$ and $x = a$, the amplitudes and their first derivatives should be continuous even at these points. This requires that

$$\psi_1 = \psi_2, \quad \text{at } x = 0,$$

$$\frac{d\psi_1}{dx} = \frac{d\psi_2}{dx}, \quad \text{at } x = 0,$$

and

$$\psi_2 = \psi_3, \quad \text{at } x = a,$$

$$\frac{d\psi_2}{dx} = \frac{d\psi_3}{dx}, \quad \text{at } x = a.$$

These conditions as applied to (6.29), (6.30) and (6.32), and (6.33)–(6.35) yield

$$A + B = C + D,$$

$$i\alpha(A - B) = i\beta(C - D),$$

$$Ce^{i\beta a} + De^{-i\beta a} = Fe^{i\alpha a},$$

$$i\beta Ce^{i\beta a} + i\beta De^{-i\beta a} = i\alpha Fe^{i\alpha a}.$$

The above equations can be written as

$$A + B = C + D, \tag{6.36}$$

$$A - B = \frac{\beta}{\alpha}(C - D), \tag{6.37}$$

$$Ce^{i\beta a} + De^{-i\beta a} = Fe^{i\alpha a}, \tag{6.38}$$

$$\frac{\beta}{\alpha}Ce^{i\beta a} - \frac{\beta}{\alpha}De^{-i\beta a} = Fe^{i\alpha a}. \tag{6.39}$$

Adding (6.36) and (6.37), we obtain

$$2A = \left(1 + \frac{\beta}{\alpha}\right)C + \left(1 - \frac{\beta}{\alpha}\right)D. \tag{6.40}$$

Subtracting (6.39) from (6.38), we obtain

$$\left(1 - \frac{\beta}{\alpha}\right)Ce^{i\beta a} + \left(1 + \frac{\beta}{\alpha}\right)De^{-i\beta a} = 0. \tag{6.41}$$

Writing μ for β/α, the last two equations, (6.40) and (6.41), can be written as

$$2A = (1 + \mu)C + (1 - \mu)D \tag{6.42}$$

and

$$(1 - \mu)Ce^{i\beta a} + (1 + \mu)De^{-i\beta a} = 0. \tag{6.43}$$

Multiplying (6.42) by $(1 + \mu)e^{-i\beta a}$ and taking all the terms on one side, we have

$$2(1 + \mu)Ae^{-i\beta a} - (1 + \mu)^2 Ce^{-i\beta a} - \left(1 - \mu^2\right)De^{-i\beta a} = 0. \tag{6.44}$$

Multiplying (6.43) by $(1 - \mu)$, we have

$$(1 - \mu)^2 C e^{i\beta a} + \left(1 - \mu^2\right) D e^{-i\beta a} = 0. \tag{6.45}$$

Adding (6.44) and (6.45), we obtain

$$2(1 + \mu) A e^{-i\beta a} - (1 + \mu)^2 C e^{-i\beta a} + (1 - \mu)^2 C e^{i\beta a} = 0,$$

or

$$\left[(1 - \mu)^2 e^{i\beta a} - (1 + \mu)^2 e^{-i\beta a}\right] C = -2(1 + \mu) A e^{-i\beta a},$$

or

$$C = \frac{-2(1 + \mu) e^{-i\beta a}}{(1 - \mu)^2 e^{i\beta a} - (1 + \mu)^2 e^{-i\beta a}} = A. \tag{6.46}$$

Problem

6.6. Show that the denominator in the above expression for C can be written as

$$2i\left[\left(1 + \mu^2\right) \sin \beta a + 2i\mu \cos \beta a\right].$$

Therefore, the expression for C takes the form

$$C = e^{-i\beta a} \frac{i(1 + \mu)}{\left(1 - \mu^2\right) \sin \beta a + 2i\mu \cos \beta a} A.$$

Similarly, we can show that

$$D = e^{i\beta a} \frac{-i(1 - \mu)}{\left(1 - \mu^2\right) \sin \beta a + 2i\mu \cos \beta a} A,$$

$$F = e^{-i\alpha a} \frac{2i\mu}{\left(1 - \mu^2\right) \sin \beta a + 2i\mu \cos \beta a} A,$$

$$B = \frac{\left(1 + \mu^2\right) \sin \beta a}{\left(1 - \mu^2\right) \sin \beta a + 2i\mu \cos \beta a} A.$$

Therefore, the absolute square of the transmitted wave is given by

$$T = |F/A|^2 = \frac{4\mu^2}{(1 - \mu^2)^2 \sin^2 \beta a + (2\mu)^2 \cos^2 \beta a}. \tag{6.47}$$

Problems

6.7. Show that (6.47) can also be written as

$$T = \left[1 + \frac{V_0^2 \sin^2 \beta a}{4E(E - V_0)}\right]^{-1}. \tag{6.48}$$

6.8. Prove that

$$R = \left[1 + \frac{4E(E - V_0)}{V_0^2 \sin^2 \beta a}\right]^{-1}. \tag{6.49}$$

6.9. Find the expression for $R = |B/A|^2$ and show that $R + T = 1$.

The following points can be noted:

1. For $\beta a = n\pi$, where $n = 1, 2, \ldots$, we have $\sin^2 \beta a = 0$. Therefore, (6.48) gives $T = 1$.
2. When E is very large compared to V_0, then (6.48) yields $T \approx 1$, as V_0/E is very small.

Problems

6.10. Show that for $E \to V_0$ from the above, (6.48) yields

$$T = \frac{1}{1 + (\alpha a/2)^2}. \tag{6.50}$$

6.11. What is the value of energy E at which the transmission coefficient T becomes unity?

6.12. Draw a graph between T and E/V_0 for $mV_0a^2/h^2 = 5$.

Thus, although classically all the electrons would be transmitted, the above analysis shows that quantum mechanically these microscopic entities will be transmitted as well as reflected.

Case 4: $E < V_0$.
When the energy of the incident particle is less than the potential energy V_0 of the barrier, as shown in figure 6.7, then classically all the particles will be reflected. To determine the quantum mechanical behaviour of this phenomenon, we write

$$\alpha = \frac{\sqrt{2mE}}{\hbar} \qquad \text{and} \qquad \gamma = \frac{\sqrt{2m(V_0 - E)}}{\hbar} \tag{6.51}$$

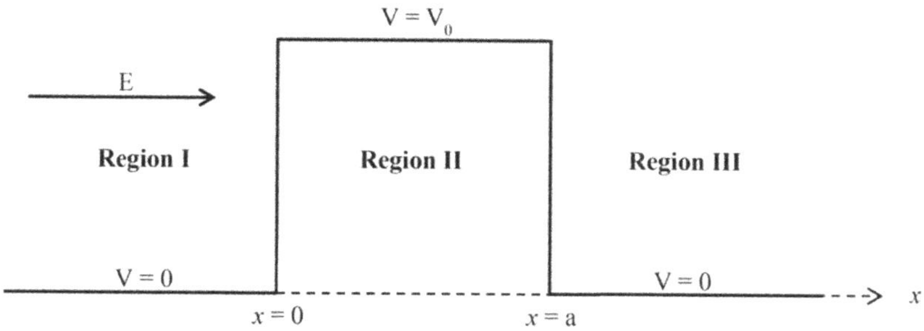

Figure 6.7. Potential barrier with $E < V_0$.

where α^2 and γ^2 are both positive. We note that $\beta^2 = -\gamma^2$. This yields $\beta = \pm \mathrm{i}\gamma$. We choose $\beta = \mathrm{i}\gamma$. The other choice gives the same values of T and R. Hence, in the results obtained in case 3, only β is to be replaced by $\mathrm{i}\gamma$ to obtain the results in case 4. The expression for T for case 3 ($E > V_0$) can be written as

$$\frac{4\left(\frac{\beta}{\alpha}\right)^2}{\left[1+\left(\frac{\beta}{\alpha}\right)^2\right]^2 \sin^2 \beta a + \left(\frac{2\beta}{\alpha}\right)^2 \cos^2 \beta a}.$$

Multiplying the numerator and the denominator by α^4/β^4, we obtain

$$\frac{4\left(\frac{\alpha}{\beta}\right)^2}{\left[1+\left(\frac{\alpha}{\beta}\right)^2\right]^2 \sin^2 \beta a + \left(\frac{2\alpha}{\beta}\right)^2 \cos^2 \beta a}.$$

Replacing β by $\mathrm{i}\gamma$, we obtain the expression for the transmission T in case 4 ($E < V_0$):

$$T = \frac{-4\left(\frac{\alpha}{\gamma}\right)^2}{\left[1-\left(\frac{\alpha}{\gamma}\right)^2\right]^2 \sin^2 \mathrm{i}\gamma a - \left(\frac{2\alpha}{\gamma}\right)^2 \cos^2 \mathrm{i}\gamma a}. \qquad (6.52)$$

Recalling that

$$\sin x = \frac{e^{ix} - e^{-ix}}{2i}, \qquad \cos x = \frac{e^{ix} + e^{-ix}}{2},$$

$$\sinh x = \frac{e^{x} - e^{-x}}{2}, \qquad \cosh x = \frac{e^{x} + e^{-x}}{2},$$

we can write

$$\sin^2 \mathrm{i}\gamma a = -\sinh^2 \gamma a,$$

$$\cos^2 \mathrm{i}\gamma a = \cosh^2 \gamma a.$$

Equation (6.52) therefore reduces to

$$T = \frac{4\left(\dfrac{\alpha}{\gamma}\right)^2}{\left[1 - \left(\dfrac{\alpha}{\gamma}\right)^2\right]^2 \sinh^2 \gamma a + \left(\dfrac{2\alpha}{\gamma}\right)^2 \cosh^2 \gamma a}. \tag{6.53}$$

This shows that there is a transmission even when the energy of the incident particles is less than the height of the potential barrier. This quantum mechanical phenomenon is against the classical expectations. As $E < V_0$, classically, no electron will be transmitted. The tunneling of a microscopic entity as predicted by quantum mechanics is experimentally observed in physical problems. In this case too, no restrictions are imposed on the energy value spectrum. Why?

Problems

6.13. Obtain by direct calculations the expression for T.

6.14. Obtain the expression for the reflection coefficient R for $E < V_0$ and show that $R + T = 1$.

6.15. Show that

$$T = \left[1 + \frac{V_0^2 \sin^2 \gamma a}{4E(V_0 - E)}\right]^{-1}.$$

6.16. Prove that

$$R = \left[1 + \frac{4E(V_0 - E)}{V_0^2 \sinh^2 \gamma a}\right]^{-1}.$$

6.17. Show that $R + T = 1$.

6.3 α-particle decay

An interesting application of the above results is in the decay of α-particles. An α-particle is a conglomeration of two protons and two neutrons—what is usually termed as the helium nucleus. These are bound so tightly that the α-particle retains its identity even within the nucleus. The nuclei of heavier elements like U^{238} decay spontaneously with the emission of α-particles. These particles are emitted from U^{238} with an energy of about 4.2 MeV while the repulsive potential inside the nucleus is about 30 MeV. Within the nucleus, two forces are acting on the α-particle. (1) The repulsive Coulomb force between the α-particle and the residual nucleus, because both are positively charged; it is a long range force which tries to throw the α-particle outside the nucleus. (2) The much stronger but short range attractive nuclear force which tries to retain the α-particle inside the nucleus. Within the nucleus, the attractive nuclear interaction dominates. The resultant force on the α-particle is attractive and the corresponding potential is much greater than its energy, therefore, classically, an α-particle inside the nucleus and of an energy of about 4.2 MeV should remain confined to the nucleus as it cannot overcome the strong attractive potential and escape. However, we have already seen that quantum mechanically a microscopic object can penetrate through a barrier and by virtue of this characteristic an α-particle can come out of the nucleus despite its low energy. This is known as the *tunnelling process*. Let us now see how it predicts the probability per second of finding the α-particle outside the nucleus. Of course, this probability will depend upon the shape of the barrier.

Let E be the energy of the α-particle within the nucleus and encountering a potential $V(x)$. The resultant potential energy within the nucleus is attractive; inside the nucleus, it is assumed to be practically constant. Just outside the periphery of the nucleus, the attractive nuclear potential becomes abruptly zero and only the Coulomb potential, which decreases with an increase in distance, plays a role. This is shown in figure 6.8. Classically, the particle of energy E cannot leave the nucleus because the attractive potential energy is about double that of the energy of the α-particle. Quantum mechanically there is a non-zero probability of the transmission of the α-particle in the nucleus into the classically forbidden region. Let us *assume* that the α-particle is bouncing back and forth with a velocity v inside the nucleus of diameter $2R$. As it traverses a distance v in one second, it will take $2R/v$ seconds to

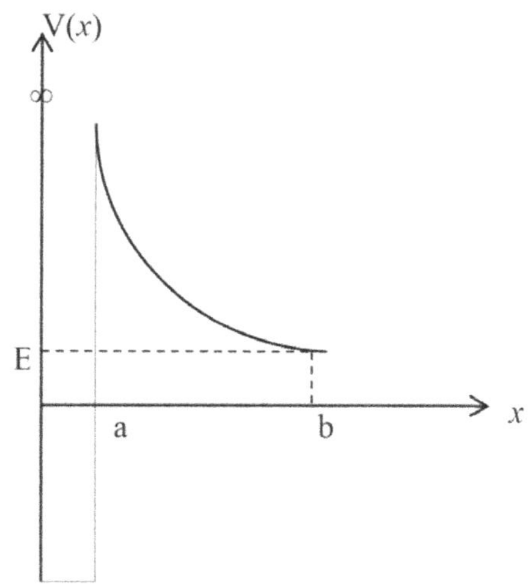

Figure 6.8. The potential barrier for α-decay.

cover a distance $2R$. It will then go back and therefore strike the boundary of the nucleus $v/2R$ times per second. We denote this number by N so that

$$N = \frac{v}{2R}.$$

Let T be the probability that an α-particle will penetrate the barrier of nuclear potential in one collision. Then TN is the probability that it will penetrate the barrier in $N = v/2R$ collisions. We denote this by λ. Thus λ is the probability per second that the nucleus will emit an α-particle.

When the energy E of the incident particle is less than the potential energy V_0 of the square potential barrier, quantum mechanically the particle can be transmitted through the square barrier of width a, the probability of transmission being given by

$$P(\equiv T) = \frac{4\left(\frac{\alpha}{\gamma}\right)^2}{\left[1 - \left(\frac{\alpha}{\gamma}\right)^2\right]^2 \sinh^2 \gamma a + \left(\frac{2\alpha}{\gamma}\right)^2 \cosh^2 \gamma a}. \tag{6.53'}$$

This result can be used to deduce the probability of transmission through a barrier of arbitrary shape. Consider a potential barrier of the shape shown in figure 6.9. Such a potential barrier can be approximated by a series of thin square potential barriers placed side by side. For a thin square barrier, the probability of transmission can be approximated to

$$P \approx \exp(-2\gamma a). \tag{6.53a}$$

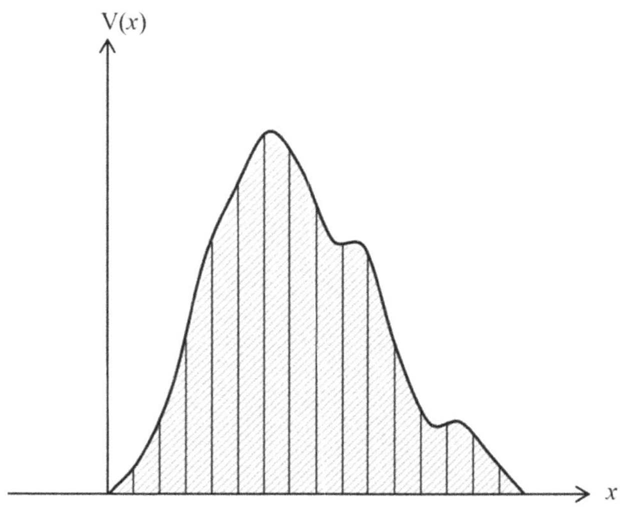

Figure 6.9. A barrier of arbitrary shape approximated as a succession of thin square barriers.

Problem

6.18. Show that for a thin square barrier (6.53′) reduces to (6.53*a*).

Thus, the probability that a particle of energy less than the potential energy of a thin square barrier of width Δx_i will be transmitted through it is:

$$P_i \approx \exp\left(-2\gamma_i \Delta x_i\right).$$

As the transmission of a particle through successive thin barriers is composed of independent events, the probability of transmission through the entire barrier is obtained by multiplying the probabilities of successive independent events:

$$P = P_1 P_2 P_3 \cdots \approx \exp\left(-2\sum_i \gamma_i \Delta x_i\right).$$

If the intervals Δx_i, the thicknesses of the barriers, are very small, then the sum over i may be replaced by an integral and we have

$$P \approx \exp\left(-2\int_0^a \gamma \mathrm{d}x\right),$$

therefore:

$$\lambda \approx \frac{v}{2R} \exp\left[-2\int_R^\infty \frac{\sqrt{2m(V_o - E)}}{\hbar} \mathrm{d}x\right].$$

We have taken the limits from R to ∞, as the region from 0 to R contributes very little to the integral. All the factors on the right-hand side of this equation except v and R are known. So if we can determine these two quantities, then λ can be calculated. The velocity v of the α-particle can be determined by *assuming* that this velocity is comparable to that possessed by the α-particle *after emission*. This can then be computed by using the value of energy $E\,(= 1/2\, mv^2)$ of the α-particle outside the nucleus on penetration through the barrier. Gamow *et al* calculated values of λ which agreed well with experiments.

6.4 The square-well potential

Let us next consider a square-well potential, that is, a potential given by

$$\begin{aligned} V(x) &= V_0, \quad \text{for } x < -a, \\ &= 0, \quad \text{for } -a < x < a, \\ &= V_0, \quad \text{for } x > a. \end{aligned} \qquad (6.54)$$

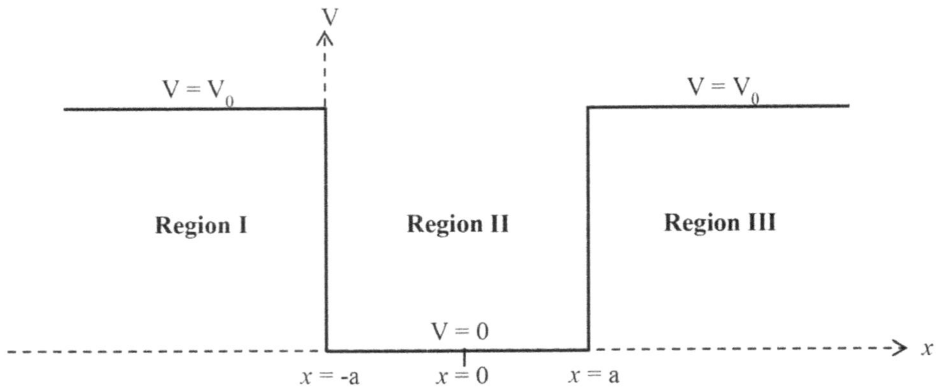

Figure 6.10. The square-well potential.

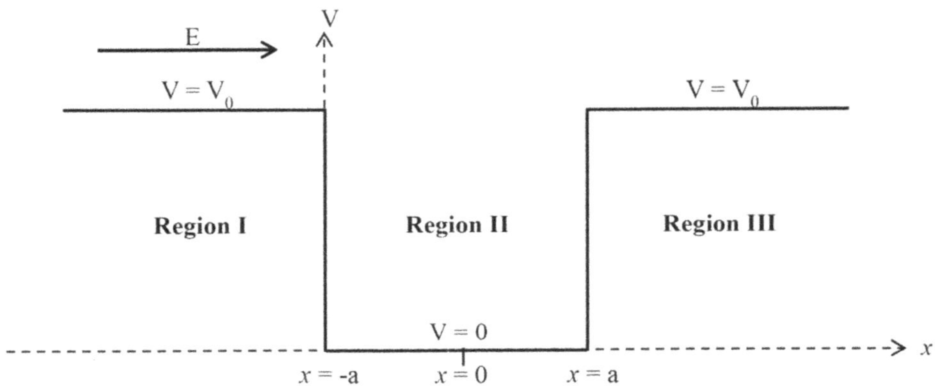

Figure 6.11. Square-well potential with $E > V_0$.

The width of the well has been taken as $2a$ and its depth is V_0. The well extends from $x = -a$ to $x = a$, as shown in figure 6.10. V_0 is a positive number. For the sake of symmetry which will facilitate the calculations, the origin, $x = 0$, has been taken at the middle of the well.

Case 5. $E > V_0$.
Suppose that a microscopic entity, say an electron of mass m and energy $E > V_0$, is incident upon the well from the left. This is shown in figure 6.11. According to classical physics, the electron cannot be reflected; in fact the attractive force at $x = -a$ will accelerate it further. But what does quantum mechanics predict about its behaviour?

The Schrödinger amplitude equation in the three regions I, II, III can be written as

$$\text{Region I:}\quad \frac{\mathrm{d}^2\psi_1}{\mathrm{d}x^2} + \frac{2m}{\hbar^2}(E - V_0)\psi_1 = 0, \quad \text{for } x < -a,$$

$$\text{Region II:} \quad \frac{d^2\psi_2}{dx^2} + \frac{2m}{\hbar^2}E\psi_2 = 0, \quad \text{for } -a < x < a,$$

$$\text{Region III:} \quad \frac{d^2\psi_3}{dx^2} + \frac{2m}{\hbar^2}(E - V_0)\psi_3 = 0, \quad \text{for } x > a.$$

The solutions of the Schrödinger equation for these three regions are:

$$\psi_1 = Ae^{i\beta x} + Be^{-i\beta x}, \tag{6.55}$$

$$\psi_2 = Ce^{i\alpha x} + De^{-i\alpha x}, \tag{6.56}$$

$$\psi_3 = Fe^{i\beta x} + Ge^{-i\beta x}, \tag{6.57}$$

where

$$\alpha = \frac{\sqrt{2mE}}{\hbar}$$

and

$$\beta = \frac{\sqrt{2m(E - V_0)}}{\hbar}.$$

Since in region III, no electrons are moving from right to left, the expression $Ge^{-i\beta x}$ must be zero. Equation (6.57) therefore reduces to

$$\psi_3 = Fe^{i\beta x}. \tag{6.58}$$

Differentiating (6.55), (6.56) and (6.58), we obtain

$$\frac{d\psi_1}{dx} = i\beta Ae^{i\beta x} - i\beta Be^{-i\beta x},$$

$$\frac{d\psi_2}{dx} = i\alpha Ce^{i\alpha x} - i\alpha De^{-i\alpha x},$$

$$\frac{d\psi_3}{dx} = i\beta Fe^{i\beta x}.$$

The boundary conditions are as follows.

For $x = -a$,

$$\psi_1 = \psi_2,$$

$$\frac{d\psi_1}{dx} = \frac{d\psi_2}{dx}.$$

For $x = a$,

$$\psi_2 = \psi_3,$$

$$\frac{d\psi_2}{dx} = \frac{d\psi_3}{dx}.$$

Applying the boundary conditions, we obtain

$$Ae^{-i\beta a} + Be^{i\beta a} = Ce^{-i\alpha a} + De^{i\alpha a},$$

$$Ce^{i\alpha a} + De^{-i\alpha a} = Fe^{i\beta a},$$

$$i\beta Ae^{-i\beta a} - i\beta Be^{i\beta a} = i\alpha Ce^{-i\alpha a} - i\alpha De^{i\alpha a},$$

$$i\alpha Ce^{i\alpha a} - i\alpha De^{-i\alpha a} = i\beta Fe^{i\beta a}.$$

Solving these equations, we obtain

$$\frac{B}{A} = e^{-2i\beta a} = \frac{i\left(1 - \dfrac{\alpha^2}{\beta^2}\right)\sin 2\alpha a}{\left(1 + \dfrac{\alpha^2}{\beta^2}\right)\sin 2\alpha a + 2i\dfrac{\alpha}{\beta}\cos 2\alpha a},$$

$$\frac{F}{A} = e^{-2i\beta a} = \frac{2i\dfrac{\alpha}{\beta}}{\left(1 + \dfrac{\alpha^2}{\beta^2}\right)\sin 2\alpha a + 2i\dfrac{\alpha}{\beta}\cos 2\alpha a},$$

therefore

$$R = |B/A|^2 = \frac{\left(1 - \dfrac{\alpha^2}{\beta^2}\right)\sin^2 2\alpha a}{\left(1 + \dfrac{\alpha^2}{\beta^2}\right)^2 \sin 2\alpha a + 4\dfrac{\alpha^2}{\beta^2}\cos^2 2\alpha a}, \tag{6.59}$$

$$T = |F/A|^2 = \frac{4\dfrac{\alpha^2}{\beta^2}}{\left(1 + \dfrac{\alpha^2}{\beta^2}\right)^2 \sin 2\alpha a + 4\dfrac{\alpha^2}{\beta^2}\cos^2 2\alpha a}. \tag{6.60}$$

Equations (6.59) and (6.60) show that there is a finite probability that the incident electron with $E > V_0$ would be *transmitted as well as reflected*. This is a quantum mechanical result. Classically this would have never occurred.

For $E \gg V_{0,}$ we have

$$\beta = \frac{\sqrt{2m(E - V_0)}}{\hbar} \approx \frac{\sqrt{2mE}}{\hbar} = \alpha.$$

When, in addition to that, $2\alpha a = n\pi$, $n = 1, 2, 3, \ldots$, then since $\sin^2 2\alpha a = \sin^2 n\pi = 0$ and $\cos^2 2\alpha a = \cos^2 n\pi = 1$, (6.59) and (6.60) give $R = 0$, $T = 1$. Hence, the electron would be transmitted even when $V_0 \neq 0$ provided that the width $2a$ of the potential well is given by

$$2\alpha a = n\pi, \quad n = 1, 2, 3, \ldots.$$

Substituting the expression $\sqrt{2mE}/\hbar$ for α, this equation transforms into $8mEa^2/\hbar^2 = n^2\pi^2$ or $E = n^2\pi^2\hbar^2/8ma^2$. Thus, whenever the incident energy is as given above, practically there is only transmission. This phenomenon has its application in the scattering of low energy electrons (~0.1 eV) by noble gas atoms. For instance, it was first observed by Ramsauer and Townsend that in the scattering by neon and argon atoms, *almost* all such electrons are transmitted. This is called *transmission resonance*. A more accurate description no doubt must involve three-dimensional analysis.

Case 6. $0 < E < V_0$.
So far, we have considered those potentials with finite discontinuities in which the motion of the microscopic entity was *not* confined to a finite region of space. And it was observed that the positive energy spectrum forms a continuum in each case. Let us now consider a situation when the electron is *confined* to a finite region of space. It is then said to be *bounded*. We will show that in this case the auxiliary conditions imposed on eigenfunctions of the Schrödinger wave equation require that the energy spectrum be discrete.

We assume that with $0 < E < V_0$, the particle is *confined to a potential well*, as shown in figure 6.12. The discussion is again very much simplified by taking the

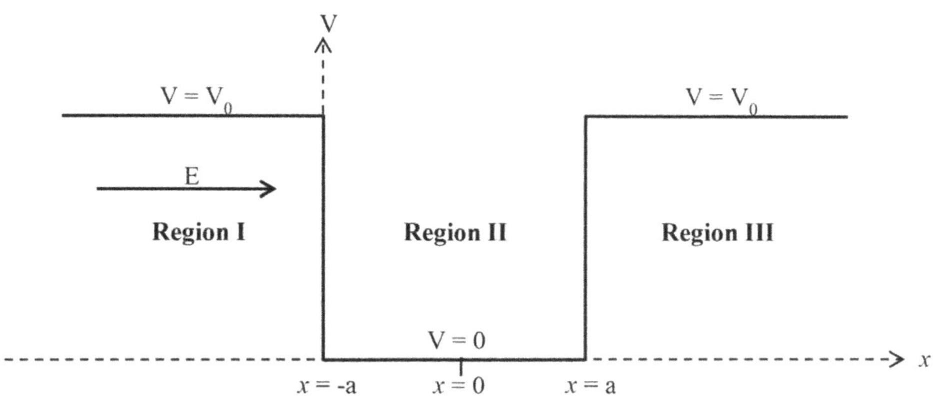

Figure 6.12. Square-well potential with $0 < E < V_0$.

origin at the centre of the well of width $2a$ so that we can take advantage of the symmetry about the centre. The potential is given by

$$\begin{aligned} V(x) &= V_0, \quad \text{for } x < -a, \\ &= 0, \quad \text{for } -a < x < a, \\ &= V_0, \quad \text{for } x > a. \end{aligned} \tag{6.50'}$$

$V(x)$ is the potential, V_0 is positive and gives the depth of the potential well. The microscopic entities, say electrons, are *classically confined to region II where the potential is zero*, the potential walls of the well having a height V_0. We will consider here the case for which the total energy E of the electron is less than V_0.

The Schrödinger wave equation in the three regions I, II, III is

$$\text{Region I:} \quad \frac{d^2\psi_1}{dx^2} + \frac{2m}{\hbar^2}(E - V_0)\psi_1 = 0, \quad \text{for } x < -a,$$

$$\text{Region II:} \quad \frac{d^2\psi_1}{dx^2} + \frac{2m}{\hbar^2}E\psi_2 = 0, \quad \text{for } -a < x < a,$$

$$\text{Region III:} \quad \frac{d^2\psi_3}{dx^2} + \frac{2m}{\hbar^2}(E - V_0)\psi_3 = 0, \quad \text{for } x > a.$$

Let us write

$$\frac{2m}{\hbar^2}E = \alpha^2, \tag{6.61}$$

$$\frac{2m}{\hbar^2}(V_0 - E) = \gamma^2. \tag{6.62}$$

Then the Schrödinger wave equation can be written as

$$\text{Region I:} \quad \frac{d^2\psi_1}{dx^2} - \gamma^2\psi_1 = 0, \quad \text{for } x < -a, \tag{6.63}$$

$$\text{Region II:} \quad \frac{d^2\psi_2}{dx^2} + \alpha^2\psi_2 = 0, \quad \text{for } -a < x < a, \tag{6.64}$$

$$\text{Region III:} \quad \frac{d^2\psi_3}{dx^2} - \gamma^2\psi_3 = 0, \quad \text{for } x > a. \tag{6.65}$$

The solutions of these differential equations are:

$$\text{Region I:}\quad \psi_1 = Ae^{\gamma x} + Be^{-\gamma x}, \tag{6.66}$$

$$\text{Region II:}\quad \psi_2 = Ce^{i\alpha x} + De^{-i\alpha x}, \tag{6.67}$$

$$\text{Region III:}\quad \psi_3 = Fe^{\gamma x} + Ge^{-\gamma x}. \tag{6.68}$$

In region I, the function $e^{-\gamma x}$ tends to infinity as $x \to -\infty$. It is therefore not physically acceptable and is discarded. Similarly, in region III, the function $e^{\gamma x}$ is not physically acceptable as it goes to infinity when $x \to \infty$. The set of above equations therefore becomes

$$\text{Region I:}\quad \psi_1 = Ae^{\gamma x}, \tag{6.69}$$

$$\text{Region I:}\quad \psi_2 = Ce^{i\alpha x} + De^{-i\alpha x}, \tag{6.70}$$

$$\text{Region III:}\quad \psi_3 = Ge^{-\gamma x}, \tag{6.71}$$

Differentiating these equations with respect to x, we obtain

$$\frac{d\psi_1}{dx^2} = \gamma Ae^{i\gamma x},$$

$$\frac{d\psi_2}{dx} = i\alpha\left(Ce^{i\alpha x} - De^{-i\alpha x}\right),$$

$$\frac{d\psi_3}{dx} = -\gamma Ge^{-\gamma x}.$$

Applying the boundary conditions that

(i) at $x = -a$,

$$\psi_1 = \psi_2$$

and

$$\frac{d\psi_1}{dx} = \frac{d\psi_2}{dx},$$

(ii) at $x = a$,

$$\psi_2 = \psi_3,$$

and

$$\frac{d\psi_2}{dx} = \frac{d\psi_3}{dx}$$

we obtain

$$Ae^{-\gamma a} = Ce^{-i\alpha a} + De^{-\alpha a},$$

$$\gamma Ae^{-\gamma a} = i\alpha\left(Ce^{-i\alpha a} - De^{-\alpha a}\right),$$

$$Ge^{-\gamma a} = Ce^{i\alpha a} + De^{-i\alpha a},$$

$$-\gamma Ge^{-\gamma a} = i\alpha\left(Ce^{i\alpha a} - De^{-i\alpha a}\right).$$

Writing P for αa and Q for γa, we obtain

$$Ae^{-Q} = Ce^{-iP} + De^{iP},$$

or

$$Ae^{-Q} - Ce^{-iP} - De^{iP} = 0 \tag{6.72}$$

$$\gamma Ae^{-Q} = i\alpha\left(Ce^{-iP} - De^{iP}\right),$$

or

$$\gamma Ae^{-Q} - i\alpha Ce^{-iP} + i\alpha De^{iP} \tag{6.73}$$

$$Ge^{-Q} = Ce^{iP} + De^{-iP},$$

or

$$Ce^{iP} + De^{-iP} - Ge^{-Q} = 0 \tag{6.74}$$

$$-\gamma Ge^{-Q} = i\alpha\left(Ce^{iP} - De^{-iP}\right),$$

or

$$i\alpha Ce^{iP} - i\alpha De^{-iP} + \gamma Ge^{-Q} = 0. \tag{6.75}$$

One of the above four constants A, C, D, G is determined by normalisation. The remaining three constants are therefore subject to the above four conditions. Consequently equations (6.61) and (6.62) *cannot hold simultaneously for arbitrary values of the energy E*. The non-trivial solution will exist only if the determinant of the coefficients of A, C, D, G in these four equations is zero:

$$\begin{vmatrix} e^{-Q} & -e^{-iP} & -e^{iP} & 0 \\ \gamma e^{-Q} & -i\alpha e^{-iP} & i\alpha e^{iP} & 0 \\ 0 & e^{iP} & e^{-iP} & -e^{-Q} \\ 0 & i\alpha e^{iP} & -i\alpha e^{-iP} & \gamma e^{-Q} \end{vmatrix} = 0 \tag{6.76}$$

The first and fourth columns both contain e^{-Q} as a common factor which can be cancelled throughout. Therefore, expanding the determinant, simplifying and replacing the value of P, we obtain

$$2\gamma\alpha \cos 2\alpha a + \left(\gamma^2 - \alpha^2\right) \sin 2\alpha a = 0,$$

or

$$2\gamma\alpha\left(\cos^2 \alpha a - \sin^2\alpha a\right)+\left(\gamma^2 - \alpha^2\right)2 \sin \alpha a \cos \alpha a = 0.$$

Dividing throughout by $2\gamma\alpha \cos^2 \alpha a$, we obtain

$$\left(1 - \tan^2\alpha a\right)+\left(\frac{\gamma}{\alpha} - \frac{\alpha}{\gamma}\right) \tan \alpha a = 0,$$

or

$$\tan^2 \alpha a - 1 - \frac{\gamma}{\alpha} \tan \alpha a + \frac{\alpha}{\gamma} \tan \alpha a = 0,$$

or

$$\left(\tan \alpha a - \frac{\gamma}{\alpha}\right)\left(\tan \alpha a + \frac{\alpha}{\gamma}\right) = 0. \tag{6.77}$$

Hence, non-trivial solutions will exist only if either

$$\tan \alpha a - \frac{\gamma}{\alpha} = 0,$$

or

$$\tan \alpha a + \frac{\alpha}{\gamma} = 0.$$

These equations can be written as

$$\gamma = \alpha \tan \alpha a \tag{6.78}$$

and

$$-\gamma = \alpha \cot \alpha a. \tag{6.79}$$

Multiplying both the equations throughout by a, we obtain

$$\gamma a = \alpha a \tan \alpha a \tag{6.80}$$

and

$$-\gamma a = \alpha a \cot \alpha a. \tag{6.81}$$

The two solutions are called *even and odd solutions* of the problem. Note that α and γ as defined by

$$\alpha^2 = \frac{2m}{\hbar 2}E, \tag{6.61'}$$

$$\gamma^2 = \frac{2m}{\hbar 2}(V_0 - E), \tag{6.62'}$$

involve the total energy E of the particle.

Problem

6.19. Show that (6.80) and (6.81) cannot hold simultaneously.

None of the transcendental equations (6.80) and (6.81) can be solved analytically. The solution to any one of these equations can only be obtained using either graphical or numerical methods. To find the values of the energy E allowed by (6.80), we proceed to solve it graphically.

Let us define two parameters ξ and η by the equations

$$\xi = \alpha a \tag{6.82}$$

and

$$\eta = \gamma a. \tag{6.83}$$

Squaring and then adding (6.82) and (6.83), we obtain

$$\begin{aligned}\eta^2 + \xi^2 &= \gamma^2 a^2 + \alpha^2 a^2 \\ &= \frac{2m}{\hbar^2}(V_0 - E)a^2 + \frac{2m}{\hbar^2}Ea^2 \\ &= \frac{2m}{\hbar^2}V_0 a^2.\end{aligned}$$

Writing

$$\frac{2m}{\hbar^2}V_0 a^2 = \lambda^2, \tag{6.85}$$

we obtain

$$\eta^2 + \xi^2 = \lambda^2. \tag{6.86}$$

This is the equation of a circle of radius $\lambda = (\sqrt{2mV_0}/\hbar)a$ in the ξ–η plane.

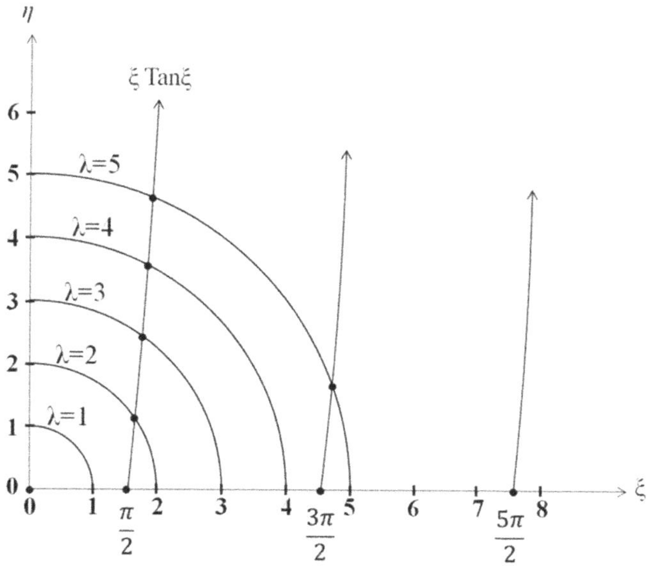

Figure 6.13. Graphical solution of $\sqrt{\lambda^2 - \xi^2} = \xi \tan \xi$.

The function η (= $\sqrt{\lambda^2 - \xi^2}$) has been plotted versus ξ in figure 6.13. As ξ (= αa) and η (= γa) are restricted to positive values, the function η is plotted against ξ in the first quadrant. Consequently, the function η as given by $\sqrt{\lambda^2 - \xi^2}$ is a quarter circle of known radius λ in the first quadrant. In this quadrant, the function $\xi \tan \xi$ has zeros at $\xi = 0, \pi, 2\pi, \ldots$ and asymptotes at $\xi = \pi/2, 3\pi/2, 5\pi/2, \ldots$. The curves for $\eta = \xi \tan \xi$ are also shown in figure 6.13. The points of intersection of two curves in the first quadrant will give us permissible energy values. Figure 6.13 shows that the *number of solutions*, indicated by the points where the curves for $\xi \tan \xi$ and $\sqrt{\lambda^2 - \xi^2}$ intersect, depends upon the radius of the quarter-circle. Each solution gives an energy eigenvalue E which is less than V_0. If $\lambda < \pi$, there exists only one energy eigenvalue solution; if $\pi \leqslant \lambda < 2\pi$, there exist two such solutions, and so on.

Problems

6.20. Find the two energy solutions for $\lambda = 4$.

6.21. Show that when the the potential well is deeper and/or broader, then there are more bound states of the system.

From (6.77) and (6.79), we have seen that one of the two solutions is given by

$$\tan \xi = \frac{\sqrt{\lambda^2 - \xi^2}}{\xi}.$$

For large values of λ, the right-hand side of the above equation becomes very large. Therefore, the left-hand side should tend to infinity. That is,

$$\tan \xi \to \infty$$

$$\frac{\sin \xi}{\cos \xi} \to \infty,$$

or

$$\cos \xi \approx 0,$$

or

$$\xi \approx \left(n + \frac{1}{2}\right)\pi, \quad n = 0, 1, 2, \ldots,$$

or

$$\xi^2 \approx \left(n + \frac{1}{2}\right)^2 \pi^2, \quad n = 0, 1, 2, \ldots,$$

or

$$\frac{2mE_n}{\hbar^2} a^2 \approx \left(n + \frac{1}{2}\right)^2 \pi^2,$$

or

$$E_n \approx \left(n + \frac{1}{2}\right)^2 \frac{\pi^2 \hbar^2}{2ma^2}, \tag{6.84}$$

where to emphasise that energy depends upon the quantum number n, we have replaced E by E_n.

Problem

6.22. Discuss the solution of

$$-\gamma a = \alpha a(\cot \alpha a).$$

Table 6.1 gives the classical as well as quantum mechanical results for (i) the potential step, (ii) the potential barrier and (iii) the square-well potential. We notice that quantum mechanics and classical theory only yield similar results for the potential step when $0 < E < V_0$. Even in this case, quantum mechanically, the transmission coefficient is zero, some waves cross the potential step and enter the classically forbidden region.

Table 6.1. Comparison of reflected and transmitted probability fluxes in classical theory and quantum mechanics.

Potential step	Classical theory	Quantum mechanics	
$E > V_0 > 0$	$T = 1$	$T \neq 0$	
	$R = 0$	$R \neq 0$	
$0 < E < V_0$	$T = 0$	$T = 0$	
	$R = 1$	$R = 1$	
Potential barrier	**Classical theory**	**Quantum mechanics**	
$E > V_0$	$T = 1$	$T \neq 0$	
	$R = 0$	$R \neq 0$	
$E < V_0$	$T = 0$	$T \neq 0$	(α-particle decay)
	$R = 1$	$R \neq 0$	
Square-well potential	**Classical theory**	**Quantum mechanics**	
$E > V_0$	$T = 1$	$T \neq 0$	
	$R = 0$	$R \neq 0$	
$E < V_0$	$T = 0$	$T \neq 0$	
	$R = 1$	$R \neq 0$	

Additional problems

6.23. Derive the expressions for reflection and transmission coefficients R and T without using the formula $\mathbf{J} = \hbar/m(\Psi^*\nabla\Psi)$.

6.24. What is α-decay? Give its quantum mechanical theory and indicate how this may be used to estimate nuclear radii.

6.25. A beam of electrons of energy $E\,(> 0)$ meets a potential barrier described by

$$V(x) = \begin{cases} 0, & x < 0, \\ V_0, & 0 < x < a, \quad \text{where } V_0 > E, \\ 0, & x > a. \end{cases}$$

Show that the ratio, T, of the number of electrons crossing unit area per unit time in the incident and transmitted beams, is

$$T = \frac{4}{4\cosh^2 \gamma a + \left(\dfrac{\gamma}{a} - \dfrac{a}{\gamma}\right)^2 \sinh^2 \gamma a}$$

where

$$\alpha\hbar = \sqrt{2mE} \qquad \text{and} \qquad \gamma\hbar = \sqrt{2m(V_0 - E)}.$$

IOP Publishing

Quantum Mechanics

Mohammad Saleem

Chapter 7

Spherically symmetric potentials

7.1 Introduction

So far, except for the free particle, we have solved the Schrödinger wave equation for only one-dimensional systems: a one-dimensional box, a linear harmonic oscillator, the potential step, etc. This was done because of the mathematical simplicity in the solution of the differential equation. We will now consider the application of quantum mechanics to a three-dimensional system. Owing to the importance of such problems in physics and chemistry, it is imperative to analyse such cases in detail. Physical systems with spherically symmetric potentials, i.e. with potentials depending only upon the magnitude of the radius vector $\mathbf{r}$ drawn from a fixed point usually taken as the origin, and not upon its orientation, play an important role in atomic and molecular physics and in the discipline of chemistry. The hydrogen atom is the most significant example of a system with such a potential. This forms the basis for the analysis of spectra of complex atomic systems, such as those of singly ionised helium and doubly ionised lithium atoms. Such atoms are called hydrogenic or hydrogen-like atoms. We shall first discuss in general the characteristics of a system with a spherically symmetric potential and then confine ourselves to hydrogen and hydrogen-like atoms. We will determine the solution of the Schrödinger wave equation for a spherically symmetric potential. This will also be the solution for the hydrogen atom which is an exquisite example, with reduced mass, of such a system. The analysis will then be easily extended to hydrogen-like atoms.

7.2 Spherically symmetric potentials

Let us consider a particle moving in a spherically symmetric potential. By definition, the potential energy of such a system depends upon r only. As it does not depend upon time explicitly, it is sufficient in this case to solve the Schrödinger amplitude equation. By virtue of its spherical symmetry, it will be convenient to solve the

doi:10.1088/978-0-7503-1206-6ch7

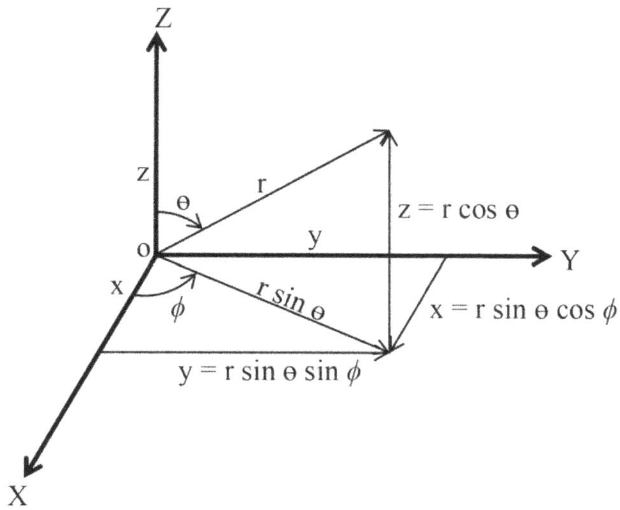

Figure 7.1. The relation between the rectangular Cartesian and spherical polar coordinates.

Schrödinger amplitude equation in spherical polar coordinates r, θ, ϕ. These are related to the rectangular Cartesian coordinates x, y, z by the following equations:

$$x = r\sin\theta\cos\phi,$$

$$y = r\sin\theta\sin\phi,$$

$$z = r\cos\theta,$$

where the ranges of r, θ, ϕ are $r \geqslant 0$, $0 \leqslant \theta \leqslant \pi$, $0 \leqslant \phi \leqslant 2\pi$. This is shown in figure 7.1.

Problem

7.1. Show that in spherical polar coordinates the Laplacian operator ∇^2 can be expressed as

$$\nabla^2 = \frac{1}{r^2}\frac{\partial}{\partial r}\left(r^2\frac{\partial}{\partial r}\right) + \frac{1}{r^2\sin\theta}\frac{\partial}{\partial\theta}\left(\sin\theta\frac{\partial}{\partial\theta}\right) + \frac{1}{r^2\sin^2\theta}\frac{\partial^2}{\partial\phi^2}. \qquad (7.1)$$

7.3 Separation of variables

For a system with spherically symmetric potential, it is convenient to use spherical polar coordinates because then the solution can be expressed as a product of a radial function $R(r)$ depending on r only and a function $Y(\theta, \phi)$ of angular coordinates alone. We proceed as follows.

Substituting the expression for ∇^2, as given in (7.1), in the Schrödinger amplitude equation,

$$\nabla^2\psi + \frac{2m}{\hbar^2}(E - V(r))\psi = 0,$$

we obtain

$$\frac{1}{r^2}\frac{\partial}{\partial r}\left(r^2\frac{\partial\psi}{\partial r}\right) + \frac{1}{r^2 \sin\theta}\frac{\partial}{\partial\theta}\left(\sin\theta\frac{\partial\psi}{\partial\theta}\right) + \frac{1}{r^2\sin^2\theta}\frac{\partial^2\psi}{\partial\phi^2} + \frac{2m}{\hbar^2}(E - V(r))\psi = 0. \quad (7.2)$$

To solve this differential equation, we suppose that the wave function $\psi(r, \theta, \phi)$ can be expressed as a product of two functions R and Y depending, respectively, on the radial and angular coordinates only. Then we may write

$$\Psi(r, \theta, \phi) = R(r)Y(\theta, \phi). \quad (7.3)$$

Differentiating with respect to r, θ, ϕ, and substituting the expressions for ψ and its derivatives in (7.2), we obtain

$$\frac{1}{r^2}\frac{\mathrm{d}}{\mathrm{d}r}\left(r^2\frac{\mathrm{d}R}{\mathrm{d}r}Y\right) + \frac{R}{r^2\sin\theta}\frac{\partial}{\partial\theta}\left(\sin\theta\frac{\partial Y}{\partial\theta}\right) + \frac{R}{r^2\sin^2\theta}\frac{\partial^2 Y}{\partial\phi^2} + \frac{2m}{\hbar^2}(E - V(r))RY = 0.$$

In accordance with the usual practice, we have used in the differential equation an ordinary derivative instead of a partial derivative wherever a function depends upon one independent variable only. This practice will be continued in the further analysis. Multiplying the above equation throughout by r^2/RY, we obtain

$$\frac{1}{R}\frac{\mathrm{d}}{\mathrm{d}r}\left(r^2\frac{\mathrm{d}R}{\mathrm{d}r}\right) + \frac{1}{Y\sin\theta}\frac{\partial}{\partial\theta}\left(\sin\theta\frac{\partial Y}{\partial\theta}\right) + \frac{1}{Y\sin^2\theta}\frac{\partial^2 Y}{\partial\phi^2} + \frac{2m}{\hbar^2}r^2(E - V(r)) = 0.$$

The terms in this differential equation may be rearranged so that we may write it as

$$\frac{1}{R}\frac{\mathrm{d}}{\mathrm{d}r}\left(r^2\frac{\mathrm{d}R}{\mathrm{d}r}\right) + \frac{2m}{\hbar^2}r^2(E - V(r)) = -\left[\frac{1}{Y\sin\theta}\frac{\partial}{\partial\theta}\left(\sin\theta\frac{\partial Y}{\partial\theta}\right) + \frac{1}{Y\sin^2\theta}\frac{\partial^2 Y}{\partial\phi^2}\right]. \quad (7.4)$$

The left-hand side of differential equation (7.4) involves the variable r only while its right-hand side is a function of angular coordinates alone. Therefore, each side can be varied independently of the other. Hence, in order that the equation will hold, it is imperative that each side be equal to the *same constant*. This constant is denoted by C. Then, we may write

$$\frac{1}{r^2}\frac{\mathrm{d}}{\mathrm{d}r}\left(r^2\frac{\mathrm{d}R}{\mathrm{d}r}\right) + \frac{2m}{\hbar^2}r^2(E - V(r)) = C \quad (7.5)$$

and

$$-\left[\frac{1}{Y\sin\theta}\frac{\partial}{\partial\theta}\left(\sin\theta\frac{\partial Y}{\partial\theta}\right)+\frac{1}{Y\sin^2\theta}\frac{\partial^2 Y}{\partial\phi^2}\right]=C. \tag{7.6}$$

Let us first solve the differential equation involving angular coordinates alone. Suppose that its solution $Y(\theta, \phi)$ can be expressed as a product of two functions P and F, the former depending on θ only and the latter depending on ϕ alone. Then we have

$$Y(\theta, \phi) = P(\theta)F(\phi). \tag{7.7}$$

Differentiating with respect to θ and ϕ and substituting the expressions for Y and its derivatives in (7.6) and simplifying, we obtain

$$-\left[\frac{1}{Q\sin\theta}\frac{\mathrm{d}}{\mathrm{d}\theta}\left(\sin\theta\frac{\mathrm{d}Q}{\mathrm{d}\theta}\right)+\frac{1}{F\sin^2\theta}\frac{\mathrm{d}^2F}{\mathrm{d}\phi^2}\right]=C.$$

Multiplying throughout by $\sin^2\theta$, we obtain

$$-\left[\frac{\sin\theta}{Q}\frac{\mathrm{d}}{\mathrm{d}\theta}\left(\sin\theta\frac{\mathrm{d}Q}{\mathrm{d}\theta}\right)+\frac{1}{F}\frac{\mathrm{d}^2F}{\mathrm{d}\phi^2}\right]=C\sin^2\theta,$$

or

$$\frac{\sin\theta}{Q}\frac{\mathrm{d}}{\mathrm{d}\theta}\left(\sin\theta\frac{\mathrm{d}Q}{\mathrm{d}\theta}\right)+C\sin^2\theta=-\frac{1}{F}\frac{\mathrm{d}^2F}{\mathrm{d}\phi^2}. \tag{7.8}$$

Arguing as before, each side must be equal to the same constant which we denote by m^2, where m is called the *magnetic quantum number* because it is most significant in problems involving atoms in magnetic fields. We can then write

$$\frac{\sin\theta}{Q}\frac{\mathrm{d}}{\mathrm{d}\theta}\left(\sin\theta\frac{\mathrm{d}Q}{\mathrm{d}\theta}\right)+C\sin^2\theta=m^2 \tag{7.9}$$

and

$$-\frac{1}{F}\frac{\mathrm{d}^2F}{\mathrm{d}\phi^2}=m^2,$$

i.e.

$$\frac{1}{F}\frac{\mathrm{d}^2F}{\mathrm{d}\phi^2}+m^2=0. \tag{7.10}$$

The solution of Schrödinger's amplitude equation (7.4) for a spherically symmetric potential $V(r)$ can therefore be written as

$$\psi(r, \theta, \phi) = R(r)Y(\theta, \phi)$$
$$= R(r)P(\theta)F(\phi).$$

If we know each of the functions $R(r)$, $P(\theta)$ and $F(\phi)$, their product will give us the solution of the problem. For this purpose, we have to solve the differential equations (7.5), (7.9) and (7.10).

7.4 Solution of the differential equation for $F(\phi)$

Let us first solve the differential equation (7.10). This can be written as

$$\frac{d^2F(\phi)}{d\phi^2} + m^2F(\phi) = 0.$$

For $m \neq 0$, there are two linearly independent solutions of this second-order ordinary differential equation. These are

$$F(\phi) = Ae^{im\phi} \qquad \text{and} \qquad F(\phi) = Be^{-im\phi}.$$

For $m = 0$, this differential equation reduces to $d^2F/d\phi^2 = 0$ and has the general solution

$$F(\phi) = C + D\phi.$$

Let us consider the solution $F(\phi) = Ae^{im\phi}$ for $m \neq 0$. Since ϕ can take the values between 0 and 2π, the function $F(\phi)$ will never become infinite. As the function $F(\phi)$ must be single-valued, we have

$$F(2\pi) = F(0),$$

or

$$Ae^{im2\pi} = Ae^{im\times 0} = A,$$

or

$$A\left(e^{im2\pi} - 1\right) = 0,$$

Therefore, either $A = 0$ or $(e^{im2\pi} - 1) = 0$. If $A \neq 0$, then we must have

$$\left(e^{im2\pi} - 1\right) = 0,$$

or

$$e^{im2\pi} = 1 = e^{\pm i2\pi} = e^{\pm i4\pi}\pi = \cdots.$$

This set of equations shows that $m = \pm 1, \pm 2, \ldots$. That is, in order that $F(\phi)$ may be a single-valued function, $m \neq 0$ must take positive or negative integral values.

If $A = 0$, then the solution $F(\phi) = Ae^{im\phi}$ will become zero for all ϕ; we do not obtain any information about the values of m. However, in this case, in the second independent solution $F(\phi) = B\mathrm{e}^{-\mathrm{i}m\phi}$, the parameter B cannot vanish. This is because if B were also zero, then $F(\phi)$ and consequently $\psi(r, \theta, \phi)$ will be zero for all ϕ; the particle will disappear! Thus for $A = 0$, the solution $F(\phi) = B\mathrm{e}^{-\mathrm{i}m\phi}$ is valid only for $B \neq 0$. This solution $F(\phi) = B\mathrm{e}^{-\mathrm{i}m\phi}$ does not give any new information about the system; as already shown m can have all negative and positive integral values. Since the two linearly independent solutions give the same information, any one of them may be retained. Thus for non-zero m, the physical solution may be written as

$$F(\phi) = A\mathrm{e}^{\mathrm{i}m\phi}, \qquad m = \pm 1, \pm 2, \ldots. \tag{7.11}$$

In order to normalise this part of the wave function, we have

$$\int_0^{2\pi} F^*(\phi)F(\phi)\,\mathrm{d}\phi = A^*A\int_0^{2\pi} \mathrm{e}^{-\mathrm{i}m\phi}\mathrm{e}^{\mathrm{i}m\phi}\,\mathrm{d}\phi = 2\pi A^2 = 1,$$

where the phase of A, being physically unimportant, has been taken as zero. This yields

$$A = \frac{1}{\sqrt{2\pi}}$$

so that for $m \neq 0$, we have

$$F(\phi) = \frac{1}{\sqrt{2\pi}}\mathrm{e}^{\mathrm{i}m\phi}, \qquad m = \pm 1, \pm 2, \ldots. \tag{7.12}$$

This is the normalised solution for the differential equation for $F(\phi)$ for $m \neq 0$.

For $m = 0$, the solution of differential equation (7.10) is

$$F(\phi) = C + D\phi. \tag{7.13a}$$

In order that $F(\phi)$ may be single-valued, we must have

$$F(2\pi) = F(0),$$

or

$$C + 2\pi D = C,$$

or

$$2\pi D = 0,$$

or

$$D = 0.$$

Equation (7.13a) therefore yields

$$F(\phi) = C, \quad \text{for } m = 0.$$

The normalisation again yields the value of the constant C as $1/\sqrt{2\pi}$. Therefore, for $m = 0$, we may write

$$F(\phi) = \frac{1}{\sqrt{2\pi}}. \tag{7.13b}$$

Combining (7.12) and (7.13*b*), we obtain

$$F(\phi) = \frac{1}{\sqrt{2\pi}} e^{im\phi}, \quad m = 0, \pm 1, \pm 2, \ldots. \tag{7.14}$$

To emphasise that $F(\phi)$ depends upon quantum number m, we write it as $F_m(\phi)$ so that

$$F_m(\phi) = \frac{1}{\sqrt{2\pi}} e^{im\phi}, \quad m = 0, \pm 1, \pm 2, \ldots. \tag{7.15}$$

This is the normalised solution of the differential equation (7.10) for $F(\phi)$. Then by (7.7) and (7.15), we have

$$Y(\theta, \phi) = P(\theta)F(\phi) = P(\theta)\frac{1}{\sqrt{2\pi}} e^{im\phi}, \quad m = 0, \pm 1, \pm 2, \ldots. \tag{7.16a}$$

A few normalised azimuthal wave functions $F_m(\phi)$ may be written as

$$F_0(\phi) = \frac{1}{\sqrt{2\pi}},$$

$$F_1(\phi) = \frac{1}{\sqrt{2\pi}} e^{i\phi},$$

$$F_{-1}(\phi) = \frac{1}{\sqrt{2\pi}} e^{-i\phi},$$

$$F_2(\phi) = \frac{1}{\sqrt{2\pi}} e^{2i\phi},$$

$$F_{-2}(\phi) = \frac{1}{\sqrt{2\pi}} e^{-2i\phi}.$$

7.5 Solution of the differential equation for $P(\theta)$

In section 7.4, we obtained the solution of the differential equation involving the variable ϕ only. Let us next solve the $P(\theta)$ differential equation, i.e.

$$\frac{\sin\theta}{P(\theta)} \frac{d}{d\theta}\left(\sin\theta \frac{dP(\theta)}{d\theta}\right) + C\sin^2\theta = m^2. \tag{7.9'}$$

If we can obtain the solution $P(\theta)$ of this differential equation, we will obtain the solution $Y(\theta, \phi)$ of differential equation (7.6) as

$$Y(\theta, \phi) = P(\theta)F(\phi).$$

Multiplying (7.9′) throughout by $P(\theta)/\sin\theta$, we obtain

$$\frac{\mathrm{d}}{\mathrm{d}\theta}\left(\sin\theta\frac{\mathrm{d}P(\theta)}{\mathrm{d}\theta}\right) + CP(\theta)\sin\theta = m^2\frac{P(\theta)}{\sin\theta}. \tag{7.16b}$$

For its solution, it is convenient for calculations to introduce a new variable z such that

$$z = \cos\theta. \tag{7.16c}$$

Then differential equation (7.16b) becomes

$$\frac{\mathrm{d}}{\mathrm{d}z}\left[(1 - z^2)\frac{\mathrm{d}P(z)}{\mathrm{d}z}\right] + \left[C - \frac{m^2}{1 - z^2}\right]P(z) = 0 \tag{7.17}$$

where the domain of z in $P(z)\{\equiv P(\cos\theta)\}$ is $-1 \leqslant z \leqslant 1$ as θ varies from 0 to π. Note that one term of this differential equation becomes infinite when z takes either of its extremum values. The points $z = \pm 1$ are therefore called the singular points of this differential equation. We will study the behaviour of the function $P(z)$ at these extreme points $z = +1$ and $z = -1$ as we proceed further.

If we try to solve this differential equation, as it stands, by expanding $P(z)$ as a power series, we obtain a recursion formula which involves more than two terms. To avoid that, we make a further substitution suggested by the theory of differential equations. We write

$$P(z) = (1 - z^2)^{|m|/2}G(z), \tag{7.18}$$

where $|m|$ is the absolute value of m. In order to avoid writing $|m|$ again and again, we write a for it and will replace it at the end. Then differentiating equation

$$P(z) = \left(1 - z^2\right)^{a/2}G(z)$$

with respect to z, we obtain

$$\frac{\mathrm{d}P(z)}{\mathrm{d}z} = \frac{a}{2}\left(1 - z^2\right)^{\frac{a}{2}-1}(-2z)G(z) + \left(1 - z^2\right)^{\frac{a}{2}}\frac{\mathrm{d}G(z)}{\mathrm{d}z},$$

or

$$P'(z) = -az\left(1 - z^2\right)^{\frac{a}{2}-1}G(z) + \left(1 - z^2\right)^{\frac{a}{2}}G'(z),$$

where $P'(z)$ and $G'(z)$ are, respectively, the first derivatives of $P(z)$ and $G(z)$ with respect to z. Then differential equation (7.17) takes the form

$$\left(1 - z^2\right)G''(z) - 2(a + 1)zG'(z) + \{C - a(a + 1)\}G(z) = 0. \tag{7.19}$$

In which $G''(z)$ represents the second derivative with respect to G. If we can solve this differential equation, we will obtain the expression for $G(z)$. Substituting that expression in (7.18), we obtain the required solution of the $P(z)$ differential equation (7.17) and consequently that of (7.9). Now, to solve the differential equation (7.19), we write $G(z)$ as a power series:

$$G(z) = a_0 + a_1 z + a_2 z^2 + a_3 z^3 + a_4 z^4 + \cdots.$$

Differentiating $G(z)$ once and twice, we obtain

$$G'(z) = a_1 + 2a_2 z + 3a_3 z^2 + 4a_4 z^3 + \cdots$$

and

$$G''(z) = 1.2a_2 + 2.3a_3 z + 3.4a_4 z^2 + \cdots.$$

Substituting the expressions for $G(z)$, $G'(z)$ and $G''(z)$ in differential equation (7.19) and simplifying, we obtain

$$\begin{aligned}\big(1.2a_2 &+ 2.3a_3 z + 3.4a_4 z^2 + 4.5a_5 z^3 + \cdots \\ &- 1.2a_2 z^2 + 2.3a_3 z^3 + \cdots \\ &- 2.1qa_1 z - 2.2qa_2 z^2 - 2.3a_3 z^3 + \cdots \\ &+ sa_0 + sa_1 z + sa_2 z^2 + sa_3 z^3 + \cdots\big) = 0,\end{aligned}$$

where $q = a + 1$, $s = C - aq$ and $a = |m|$. Since this equation must hold for all values of z varying between -1 and 1, the coefficient of each power of z must vanish separately. Thus, we have

$$\begin{aligned}&1.2a_2 + sa_0 = 0, \\ &2.3a_3 + (s - 2.1q)a_1 = 0, \\ &3.4a_4 + (s - 2.2q - 1.2)a_2 = 0, \\ &4.5a_5 + (s - 2.3q - 2.3)a_3 = 0, \\ &\cdots\end{aligned}$$

Consequently, for the coefficient of z^ν, we have

$$(\nu + 1)(\nu + 2)a_{\nu+2} + \left[s - 2\nu q - \nu(\nu - 1)\right]a_\nu = 0.$$

This leads to the recursion formula

$$a_{\nu+2} = -\frac{s - 2\nu q - \nu(\nu - 1)}{(\nu + 2)(\nu + 1)}a_\nu, \quad \nu = 0, 1, 2, \ldots. \tag{7.20}$$

This formula enables us to calculate $a_2, a_4, a_6, \ldots$ in terms of a_0, and $a_3, a_5, a_7, \ldots$ in terms of a_1. The coefficients a_0 and a_1 are arbitrary. This is what it must be! The general solution of a second-order ordinary differential equation must contain two arbitrary constants. The recursion formula shows that a with *even and odd indices are not related.* The series $G(z)$ can therefore be written as a sum of two linearly independent series:

$$G(z) = \left(a_0 + a_2z^2 + a_4z^4 + \cdots\right) + \left(a_1z + a_3z^3 + a_5z^5 + \cdots\right)$$
$$= f(z) + g(z).$$

That is, if a_0 and a_1 are known, all the a can be calculated and the solution $G(z)$ determined. Proceeding along the same lines as for the harmonic oscillator, it can be shown that the two infinite power series for $G(z)$, whose coefficients are related among themselves by the above recursion formula, diverge for $z = \pm 1$. The solution is therefore not physically acceptable. In order to make the solution acceptable, we have to make one of the functions $f(z)$ and $g(z)$ terminate after certain terms by taking the numerator in the recursion formula equal to zero for some value u of v and by making the other function vanish by choosing its first a equal to zero. That is, for any value of u, we have to discard one of the two functions. The other function by a specific choice of the numerator in the recursion formula converges to a polynomial. Hence, for a physically acceptable solution, we must have either (i) $a_1 = 0$ and $s - 2uq - u(u-1) = 0$ for even u or (ii) $a_0 = 0$ and $s - 2uq - u(u-1) = 0$ for odd u. Thus, we have

$$s - 2uq - u(u - 1) = 0$$

where u is even for (i) and odd for (ii). Substituting the expressions for q $(= a + 1)$ and s $(= C - aq = C - a(a + 1))$ in the above equation, we obtain

$$C - a(a + 1) = 2u(a + 1) + u(u - 1),$$

or

$$C = a(a + 1) + u(a + 1) + u(a + 1) + u(u - 1),$$

or

$$C = (a + 1)(a + u) + u(a + u),$$

or

$$C = (a + u)(a + u + 1).$$

Writing $|m|$ for a, we obtain

$$C = (|m| + u)(|m| + u + 1).$$

Writing ℓ for $(|m| + u)$, we obtain

$$C = \ell(\ell + 1). \tag{7.21}$$

Hence, $C = \ell(\ell + 1)$ where $\ell = (|m| + u)$ such that $|m| = 0, 1, 2, \ldots$ and $u = 0, 1, 2, \ldots$. This means that ℓ can have the values $0, 1, 2, \ldots$. ℓ is called the *orbital* or *azimuthal quantum number* since θ is the azimuthal angle. As $\ell = |m| + u$, for given ℓ and m, the value of u and as a consequence of it that polynomial which is a solution of the above differential equation is uniquely specified. The values of the quantum number ℓ are frequently associated with the *spectroscopic notation*; for instance, the state with $\ell = 0$ is called an s-state, the $\ell = 1$ state is denoted by p, the $\ell = 2$ state by d and the $\ell = 3$ state by f. The above analysis shows that in order for differential equation (7.17) to have a physically acceptable solution, it must be of the form

$$\frac{\mathrm{d}}{\mathrm{d}z}\left[(1 - z^2)\frac{\mathrm{d}P(z)}{\mathrm{d}z}\right] + \left[\ell(\ell + 1) - \frac{m^2}{1 - z^2}\right]P(z) = 0 \qquad (7.22)$$

where $\ell = 0, 1, 2, \ldots$ and $m = 0, \pm1, \pm2, \ldots$.

7.6 Legendre polynomials and associated Legendre functions

A solution of differential equation (7.22) for $m = 0$ is called the *Legendre polynomial* or *Legendre function* of degree ℓ and is denoted by $P_\ell(z)$ so that $P_\ell(z) \equiv P(z)$ for $m = 0$. The solution of the same differential equation for $m \neq 0$ is called the *associated Legendre function* or *polynomial* of degree ℓ and *order* $|m|$ and is denoted by $P_\ell^{|m|}(z)$ so that $P_\ell^{|m|}(z) = P(z)$ for $m \neq 0$.

The functions $P(z)$ which satisfy the differential equation (7.22) were known to mathematicians long before the advent of quantum mechanics. The properties of the functions $P(z)$ required for the computation of physical quantities can be obtained by using the recursion formula. However, this process is very cumbersome and therefore will not be used. Instead, we will define these functions by using differential expressions or generating functions and obtaining their various characteristics.

The associated Legendre functions $P_\ell^{|m|}(z)$ of degree ℓ and order $|m|$ are related to Legendre functions $P_\ell(z)$ by the equation

$$P_\ell^{|m|}(z) = (1 - z^2)^{|m|/2}\frac{\mathrm{d}^{|m|}}{\mathrm{d}z^{|m|}}P_\ell(z), \qquad (7.23)$$

where the Legendre polynomials $P_\ell(z)$ are defined by means of the following generating function:

$$\sum_{\ell=0}^{\infty} P_\ell(z)t^\ell = \frac{1}{\sqrt{1 - 2tz + t^2}}, \qquad |t| < 1. \qquad (7.24)$$

Problems

7.2. Show that the functions $P_\ell(z)$, as defined by (7.24), satisfy the differential equation

$$\frac{\mathrm{d}}{\mathrm{d}z}\left[\left(1 - z^2\right)\frac{\mathrm{d}P_\ell(z)}{\mathrm{d}z}\right] + \ell(\ell + 1)P_\ell(z) = 0, \quad \ell = 0, 1, 2, \ldots. \qquad (7.25)$$

7.3. Show that the Legendre polynomials are given by

$$P_\ell(z) = \frac{1}{2^\ell\,\ell!}\frac{\mathrm{d}^\ell}{\mathrm{d}z^\ell}\left(z^2 - 1\right)^\ell. \tag{7.26}$$

The first few Legendre polynomials $P_\ell(z)$ are

$$\begin{aligned}
P_0(z) &= 1,\\
P_1(z) &= z,\\
P_2(z) &= \frac{1}{2}\left(3z^2 - 1\right),\\
P_3(z) &= \frac{1}{2}\left(5z^3 - 3z\right),\\
P_4(z) &= \frac{1}{8}\left(35z^4 - 30z^2 + 3\right),\\
P_5(z) &= \frac{1}{8}\left(63z^5 - 70z^3 + 15z\right).
\end{aligned} \tag{7.27}$$

Now, on differentiating equation (7.25) $|m|$ times, we obtain

$$\left(1 - z^2\right)\frac{\mathrm{d}^{|m|+2}}{\mathrm{d}z^{|m|+2}}P_\ell(z) - 2(|m| + 1)z\frac{\mathrm{d}^{|m|+1}}{\mathrm{d}z^{|m|+1}}P_\ell(z)$$

$$+ \ell(\ell + 1) - |m|(|m| + 1)\frac{\mathrm{d}^{|m|}}{\mathrm{d}z^{|m|}}P_\ell(z) = 0. \tag{7.28}$$

Problem

7.4. Using (7.25), show that differential equation (7.28) can be transformed into

$$\frac{\mathrm{d}}{\mathrm{d}z}\left[\left(1 - z^2\right)\frac{\mathrm{d}P_\ell^{|m|}(z)}{\mathrm{d}z}\right] + \left[\ell(\ell + 1) - \frac{m^2}{1 - z^2}\right]P_\ell^{|m|}(z) = 0. \tag{7.29}$$

Since this is a homogeneous linear differential equation, a constant multiple of the associated Legendre function $P_\ell^{|m|}(z)$ is also a solution of this differential equation. A comparison of differential equations (7.22) and (7.29) shows that, the differential equations being identical, the solution of one equation must be a constant multiple of the solution of the other equation. Hence, we may write

$$P_\ell(z) = \text{const}\cdot P_\ell^{|m|}(z). \tag{7.30}$$

Problem

7.5. Show that if the polynomials $P_\ell(z)$ are to be normalised, then the constant of normalisation should be equal to

$$\in\sqrt{\frac{2\ell+1}{2}\frac{(\ell-|m|)!}{(\ell+|m|)!}},$$

where

$$\in=(-1)^m, \quad \text{for } m>0,$$

$$\in=1, \quad \text{for } m\leqslant 0.$$

It can easily be shown that $\in$ can also be written as

$$\in=(-1)^{1/2}(m+|m|).$$

Hint: remember that the range of integration for θ is from 0 to π.

The normalised solution $P_\ell(z)$ of differential equation (7.22) is given by

$$P_\ell(z)=\sqrt{\frac{2\ell+1}{2}\frac{(\ell-|m|)!}{(\ell+|m|)!}}P_\ell^{|m|}(z). \tag{7.31}$$

The first few associated Legendre functions $P_\ell^{|m|}(z)$ are

$$P_1^1(z)=\left(1-z^2\right)^{1/2},$$

$$P_2^1(z)=3\left(1-z^2\right)^{1/2}z,$$

$$P_2^2(z)=3\left(1-z^2\right),$$

$$P_3^1(z)=\frac{3}{2}\left(1-z^2\right)^{1/2}\left(5z^2-1\right),$$

$$P_3^2(z)=15z\left(1-z^2\right),$$

$$P_3^3(z)=15\left(1-z^2\right)^{3/2}.$$

7.7 Spherical harmonics

The solutions $Y_{\ell m}(\theta,\phi)$ of the angular part of the Schrödinger wave equation are defined by

$$Y_{\ell m}(\theta,\phi)=(-1)^{\frac{1}{2}(m+|m|)}P_{\text{norm}}(\theta)F_{\text{norm}}(\phi),$$

are known as *spherical harmonics* and are given by

$$Y_{\ell m}(\theta, \phi) = (-1)^{\frac{1}{2}(m+|m|)} \sqrt{\frac{2\ell+1}{2} \frac{(\ell-|m|)!}{(\ell+|m|)!}} P_\ell^{|m|}(\cos\theta) \frac{1}{\sqrt{2\pi}} e^{im\phi}$$

$$= (-1)^{\frac{1}{2}(m+|m|)} \sqrt{\frac{2\ell+1}{4\pi} \frac{(\ell-|m|)!}{(\ell+|m|)!}} P_\ell^{|m|}(\cos\theta) e^{im\phi}.$$

These functions describe the angular momentum properties of the system when the quantum numbers ℓ and m have integral values. The requirement of integral values for m stems from the constraint of single-valuedness on the solution of the differential equation (7.10). The quantum number ℓ has integral values due to the constraint that the solution of differential equation (7.9) is convergent in the region $\theta = 0$ to π. These are independent of energy E and potential $V(r)$.

The first few spherical harmonics are:

$$Y_{00} = \sqrt{\frac{1}{4\pi}},$$

$$Y_{10} = \sqrt{\frac{3}{4\pi}} \cos\theta,$$

$$Y_{20} = \sqrt{\frac{5}{16\pi}} \left(3\cos^2\theta - 1\right),$$

$$Y_{11} = -\sqrt{\frac{3}{8\pi}} \sin\theta e^{i\phi},$$

$$Y_{1,-1} = \sqrt{\frac{3}{8\pi}} \sin\theta e^{-i\phi},$$

$$Y_{21} = -\sqrt{\frac{15}{8\pi}} \sin\theta \cos\theta e^{i\phi},$$

$$Y_{2,-1} = \sqrt{\frac{15}{8\pi}} \sin\theta \cos\theta e^{-i\phi},$$

$$Y_{22} = \sqrt{\frac{15}{32\pi}} \sin^2\theta e^{2i\phi},$$

$$Y_{33} = -\sqrt{\frac{35}{64\pi}} \sin^3\theta e^{3i\phi}.$$

The choice of the phase factor $(-1)^{\frac{1}{2}(m+|m|)}$ affects the sign of some of the spherical harmonics. But it does not influence the physical results which always involve $Y^*_{\ell m} Y_{\ell m}$. It has been chosen in accordance with one of the conventions used in the literature.

It is important to note that *for a spherically symmetric potential*, only the equations involving angular coordinates can be solved without any further knowledge about the classical expression for the potential energy because in this case both the differential equations involving angular coordinates are independent of the potential energy. This is the furthest we can go without knowledge of the expression for the spherically symmetric potential $V(r)$. However, if the expression for the spherically symmetric potential $V(r)$ for any physical system is known, then an attempt can be made to solve the radial differential equation. That is what we are going to do now.

The last part of the spherically symmetric potential problem involves the solution of radial differential equation (7.5). In view of the fact that C has been determined as $\ell(\ell + 1)$, this differential equation can be written as

$$\frac{1}{r^2}\frac{\mathrm{d}}{\mathrm{d}r}\left(r^2\frac{\mathrm{d}R}{\mathrm{d}r}\right) + \left[\frac{2m}{\hbar^2}r^2(E - V(r)) - \frac{l(l+1)}{r^2}\right]R = 0. \tag{7.32}$$

It can be solved only if we know the classical expression for the spherically symmetric potential of the physical system. We will solve it for hydrogen and hydrogen-like atoms. Once we have solved it, then a complete solution of the Schrödinger wave equation for spherically symmetric potential $V(r) = -Ze^2/r$ is obtained:

$$\Psi(r, \theta, \phi) = R(r)Y(\theta, \phi).$$

7.8 Hydrogen and hydrogenic atoms

Let us consider a physical system in which a negatively charged electron is moving under the influence of a positively charged nucleus. This is the first *more than one-particle system* which we have encountered in the solution of the Schrödinger amplitude equation. If the nucleus consists of a single proton, the system is a hydrogen atom and its potential energy is $-e^2/r$, where $-e$ is the charge on the electron and r is the distance between the electron and the nucleus. If the motion of the nucleus is taken into consideration, then it is well known that the two-body problem can be reduced to a one-body problem by replacing m by μ, where

$$\frac{1}{\mu} = \frac{1}{m} + \frac{1}{M}.$$

Here m is the mass of the electron, M is the mass of the proton and μ is the *reduced mass* of the system. If a nucleus consists of Z protons and all but one of its electrons have been removed by ionisation, such as a singly ionised helium atom or a doubly ionised lithium atom, then, as already mentioned, the system is said to be a hydrogen-like or hydrogenic atom and its potential energy is $-Ze^2/r$. Z is the *atomic number* of the atom. Thus, the hydrogen atom is a special case of a hydrogenic atom with $Z = 1$. We will now solve the Schrödinger amplitude equation for a hydrogenic atom. Since the classical expression for the potential energy of such an atom is

$$V(r) = -\frac{Ze^2}{r},$$

it is spherically symmetric. The angular part being independent of this expression will be the same as for any spherically symmetric potential. The differential equations corresponding to angular coordinates have already been solved and the solutions are equally valid for any hydrogenic atom.

7.9 The solution of the radial equation

The hydrogenic atom is one of those few physical problems of immediate interest for which the Schrödinger amplitude equation can be solved exactly. We have already shown that in spherical polar coordinates, the solution of the Schrödinger wave equation for a system with a spherically symmetric potential can be expressed as a product of two functions, one dependent upon angular coordinates only, while the other depends upon the variable r alone. The differential equation involving angular coordinates does not depend upon $V(r)$, which is a function of r alone, and therefore the (θ, ϕ) differential equation can be solved without explicit knowledge of $V(r)$. This has already been done. Let u now solve the radial differential equation. For a hydrogenic atom, the potential energy is

$$V(r) = -Ze^2/r$$

where $+Ze$ is the charge on the atomic nucleus and $-e$ is the charge on the only remaining electron when the atom has been ionised. It may be noted that as the mass of the atomic nucleus is much greater than the mass of the electron, the reduced mass is only slightly smaller than the electronic mass.

We will solve the corresponding radial equation for the hydrogenic atom as detailed below. For the hydrogenic atom, $V(r) = -Ze^2/r$. The radial equation then becomes

$$\frac{1}{r^2}\frac{\mathrm{d}}{\mathrm{d}r}\left(r^2\frac{\mathrm{d}R(r)}{\mathrm{d}r}\right) + \left[\frac{2\mu}{\hbar^2}\left(E + \frac{Ze^2}{r}\right) - \frac{\ell(\ell+1)}{r^2}\right]R(r) = 0. \tag{7.33}$$

This differential equation shows that

$$V(r) = -\frac{Ze^2}{r} + \frac{\ell(\ell+1)}{r^2}\frac{\hbar^2}{2\mu} \tag{7.34}$$

is the *effective potential*. The term $\ell(\ell + 1)/r^2\hbar^2/2\mu$ is called the *effective centrifugal potential* or *centrifugal barrier*. This additional potential is repulsive. The radial differential equation (7.33) which determines the energy levels of the system does not contain the magnetic quantum number m. It can be noted that, at large distances, the effective centrifugal potential

$$\frac{\ell(\ell+1)}{r^2}\frac{\hbar^2}{2\mu}$$

is negligible in magnitude as compared to

$$V(r) = -\frac{Ze^2}{r}.$$

However, it dominates at small distances.

We first consider the case for *negative total energy* E (i.e. $E < 0$), i.e. when the kinetic energy is less than the potential energy. In this case, the total energy is insufficient to ionise the atom. Hence, we will be concentrating on the bound states.

Remarks

The radial equation can be simplified by introducing a new function χ such that

$$\chi(r) = rR(r).$$

That is:

$$R(r) = \frac{\chi(r)}{r}.$$

Differentiating with respect to r, substituting the expression for $R(r)$ and its derivatives with respect to r in (7.33) and simplifying, we obtain

$$\frac{\mathrm{d}^2\chi}{\mathrm{d}r^2} + \frac{2\mu}{\hbar^2}\left[E + \frac{Ze^2}{r} - \frac{\ell(\ell+1)}{r^2}\frac{\hbar^2}{2\mu}\right]\chi = 0.$$

Equation $R(r) = \chi(r)/r$ shows that the function $\chi(r)$ must be zero at the origin as otherwise the solution $R(r)$ will blow up at $r = 0$. This does not mean that $R(r)$ should also vanish at the origin because $\chi(r)/r$ may not be zero. But we will not adopt this procedure. It may be noted that the centrifugal barrier becomes large for small values of r. It prevents a particle with non-zero angular momentum from getting too close to the origin.

We simplify (7.33*a*) by writing

$$\rho = 2\sqrt{-\frac{2\mu}{\hbar^2}E}\,r \qquad \text{and} \qquad \lambda = \frac{Ze^2\mu}{\hbar^2}\frac{2r}{\rho} \tag{7.35}$$

so that the new variable ρ also varies from zero to infinity. The radial differential equation then becomes

$$\frac{1}{\rho^2}\frac{\mathrm{d}}{\mathrm{d}\rho}\left(\rho^2\frac{\mathrm{d}R}{\mathrm{d}\rho}\right) + \left[-\frac{1}{4} - \frac{\ell(\ell+1)}{\rho^2} + \frac{\lambda}{\rho}\right]R = 0,$$

where, as is customary in the literature, instead of using a different letter S for the function of the variable ρ, we have used the same letter R to denote the function of ρ. This differential equation does not involve the quantum number m. Therefore, its solution cannot depend upon m. Thus, for a given value of quantum number ℓ, there will be $(2\ell + 1)$-fold degeneracy because all permissible values of m, ranging in unit steps from $-\ell$ to ℓ, will have the same energy. This differential equation can be written as

$$\frac{d^2R}{d\rho^2} + \frac{2}{\rho}\frac{dR}{d\rho} + \left[-\frac{1}{4} - \frac{\ell(\ell+1)}{\rho^2} + \frac{\lambda}{\rho}\right] + R = 0. \tag{7.36}$$

We may emphasise that *only negative energy states of the hydrogenic atom* are being considered here. Following the procedure adopted in the case of a harmonic oscillator, we first find the asymptotic solution of this differential equation. For large values of ρ, this reduces to

$$\frac{d^2R}{d\rho^2} - \frac{1}{4}R = 0. \tag{7.37}$$

There are two linearly independent solutions of this second-order differential equation. These are

$$R \sim \exp\left(\frac{\rho}{2}\right)$$

and

$$R \sim \exp\left(-\frac{\rho}{2}\right).$$

The first solution tends to infinity as $\rho \to \infty$ and is therefore not physically acceptable. The only physically acceptable solution of differential equation (7.37) and consequently the physically acceptable asymptotic solution of differential equation (7.36) is $R \sim \exp(-\rho/2)$. Following the procedure in solving differential equations, we may now assume that the general solution of differential equation (7.36) is

$$R(\rho) = \exp\left(-\frac{\rho}{2}\right)F(\rho), \tag{7.38}$$

where $F(\rho)$ is a power series in ρ. Differentiating with respect to ρ, we obtain

$$\frac{dR}{d\rho} = -\frac{1}{2}\exp\left(-\frac{\rho}{2}\right)F(\rho) + \exp\left(-\frac{\rho}{2}\right)F'(\rho).$$

Differentiating again with respect to ρ, we obtain

$$\frac{d^2R}{d\rho^2} = \frac{1}{4}\exp\left(-\frac{\rho}{2}\right)F(\rho) - \frac{1}{2}\exp\left(-\frac{\rho}{2}\right)F'(\rho)$$

$$-\frac{1}{2}\exp\left(-\frac{\rho}{2}\right)F'(\rho) + \exp\left(-\frac{\rho}{2}\right)F''(\rho),$$

or

$$\frac{d^2R}{d\rho^2} = \left[F''(\rho) - F'(\rho) + \frac{1}{4}F(\rho)\right]\exp\left(-\frac{\rho}{2}\right).$$

Substituting these expressions in (7.36) and cancelling the common factor $\exp(-\rho/2)$, we obtain

$$\left[F''(\rho) - F'(\rho) + \frac{1}{4}F(\rho)\right] + \frac{2}{\rho}\left[-\frac{1}{2}F(\rho) + F'(\rho)\right] + \left[-\frac{1}{4} - \frac{\ell(\ell+1)}{\rho^2} + \frac{\lambda}{\rho}\right]F(\rho) = 0,$$

or

$$F''(\rho) + \left(\frac{2}{\rho} - 1\right)F'(\rho) + \left[-\frac{\ell(\ell+1)}{\rho^2} + \frac{\lambda - 1}{\rho}\right]F(\rho) = 0. \qquad (7.39)$$

Let us sum up to this point the analysis of the radial wave equation. This equation has the solution

$$R(\rho) = \exp\left(-\frac{\rho}{2}\right)F(\rho) \qquad (7.40)$$

where $F(\rho)$ satisfies differential equation (7.39). If we can determine $F(\rho)$, the solution $R(\rho)$ of the radial wave equation will be known. To solve differential equation (7.39), we note that the coefficients of $F(\rho)$ and $F'(\rho)$ are not analytical at $\rho = 0$ and are of the form $\phi(\rho)/\rho$ and $\psi(\rho)/\rho^2$. We may therefore use the Forbenius method and write $F(\rho)$ as a power series such that:

$$F(\rho) = \rho^s\left(a_0 + a_1\rho + a_2\rho^2 + a_3\rho^3 + a_4\rho^4 + \cdots\right), \quad a_0 \neq 0, \quad s \geqslant 0,$$

$$\equiv \rho^s L(\rho), \qquad (7.41a)$$

where

$$L(\rho) = \left(a_0 + a_1\rho + a_2\rho^2 + a_3\rho^3 + a_4\rho^4 + \cdots\right), \quad a_0 \neq 0. \qquad (7.41b)$$

Differentiating (7.41a) with respect to ρ, we obtain

$$F'(\rho) = s\rho^{s-1}L(\rho) + \rho^s L'(\rho).$$

Differentiating again with respect to ρ, we obtain

$$\begin{aligned} F''(\rho) &= s(s-1)\rho^{s-2}L(\rho) + s\rho^{s-1}L'(\rho) + s\rho^{s-1}L'(\rho) + \rho^s L''(\rho) \\ &= \rho^s L''(\rho) + 2s\rho^{s-1}L'(\rho) + s(s-1)\rho^{s-2}L(\rho). \end{aligned}$$

Substituting these expressions for $F(\rho)$, $F(\rho)$ and $F''(\rho)$ in (7.39) and simplifying, we obtain

$$\begin{aligned} &\rho^2 L''(\rho) + \rho[2(s+1) - \rho]L'(\rho) \\ &\quad + [\rho(\lambda - s - 1) + s(s+1) - \ell(\ell+1)]L(\rho) = 0. \end{aligned} \tag{7.42}$$

Equation (7.41b) shows that $L(\rho)$ has been taken as a power series:

$$L(\rho) = a_0 + a_1\rho + a_2\rho^2 + a_3\rho^3 + a_4\rho^4 + \cdots.$$

Differentiating $L(\rho)$, as given by the above equation, once and twice, we obtain

$$L'(\rho) = a_1 + 2a_2\rho + 3a_3\rho^2 + 4a_4\rho^3 + \cdots$$

and

$$L''(\rho) = 2.1a_2 + 3.2a_3\rho + 4.3a_4\rho^2 + \cdots.$$

Substituting the expressions for $L(\rho)$, $L'(\rho)$ and $L''(\rho)$ in differential equation (7.42), we obtain

$$\begin{aligned} &\rho^2\left(2.1a_2 + 3.2a_3\rho + 4.3a_4\rho^2 + 5.4a_5\rho^3 + \cdots\right) \\ &+ \rho[2(s+1) - \rho]\left(a_1 + 2a_2\rho + 3a_3\rho^2 + 4a_4\rho^3 + \cdots\right) \\ &+ \left[\rho(\lambda - s - 1) + s(s+1) - \ell(\ell+1)\right] \\ &\left(a_0 + a_1\rho + a_2\rho^2 + a_3\rho^3 + a_4\rho^4 + \cdots\right) = 0. \end{aligned}$$

Combining different sets of elements with the same power of ρ, we obtain

$$\begin{aligned} &\left[s(s+1) - \ell(\ell+1)\right]a_0 \\ &+ \left[(\lambda - s - 1)a_0 + \{s(s+1) - \ell(\ell+1)\}a_1 + 2(s+1)a_1\right]\rho \\ &+ \left[(\lambda - s - 1)a_1 + \{s(s+1) - \ell(\ell+1)\}a_2 + \left\{2(s+1)2a_2 - a_1\right\} + 2.1a_2\right]\rho^2 \\ &+ \left[(\lambda - s - 1)a_2 + \{s(s+1) - \ell(\ell+1)\}a_3 + \left\{2(s+1)3a_3 - 2a_2\right\} + 3.2a_3\right]\rho^3 \\ &+ \cdots = 0. \end{aligned}$$

Since this is an identity in the variable ρ and its various powers, the coefficient of each power of ρ must be zero. We therefore have

$$[s(s+1) - \ell(\ell+1)]a_0 = 0,$$
$$\left[(\lambda - s - 1)a_0 + \{s(s+1) - \ell(\ell+1)\}a_1 + 2(s+1)a_1\right] = 0,$$
$$\left[(\lambda - s - 1)a_1 + \{s(s+1) - \ell(\ell+1)\}a_2 + \left\{2(s+1)2a_2 - a_1\right\} + 2.1a_2\right] = 0,$$
$$\left[(\lambda - s - 1)a_2 + \{s(s+1) - \ell(\ell+1)\}a_3 + \left\{2(s+1)3a_2 - 2a_2\right\} + 3.2a_3\right] = 0,$$
$$\cdots. \tag{7.43}$$

The first equation of this set gives

$$[s(s+1) - \ell(\ell+1)]a_0 = 0.$$

But it has been assumed that $a_0 \neq 0$. Therefore, we must have

$$s(s+1) - \ell(\ell+1) = 0.$$

This is a quadratic equation in s, i.e. $s^2 + s - \ell(\ell+1) = 0$, and yields two values, $s = \ell$ and $s = -(\ell+1)$. However (7.41a) shows that a negative value of s, i.e. $-\ell(\ell+1)$, is not physically acceptable as it will make the function $F(\rho) = \rho^s L(\rho)$ and consequently the radial solution (7.40), i.e.

$$R(\rho) = \exp\left(-\frac{\rho}{2}\right)F(\rho) = \exp\left(-\frac{\rho}{2}\right)\rho^s L(\rho),$$

infinite at the origin. Thus, one of the two solutions of the second-order radial differential equation is discarded because it blows up at the origin. For this reason, terminating the series after a finite number of terms does not help us here. This is because here the infinite value occurs at the origin. Hence, for a physically acceptable solution, we must have $s = \ell$. Equation (7.41a) therefore yields

$$F(\rho) = \rho^\ell L(\rho) \tag{7.44a}$$

and the set of equations (7.43) reduces to

$$(\lambda - \ell - 1)a_0 + 2(\ell+1)a_1 = 0,$$
$$(\lambda - \ell - 1)a_1 + 2(\ell+1)2a_2 - a_1 + 2.1a_2 = 0,$$
$$(\lambda - \ell - 1)a_2 + 2(\ell+1)3a_2 - 2a_2 + 3.2a_3 = 0,$$
$$\cdots,$$

or

$$\rho^1: (\lambda - \ell - 1)a_0 + (2\ell + 2)a_1 = 0,$$
$$\rho^2: (\lambda - \ell - 2)a_1 + (1 + 2\ell + 2)2a_2 = 0,$$
$$\rho^3: (\lambda - \ell - 3)a_2 + (2 + 2\ell + 2)3a_3 = 0,$$
$$\cdots.$$

For $\rho^{\nu+1}$, $\nu = 1, 2, 3$, we have

$$[\lambda - \ell - (\nu + 1)]a_\nu + (\nu + 2\ell + 2)(\nu + 1)a_{\nu+1} = 0,$$

or

$$a_{\nu+1} = -\frac{\lambda - \ell - (\nu + 1)}{(\nu + 2\ell + 2)(\nu + 1)}a_\nu, \quad \nu = 1, 2, 3, \ldots,$$

or

$$a_{\nu+1} = \frac{\nu + \ell + 1 - \lambda}{(\nu + 1)(\nu + 2\ell + 2)}a_\nu, \quad \nu = 1, 2, 3, \ldots. \tag{7.44b}$$

This is the recursion formula (or the recursive formula) for the coefficients. It shows that *for large* ν, the ratio of the coefficients of successive terms behaves as $1/\nu$. This *ratio is the same* as that of successive terms in the series expansion of e^ρ *for large* ν. Therefore, the asymptotic behaviour of the two series is the same. Hence, the asymptotic behaviour of $R(\rho) = e^{-\rho/2}F(\rho) = e^{-\rho/2}\rho^\ell L(\rho)$ is similar to that of $e^{-\rho/2}\rho^\ell e^\rho$, i.e. to $e^{\rho/2}\rho^\ell$. This leads to infinity with increasing ρ. As the function $R(\rho)$ diverges asymptotically, this solution is not physically permissible.

7.10 Physically acceptable solutions for the radial equation and discrete energy values

A physically acceptable solution is obtained if the series for L terminates. In that case, we have $L(\rho)$ as a polynomial in ρ. Then, for large ρ, the factor $e^{-\rho/2}$ dominates the behaviour of $R(\rho) = e^{-\rho/2}F(\rho) = e^{-\rho/2}\rho^\ell L(\rho)$ which converges as $\rho \to \infty$. The termination of the series can be achieved if the numerator in the recursion formula becomes zero for some value of ν, say n_r. Then we may write

$$n_r + \ell + 1 - \lambda = 0,$$

or

$$\lambda = n_r + \ell + 1.$$

It is customary in the literature to write n in place of $n_r + \ell + 1$. The above equation then takes the form $n = \lambda = (n_r + \ell + 1)$. Differential equation (7.39) is uniquely specified by λ and ℓ. For given λ and ℓ, the number n_r has a unique value and hence the solution is unique. Since each one of ℓ and n_r can either be zero or a positive integer, n can have only positive integral values such that $n \geqslant (\ell + 1)$. This relation shows that the maximum value of ℓ is $n - 1$. n_r is called the *radial quantum number* while n is called the *total* or *principal quantum number*. The relation $\lambda = n$ and relations (7.35*a*) which express λ in terms of energy, give

$$-\frac{Z^2e^4\mu}{2E\hbar^2} = n^2,$$

or

$$E = -|E_n| = -\frac{Z^2 e^4 \mu}{2n^2 \hbar^2}, \quad n = 1, 2, 3, \ldots, \tag{7.45a}$$

where to emphasise that energy depends on n, we have used $|E_n|$ for the magnitude of energy. This is in agreement with experiments. The minus sign indicates that we are considering only those states for which the total energy is negative. Equation (7.45a) itself shows that a hydrogenic atom possesses discrete energies. The energy values are independent of the orbital and magnetic quantum numbers ℓ and m, and depend only on total quantum number n. The principal quantum number n characterises the energy level E_n. For $Z = 1$, (7.45a) is Bohr's formula for the hydrogen atom energy spectrum. If we introduce the dimensionless constant α defined by

$$\alpha = \frac{e^2}{\hbar c}$$

equation (7.45a) reduces to

$$E = -\frac{Z^2 \alpha^2 \mu c^2}{2n^2}, \quad n = 1, 2, 3, \ldots. \tag{7.45b}$$

The constant α is called the *fine structure constant* and is nearly equal to 1/137. The quantity μc^2 is the energy released if an electron of reduced mass μ (actually the mass of the electron is m and is related to the reduced mass μ by the equation $1/\mu = 1/m + 1/M$, where M is the mass of the nucleus) is completely converted into energy. It is approximately equal to 0.51 MeV. If we denote the quantity $\hbar^2/\mu e^2$ by a_0, then we can write (7.45a) as

$$E = -\frac{Z^2 e^2}{2n^2 a_0}, \quad n = 1, 2, 3, \ldots. \tag{7.45c}$$

The constant $a_0 = \hbar^2/\mu e^2$ has the dimension of length and is equal to the radius of the first orbit of the electron in the hydrogen atom in the Bohr model. It is therefore called the *Bohr radius* for the hydrogen atom and is given by

$$a_0 = \frac{\hbar^2}{\mu e^2} = 5.29 \times 10^{-9}\,\text{cm} = 0.529\,\text{Å}$$

It may be emphasised that although the energy levels for the hydrogen atom as derived in quantum mechanics are identical with those derived by Bohr's theory, the two theories are *not identical*. Quantum mechanically, the ground state of the hydrogen atom is specified by $\ell = 0$, while the old quantum theory predicts the corresponding value of the orbital angular momentum quantum number $\ell = 1$.

Remarks

We may take this opportunity to remark that *both the solutions* of the Schrödinger amplitude equation *for a linear harmonic oscillator* are retained after imposing the constraint that the two linearly independent infinite series, the sum of which constitutes the general mathematical solution of the second-order ordinary differential equation, should be terminated after a finite number of terms. The physically acceptable solutions which are Hermite polynomials involve one of the terminated series when n is even and the other terminated series when n is odd. Thus, a Hermite polynomial $H_n(z)$, $z = \sqrt{\alpha}x$, of degree n consists of either only even or only odd powers of the variable z, depending upon whether n is even or odd. A similar conclusion is true for the Legendre polynomials $P_\ell(z)$, $z = \cos\theta$, solutions of the differential equation for the variable θ in the hydrogen atom. However, in the case of the radial equation, one of the solutions is to be *discarded* as it becomes *infinite at the origin*; it cannot be made finite. Only *one* of the two solutions of the radial wave equation is physically acceptable.

7.11 The parity of a particle in a spherically symmetric potential

We have already introduced the concept of the parity of a particle moving in a one-dimensional symmetric potential. In terms of spherical polar coordinates, this idea can be easily extended to a three-dimensional space. In a Cartesian system of coordinates, the spatial reflection is defined by

$$\mathbf{r} \to \mathbf{r}' = -\mathbf{r}.$$

Let (r, θ, ϕ) be the spherical polar coordinates of a particle. If the axes are reflected to (x', y', z') and (r', θ', ϕ') are defined conventionally with respect to the primed Cartesian coordinates x', y', z', then from figure 7.2, it can be easily seen that

$$r \to r' = r,$$

$$\theta \to \theta' = \pi - \theta,$$

$$\phi \to \phi' = \pi + \phi.$$

It should be noted that reflection is *not* an operation which can be carried out physically but it can be given a physical meaning.

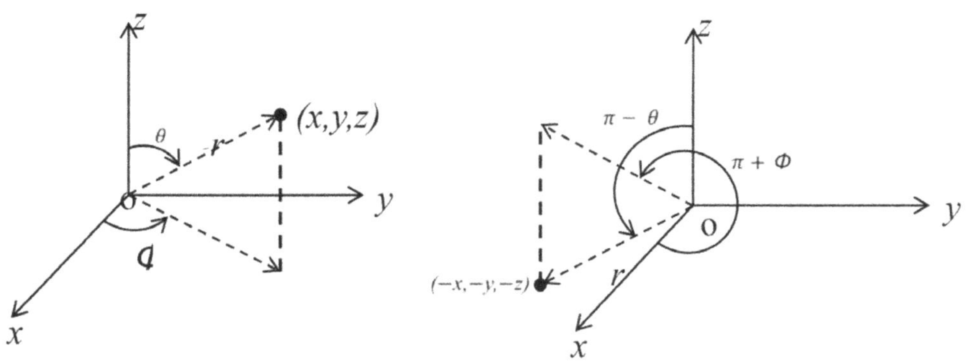

Figure 7.2. Space reflection in terms of spherical polar coordinates.

Consider a particle moving in a spherically symmetric potential in an angular momentum state $Y_{\ell m}(\theta, \phi)$. Then the state $Y_{\ell m}(\theta', \phi')$ obtained under space reflection is given by

$$\begin{aligned}
Y_{\ell m}(\theta', \phi') &= \text{const}\ P_\ell^{|m|}(\cos\theta')\mathrm{e}^{\mathrm{i}m\phi} \\
&= \text{const}\ P_\ell^{|m|}[\cos(\pi - \theta)]\mathrm{e}^{\mathrm{i}m(\pi+\phi)} \\
&= \text{const}\ P_\ell^{|m|}(-\cos\theta')\mathrm{e}^{\mathrm{i}m\phi}(-1)^{|m|} \\
&= \text{const}\ (-1)^{\ell-|m|}P_\ell^{|m|}(\cos\theta)\mathrm{e}^{\mathrm{i}m\phi}(-1)^{|m|} \\
&= \text{const}\ (-1)^{\ell-2|m|}P_\ell^{|m|}(\cos\theta)\mathrm{e}^{\mathrm{i}m\phi} \\
&= \text{const}\ (-1)^{\ell}P_\ell^{|m|}(\cos\theta)\mathrm{e}^{\mathrm{i}m\phi} \\
&= (-1)^{\ell}\, Y_{\ell m}(\theta, \phi).
\end{aligned}$$

If $R(r)$ denotes the radial part of the solution, the state $\psi(r, \theta, \phi)$ is determined by $Y_{\ell m}(\theta, \phi)R(r)$. But the radial part is unaffected by space reflection, i.e. by $r \rightarrow r' = r$. Therefore, the parity of such a particle is determined by $(-1)^{\ell}$. It is therefore positive or negative according to whether ℓ is zero/even or odd.

7.12 Comparison of the spectral series of hydrogen atom with experiments

It is well known that when current is passed through hydrogen gas sealed in a discharge tube, it becomes illuminated and produces a dull red glow. When this light is passed through a prism, four brightly coloured lines superimposed on a dark background are seen in the visible region. These lines of the spectrum constitute what is called the *Balmer series*. Figure 7.3 shows how the various series of line

Figure 7.3. The various spectral series formed by the radiative transitions of the hydrogen atom.

spectra are formed in the hydrogen gas. When the electrons fall from the states with total quantum number $n = 3, 4, 5$, and 6 to the first excited state with $n = 2$, we obtain the Balmer series. In addition to that, we have the Lyman series, Paschen series and Brackett series produced as shown in the same figure. These last three series arise as electrons make transitions from higher states to lower states with $n = 1$, 3, 4, and 5, respectively. The Lyman series lies in the ultraviolet region, the Balmer series lies in the visible region while the remaining two series lie in the infrared region.

The following table demonstrates how the spectral lines for a hydrogen atom in a magnetic field are formed:

Table 7.1.

	Higher states (n)	Lower state (n)
Lyman series	2, 3, 4, 5, 6	1
Balmer series	3, 4, 5, 6	2
Paschen series	4, 5, 6	3
Bracket series	5, 6	4

To compute the wavelengths of various lines in the Balmer series, we proceed as follows. The energy levels in the hydrogen atom are given by

$$E_n = -|E_n| = -\frac{e^4\mu}{2n^2\hbar^2}.$$

For $n = 2$ and 3, we have

$$E_2 = -\frac{e^4\mu}{8\hbar^2},$$

$$E_3 = -\frac{e^4\mu}{18\hbar^2}.$$

Therefore, the difference in energy levels is given by

$$\Delta E = E_3 - E_2 = -\frac{5e^4\mu}{72\hbar^2}.$$

Problem

7.4. Calculate the wavelengths of the four lines in the Balmer series.

7.13 The radial wave function

We have shown that the solution $R(\rho)$ of the radial equation is given by

$$R(\rho) = \mathrm{e}^{-\rho/2}F(\rho) = \mathrm{e}^{-\rho/2}\rho^{\ell}L(\rho). \tag{7.46}$$

It is physically acceptable only if $L(\rho)$ is a polynomial which satisfies differential equation (7.42) with $s = \ell$, i.e. if $L(\rho)$ satisfies the differential equation

$$\rho L''(\rho) + [2(\ell + 1) - \rho]L'(\rho) + (\lambda - \ell - 1)L(\rho) = 0. \tag{7.42'}$$

It had already been known to mathematicians for a long time that *associated Laguerre polynomials* $L_q^p \equiv L$ of degree $q - p$ and order p satisfy the differential equation

$$\rho L''(\rho) + [p + 1 - \rho]L'(\rho) + (q - p)L(\rho) = 0. \tag{7.47}$$

A comparison of (7.42′) and (7.47) shows that the desired solutions are constant multiples of the associated Laguerre functions $L_q^p(\rho)$ provided $p = 2\ell + 1$ and $q = \lambda + \ell$.

The associated Laguerre polynomial $L_q^p(\rho)$ of degree $q - p$ and order p is also defined by the equation

$$L_q^p(\rho) = \frac{\mathrm{d}^p}{\mathrm{d}\rho^p} L_q(\rho),$$

where $L_q(\rho)$ is the *Laguerre polynomial* and is defined by the following generating function (for $t < 1$):

$$\sum_{q=0}^{\infty} L_q(\rho) \frac{t^q}{q!} = \frac{1}{1 - t} \exp\left[-\rho \frac{t}{1 - t}\right].$$

It can be deduced from this equation that

$$L_q(\rho) = \mathrm{e}^{\rho} \frac{\mathrm{d}^q}{\mathrm{d}\rho^q}\left(\rho^q \mathrm{e}^{-\rho}\right).$$

The first five Laguerre polynomials are:

$$L_0 = 1,$$

$$L_1 = 1 - \rho,$$

$$L_2 = 2 - 4\rho + \rho^2,$$

$$L_3 = 6 - 18\rho + 9\rho^2 - \rho^3,$$

$$L_4 = 24 - 96\rho + 72\rho^2 - 16\rho^3 + \rho^4.$$

Problem

7.5. Show that the *normalised solution* $R_{n\ell}(r)$ of the radial wave equation for hydrogenic atoms is given by

$$R_{n\ell}(r) = -\left(\frac{2Z}{na_0}\right)^{3/2} \sqrt{\frac{(n - \ell - 1)!}{2n\{(n + \ell)!\}^3}}\, \mathrm{e}^{-\rho/2} \rho^{\ell} L_{n+1}^{2\ell+1}(\rho),$$

where

$$\rho = \frac{2Ze^2\mu}{n\hbar^2}r = \frac{2Z}{na_0}r.$$

The first few hydrogenic radial wave functions are:

$$R_{10}(r) = \left(\frac{Z}{a_0}\right)^{3/2} 2\exp(-\rho/2), \quad \rho = \frac{2Z}{na_0}r,$$

$$R_{20}(r) = \left(\frac{Z}{a_0}\right)^{3/2} \frac{1}{2\sqrt{2}}(2-\rho)\exp(-\rho/2),$$

$$R_{21}(r) = \left(\frac{Z}{a_0}\right)^{3/2} \frac{1}{2\sqrt{2}}\rho\exp(-\rho/2),$$

$$R_{30}(r) = \left(\frac{Z}{a_0}\right)^{3/2} \frac{1}{9\sqrt{3}}\left(6-6\rho+\rho^2\right)\exp(-\rho/2),$$

$$R_{31}(r) = \left(\frac{Z}{a_0}\right)^{3/2} \frac{1}{9\sqrt{6}}\rho(4-\rho)\exp(-\rho/2),$$

$$R_{32}(r) = \left(\frac{Z}{a_0}\right)^{3/2} \frac{1}{9\sqrt{30}}\rho^2\exp(-\rho/2),$$

where $\rho = \frac{2Z}{na_0}r$.

It is important to note that in quantum mechanics, the ground state of the hydrogen atom corresponds to $n = 1$, $\ell = 0$. The corresponding radial wave function $R_{10}(r)$ is given by

$$R_{10}(r) = \left(\frac{1}{a_0}\right)^{3/2} 2\exp\left(-\frac{r}{a_0}\right).$$

For $r = 0$,

$$R_{10}(r=0) = 2\left(\frac{1}{a_0}\right)^{3/2}.$$

The non-zero value of $R_{10}(r)$ at the origin might create the impression that in the ground state ($n = 1, \ell = 0$), the electron passes through the nucleus. This is not true. $R_{10}(r)$ is only a mathematical function which is related to the physical quantities only through $r^2R^*_{10}R_{10}(r)$ or the expectations values. Therefore, the radial function itself does not describe the physical motion. We take this opportunity to assert again that the classical concept of an orbit which is determined by simultaneous precise measurements of position and momentum at all times has no meaning when quantum mechanical effects are important. It has been well said that the classical concepts which have stuck in our minds but lost their meaning in quantum mechanics should not be allowed to haunt us any more! In fact, for a radial solution $R_{n\ell}(r)$, the probability of finding the electron in the distance dr between r and r + dr is proportional to $R^*_{n\ell}(r)R_{n\ell}(r)r^2\,\mathrm{d}r$ and *not* $R^*_{n\ell}(r)R_{n\ell}(r)\,\mathrm{d}r$. This is because in spherical polar coordinates, the volume $\mathrm{d}\tau\,(\equiv \mathrm{d}x\mathrm{d}y\mathrm{d}z) = r^2\mathrm{d}r\ \sin\theta\,\mathrm{d}\theta\mathrm{d}\phi$. Consequently, the radial probability density will be $r^2R^*_{n\ell}(r)R_{n\ell}(r)$. It should be remarked that, as is customary in the literature on quantum mechanics, we have used the words 'at the origin', etc. Strictly speaking, one should state: the probability of the electron being found in the distance dr between r and r + dr at time t or in the volume dτ around the point x, y, z at time t.

7.14 The spectroscopic notation

It is appropriate to mention here that in spectroscopy the values of the quantum number ℓ, i.e. 0, 1, 2, 3, 4, …, are given, respectively, the letter symbols s, p, d, f, g, …. The values of quantum number n are either indicated as these are, i.e. 1, 2, 3, 4, …, or are, respectively, denoted by capital letter symbols K, L, M, N, …. Thus, the radial function $R_{10}(r)$ is referred to as the 1s orbital of the K shell, $R_{21}(r)$ is referred to as the 2p orbital of the L shell and so on. The wave function ψ_{20m} is designated as the 2s state, ψ_{32m} is labelled as the 3d state, and so on.

7.15 The normalised solution for the hydrogenic atom

We can write the normalised solution $\psi_{n\ell m}(r, \theta, \phi)$ of the Schrödinger amplitude equation for a hydrogen-like atom. In view of the fact that $R(r)$, $P(\theta)$ and $F(\phi)$ depend on the quantum numbers n, ℓ, ℓ, m and m, respectively, we use them as subscripts when writing the solution of the Schrödinger wave equation. Of course, the function $\psi_{n\ell m}(r, \theta, \phi)$ depends on all three quantum numbers, ℓ, m, n. Thus, we have

$$\psi_{n\ell m}(r, \theta, \phi) = R_{n\ell}(r)P_{\ell m}(\theta)F_m(\phi),$$

where

$$R_{n\ell}(r) = \left(\frac{2Z}{na_0}\right)^{3/2}\sqrt{\frac{(n-\ell-1)!}{2n\{(n+\ell)!\}^3}}\,\mathrm{e}^{-\rho/2}\rho^{\ell}L^{2\ell+1}_{n+1}(\rho)$$

with

$$\rho = \frac{2Ze^2\mu}{n\hbar^2}r = \frac{2Z}{na_0}r,$$

$$P_{\ell m}(\theta) = \sqrt{\frac{2\ell+1}{2}\frac{(\ell-|m|)!}{(\ell+|m|)!}}\,P_\ell^{|m|}(\cos\theta)$$

and

$$F_m(\phi) = \frac{1}{\sqrt{2\pi}}\mathrm{e}^{\mathrm{i}m\phi}.$$

The quantum numbers n, ℓ, m can take the values as

$$n = 1, 2, 3, \ldots,$$

$$\ell = 0, 1, 2, \ldots, n-1,$$

$$m = -\ell, -\ell+1, \ldots, \ell.$$

The wave functions written above are normalised as

$$\int_0^\infty \int_0^\pi \int_0^{2\pi} \psi^*_{n\ell m}(r, \theta, \phi)\psi_{n\ell m}(r, \theta, \phi)r^2 \sin\theta \,\mathrm{d}r\mathrm{d}\theta\mathrm{d}\phi = 1.$$

This is because the functions $R_{n\ell}(r)$, $P_{\ell m}(\theta)$ and $F_m(\phi)$, such that $\psi_{n\ell m}(r, \theta, \phi) = R_{n\ell}(r)P_{\ell m}(\theta)F_m(\phi)$, have already been normalised according to

$$\int_0^\infty R^*_{n\ell}(r)R_{n\ell}(r)r^2\,\mathrm{d}r = 1,$$

$$\int_0^\pi P^*_{\ell m}(\theta)P_{\ell m}(\theta)\sin\theta\,\mathrm{d}\theta = 1,$$

$$\int_0^{2\pi} F^*_m(\phi)F_m(\phi)\,\mathrm{d}\phi = 1.$$

The wave functions are also mutually orthogonal:

$$\int_0^\infty \int_0^\pi \int_0^{2\pi} \psi^*_{n\ell m}(r, \theta, \phi)\psi_{n'\ell' m'}(r, \theta, \phi)r^2 \sin\theta \,\mathrm{d}r\mathrm{d}\theta\mathrm{d}\phi = 0$$

except for $n = n'$, $\ell = \ell'$ and $m = m'$.

The first few wave functions for a hydrogenic atom are given by

$$\psi_{100} = \frac{1}{\sqrt{\pi}}\left(\frac{Z}{a_0}\right)^{3/2} \mathrm{e}^{-\sigma}, \quad \sigma = \frac{Z}{a_0}r,$$

$$\psi_{200} = \frac{1}{4\sqrt{2\pi}}\left(\frac{Z}{a_0}\right)^{3/2} (2 - \sigma)\mathrm{e}^{-\sigma/2},$$

$$\psi_{210} = \frac{1}{4\sqrt{2\pi}}\left(\frac{Z}{a_0}\right)^{3/2} \sigma\mathrm{e}^{-\sigma/2}\cos\theta,$$

$$\psi_{211} = \frac{1}{8\sqrt{\pi}}\left(\frac{Z}{a_0}\right)^{3/2} \sigma\mathrm{e}^{-\sigma/2}\sin\theta\,\mathrm{e}^{\mathrm{i}\phi},$$

$$\psi_{21,-1} = \frac{1}{8\sqrt{\pi}}\left(\frac{Z}{a_0}\right)^{3/2} \sigma\mathrm{e}^{-\sigma/2}\sin\theta\,\mathrm{e}^{-\mathrm{i}\phi}.$$

It may be noted that in general a wave function Ψ for the hydrogen atom is given by

$$\Psi = \sum\nolimits_n \sum\nolimits_\ell \sum\nolimits_m a_{n\ell m}\psi_{n\ell m}(\mathbf{r})\mathrm{e}^{-\mathrm{i}Et/\hbar}.$$

7.16 Stationary states

One important point deserves careful consideration. The wave function $\Psi_n(\mathbf{r}, t)$ describing an eigenstate of a hydrogenic atom is given by

$$\Psi_n(\mathbf{r}, t) = A_n\psi_n(\mathbf{r})\mathrm{e}^{-\mathrm{i}E_n t/\hbar},$$

where A_n is the normalisation constant and E_n is the definite energy corresponding to this state. The probability density that the electron will be found in this state is

$$\Psi_n^*(\mathbf{r}, t)\Psi_n(\mathbf{r}, t) = A_n^* A_n\psi_n^*(\mathbf{r})\psi_n(\mathbf{r}).$$

This is independent of time. Such a state is said to be a stationary state of the hydrogenic atom. Since the probability of finding the electron in this stationary state is independent of time, once it is in this state it should remain in this state indefinitely. However, it is known that a hydrogen atom in an excited state does not remain in that state indefinitely. It goes to the ground state by radiating energy. The question arises: how does this happen? The answer to this question stems from the fact that $\Psi(\mathbf{r}, t)$ represents a stationary state of the electron only when it is under the influence of the Coulomb field of the nucleus. It *ignores* the interaction of the electron with any other particle or field. If that is included, $\Psi(\mathbf{r}, t)$ would *not* be an eigenstate of the entire system including interactions. It is this interaction which causes the electron in the excited state to fall to the ground state and emit radiation.

Example 7.1.

Problem

Show that, for a hydrogen atom in the ground state, the most probable distance of the electron from the nucleus is equal to $a_0 = \hbar^2/(me^2)$, the radius of Bohr's orbit for the hydrogen atom. (This is known as the Bohr radius because it was introduced by Bohr in his model for the hydrogen atom.)

Solution

The wave function ψ_{100} for the hydrogen atom, for which $Z = 1$, in the ground state is

$$\psi_{100} = \frac{1}{\sqrt{\pi a_0^3}} \mathrm{e}^{-r/a_0}.$$

Therefore, the probability distribution function ψ_{100}^* of the electron relative to the nucleus is given by

$$\psi_{100}^* \psi_{100} = \frac{1}{\pi a_0^3} \mathrm{e}^{-2r/a_0}.$$

This expression is independent of coordinates θ and ϕ. (What does this indicate about the shape of the hydrogen atom (in the ground state)?) We therefore have to determine those values of r for which the probability of finding an electron within a spherical shell of radius r and width dr is a maximum.

The probability of finding the electron in a volume dτ (= r^2 dr sin θ dθdϕ) is

$$\psi_{100}^* \psi_{100}\, \mathrm{d}\tau = \frac{1}{\pi a_0^3} \mathrm{e}^{-2r/a_0} r^2\, \mathrm{d}r \sin\theta\, \mathrm{d}\theta \mathrm{d}\phi.$$

Then the probability of finding the electron in the hydrogen atom is obtained by integrating over the entire volume:

$$\int \psi_{100}^* \psi_{100}\, \mathrm{d}\tau = \frac{1}{\pi a_0^3} \int_0^\infty \int_0^\pi \int_0^{2\pi} \mathrm{e}^{-2r/a_0} r^2 \mathrm{d}r \sin\theta\, \mathrm{d}\theta \mathrm{d}\phi.$$

By integrating over θ and ϕ, we obtain

$$\int \psi_{100}^* \psi_{100}\, \mathrm{d}\tau = \frac{4}{a_0^3} \int_0^\infty \mathrm{e}^{-2r/a_0} r^2\, \mathrm{d}r.$$

Therefore, the expression

$$\frac{4}{a_0^3} \mathrm{e}^{-2r/a_0} r^2\, \mathrm{d}r$$

gives the probability that the electron lies *between* the distances r and r + dr from the nucleus. This is shown in figure 7.4. The probability that the electron lies in the spherical shell between r and r + dr will have its maximum value

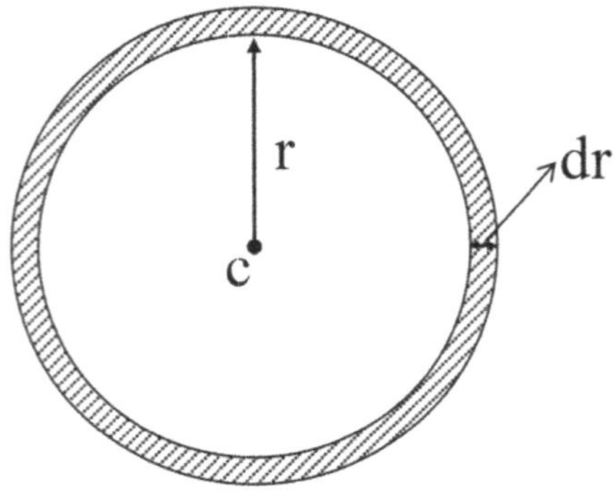

Figure 7.4. A thin spherical shell of radius r and thickness dr.

when $r^2 e^{-2r/a_0}$ is maximum. To obtain the value of r which will make $r^2 e^{-2r/a_0}$ maximum, we must put its derivative with respect to r equal to zero:

$$\frac{d}{dr}\left(r^2 e^{-2r/a_0}\right) = 0.$$

This yields

$$2re^{-2r/a_0} + r^2\left(-\frac{2}{a_0}\right)e^{-2r/a_0} = 0,$$

or

$$r = a_0.$$

This shows that the most probable value of r from the nucleus is a_0, the Bohr radius of the hydrogen atom. In this sense, the electron may be considered as if it were somewhere localised at a distance a_0 from the nucleus.

Problems

7.7. Determine the average distance of the electron from the nucleus in the hydrogen atom in the ground state.

7.8. A hydrogen atom is in the eigenstate $\psi_{n\ell m}$. Find the average distance of the electron from the nucleus.

7.9. For $\ell = 0$, find out how the size of the hydrogen atom changes with the total quantum number n. Comment on the statement that for $\ell \neq 0$, this size increases in approximately the same way.

7.10. Find the expectation values of r^2, $1/r$ and $1/r^2$ for the hydrogen atom in the nth state.

7.11. Find the expectation values of the potential and kinetic energies of a hydrogen atom in the nth state.

7.12. Show that for the hydrogen atom in the nth state, the expectation value of the square of the velocity $v_{n\ell m}$ is given by

$$\left\langle v_{n\ell m}^2 \right\rangle = \left(\frac{e^2}{nh}\right)^2.$$

Example 7.2.

Problem

The normalised energy eigenfunction for the ground state of a hydrogen-like atom is

$$\psi_{100} = \frac{1}{\sqrt{\pi}}\left(\frac{Z}{a_0}\right)^{3/2} e^{-Zr/a_0}.$$

Calculate the expectation value of $1/r$ and hence that of the potential energy. Show that the average value of the kinetic energy is given by

$$\langle T \rangle_{100} = \frac{Ze^2}{2a_0}.$$

It may be assumed that

$$\int_0^\infty e^{-x} x^n \, dx = n!.$$

Solution

The expectation value of $1/r$ for the ground state of a hydrogen-like atom is given by

$$\left\langle \frac{1}{r} \right\rangle_{100} = \int_0^\infty \int_0^\pi \int_0^{2\pi} \psi_{100}^*(r, \theta, \phi)\frac{1}{r}\psi_{100}(r, \theta, \phi) r^2 \sin\theta \, dr d\theta d\phi$$

$$= \int_0^\infty \int_0^\pi \int_0^{2\pi} \frac{1}{\pi}\left(\frac{Z}{a_0}\right)^3 e^{-2Zr/aa_0} r \sin\theta \, dr d\theta d\phi. \tag{7.49}$$

Let us write

$$\frac{2Z}{a_0} r = u.$$

Then, by differentiating the above equation, we obtain

$$\frac{2Z}{a_0} dr = du.$$

By a change in variable from r to u as defined above, (7.49) takes the form

$$\left\langle \frac{1}{r} \right\rangle_{100} = \int_0^\infty \int_0^\pi \int_0^{2\pi} \frac{1}{\pi}\left(\frac{Z}{a_0}\right)^3 e^{-u}\left(\frac{a_0}{2Z}\right) u \sin\theta \, d\theta d\phi \left(\frac{a_0}{2Z}\right) du$$

$$= \left(\frac{Z}{4\pi a_0}\right) \int_0^\infty \int_0^\pi \int_0^{2\pi} u e^{-u} \sin\theta \, d\theta d\phi du.$$

Integrating with respect to θ and ϕ, and simplifying, we obtain

$$\left\langle \frac{1}{r} \right\rangle_{100} = \left(\frac{Z}{a_0}\right) \int_0^\infty u e^{-u} \, du = \frac{Z}{a_0},$$

where we have made use of the relation

$$\int_0^\infty x^n e^{-x} \, dx = n!.$$

The potential energy can then be computed as given by

$$\langle V \rangle_{100} = \int_0^\infty \int_0^\pi \int_0^{2\pi} \psi_{100}^* \left(-\frac{Ze^2}{r}\right) \psi_{100} r^2 \sin\theta \, dr d\theta d\phi$$

$$= \left(-Ze^2\right)\left\langle \frac{1}{r} \right\rangle_{100} = \left(-Ze^2\right)\frac{Z}{a_0} = -\frac{Z^2 e^2}{a_0}.$$

Since $a_0 = \hbar^2/(\mu e^2)$, the expectation value for the potential energy takes the form

$$\langle V \rangle_{100} = -\frac{\mu Z^2 e^4}{\hbar^2}.$$

The expectation value $\langle T \rangle$ for the kinetic energy is easily obtained by using the virial theorem:

$$\langle T \rangle = \frac{1}{2}\langle \mathbf{r} \cdot \nabla V \rangle = \frac{1}{2}\left\langle \mathbf{r} \cdot \nabla\left(-\frac{Ze^2}{r}\right)\right\rangle$$

$$= -\frac{Ze^2}{2}\left\langle \mathbf{r} \cdot \left(-\frac{\mathbf{r}}{r^2}\right)\right\rangle = -\frac{Ze^2}{r}\left\langle -\frac{1}{r} \right\rangle$$

$$= -\frac{1}{2}\left\langle -\frac{Ze^2}{r} \right\rangle$$

$$= -\frac{1}{2}\langle V \rangle.$$

Hence

$$\langle T\rangle_{100} = -\frac{1}{2}\langle V\rangle_{100} = \frac{\mu Z^2 e^4}{2\hbar^2} = \frac{Z^2 e^2}{2a_0}, \quad \text{as } a_0 = \hbar^2/\left(\mu e^2\right).$$

Example 7.3.

Problem

The normalised energy eigenfunction for the ground state of the hydrogen-like atom is given by

$$\psi_{100} = \frac{1}{\sqrt{\pi}}\left(\frac{Z}{a_0}\right)^{3/2} \mathrm{e}^{-Zr/a_0}.$$

Show that the probability of finding the electron at a distance greater than its 'classical' distance $2a_0/Z$ from the nucleus is ≈0.24.

Solution

Since we are interested in determining the probability P that the electrons can be found beyond the classical distance $2a_0/Z$, we can first calculate the probability P_c of finding the electron in the range $2a_0/Z$ and then subtract it from unity. Now

$$P_e = \int \psi_{100}^* \psi_{100}\, \mathrm{d}\tau = \frac{Z^3}{\pi a_0^3}\int_0^{2a_0/Z}\int_0^{\pi}\int_0^{2\pi} \mathrm{e}^{-2Zr/a_0} r^2\, \mathrm{d}r \sin\theta\, \mathrm{d}\theta \mathrm{d}\phi$$

because the radial limits are from zero to $2a_0/Z$. Integrating with respect to θ and ϕ, we obtain

$$P_e = \frac{Z^3}{\pi a_0^3}[-\cos\theta]\int_0^{\pi}[\phi]\int_0^{2\pi}\int_0^{2a_0/Z} \mathrm{e}^{-2Zr/a_0} r^2\, \mathrm{d}r$$

$$= \frac{4Z^3}{a_0^3}\int_0^{2a_0/Z} \mathrm{e}^{-2Zr/a_0} r^2\, \mathrm{d}r. \tag{7.50}$$

Let us introduce a variable u according to the following relation:

$$\frac{2Z}{a_0} r = u$$

so that

$$\frac{2Z}{a_0}\mathrm{d}r = \mathrm{d}u.$$

Then (7.50) becomes

$$P_e = \frac{4Z^3}{a_0^3} \int_0^4 \frac{a_0^2}{4Z^2} e^{-u} u^2 \frac{a_0}{2Z} \, du$$

$$= \frac{1}{2} \int_0^4 u^2 e^{-u} \, du.$$

Integrating by parts and simplifying, we obtain

$$P_e = 1 = 13e^{-4}.$$

This gives us the probability of finding the electron between zero and $2a_0/Z$.

Hence, the probability of finding the electron beyond the range $2a_0/Z$ is

$$1 - P_e = 13e^{-4} \approx 0.24.$$

This is a very important quantum mechanical result. Just as in the case of a potential barrier and a harmonic oscillator, there is a finite probability of finding the electron in a hydrogenic atom in the classically inaccessible region. In the ground state of the hydrogenic atom, there is about a 24% chance of finding the electron outside the classical limit.

Remarks about degeneracy

The wave function $\psi_{n\ell m}$ for the hydrogen atom depends upon quantum numbers n (= 1, 2, 3, ...), ℓ (= 0, 1, 2, ..., $n-1$) and m (= $-\ell$, $-\ell+1$, ..., ℓ), while the corresponding energy value E_n depends upon the total quantum number n only and is independent of the orbital quantum number ℓ and the magnetic quantum number m. Thus, for the ground state for which $n = 1$ ($\ell = 0$, $m = 0$), the energy eigenvalue is non-degenerate. The energy eigenvalue is degenerate for all other values of n. The first excited state, with $n = 2$, is four-fold degenerate. The corresponding values of ℓ are 0, 1, and for each ℓ, the corresponding values of m are 0 and −1, 0, 1. The eigenfunctions ψ_{200}, ψ_{210}, ψ_{211}, $\psi_{21,-1}$ correspond to the same energy value E_2.

For a given value of n, there are n values of ℓ and for each value of ℓ, there are $(2\ell + 1)$ values of m. Hence, the number of linearly independent functions corresponding to a single value of n is

$$\sum_{\ell=0}^{n-1}(2\ell + 1) = 1 + 3 + 5 + \cdots + (2n - 1) \quad (n \text{ terms})$$

$$= \frac{n}{2}[2 + (n - 1) \cdot 2]$$

$$= \frac{n}{2} \cdot 2n = n^2.$$

This shows that the energy eigenvalues of the hydrogen atom are n^2-fold degenerate. That is, for a given value of n, there are n^2 linearly independent eigenfunctions corresponding to the same value of energy.

Now, how does a degeneracy arise? It is said that it is *often* due to some spatial symmetry of the system. For instance, in the case of a hydrogen atom, for a given ℓ, the degeneracy corresponding to m is generated by the spherical symmetry of the potential. If this is true, then whenever the spherical symmetry of the potential breaks down, the degeneracy of degree $2\ell + 1$ with respect to m will be removed. This does occur. When an external magnetic field is switched on and this breaks down the spherical symmetry by introducing a ϕ-dependent energy term in the Hamiltonian of the system, the degeneracy with respect to m is removed. The new formula for energy involves m and therefore the states with different values of m possess different energies.

But what about the degeneracy with respect to different values of ℓ? What is it due to and how can it be removed? Is this a characteristic of the Coulomb field? If this is so, then breaking the Coulomb symmetry should remove the ℓ-degeneracy. This can easily be checked. The Coulomb potential may be modified by adding a term inversely proportional to r^2. For convenience in computations, we may write it as $(\hbar^2/2\mu)g^2/r^2$ so that

$$V(r) = -\frac{Ze^2}{r} + \frac{\hbar^2}{2\mu}\frac{g^2}{r^2}.$$

By this addition to the potential, the Coulomb symmetry breaks down but the potential is still spherically symmetric. If we make this substitution for $V(r)$ in the radial equation, $\ell(\ell + 1)$ is replaced by $\ell(\ell + 1) + g^2$; the rest remains unchanged. Writing this as $\ell'(\ell' + 1)$, we obtain

$$\ell' = -\frac{1}{2} + \sqrt{(\ell + 1/2)^2 + g^2}.$$

The expression for the energy then takes the form

$$E = -\frac{1}{2}\mu c^2 \frac{(Z\alpha)^2}{\left[n_r + \frac{1}{2} + \sqrt{\left(\ell + \frac{1}{2}\right)^2 + g^2}\right]}.$$

It may be recalled that n_r is that value of ν for which the numerator in the recursion formula (7.44*b*) becomes zero. It is evident from this formula that for a given value of n, say $n = 3$, the eigenfunctions corresponding to different values of ℓ, for instance ($n_r = 1$, $\ell = 2$) and ($n_r = 2$, $\ell = 1$), are no longer degenerate. By breaking the Coulomb symmetry but without breaking the spherical symmetry, the ℓ degeneracy has been removed. This shows that the Coulomb symmetry is associated with the ℓ degeneracy. A very important but unanswered question is: what is the geometrical nature of the symmetry arising due to Coulmb interaction? Owing to the non-availability of the answer to this question, ℓ-degeneracy is called the *accidental degeneracy*. A thorough analysis of such situations as confronted in various physical phenomena may ultimately lead to some interesting conclusion about the geometrical structure of the symmetry causing so-called accidental degeneracy.

To sum up: with a spherically symmetric potential, the solution of the Schrödinger wave equation in spherical polar coordinates can be expressed as a product of two functions $Y(\theta, \phi)$ and $R(r)$, the former depending upon the angular coordinates θ, ϕ only while the latter $R(r)$ depends only on r. The differential equation for $Y(\theta, \phi)$ does not depend upon the potential energy of the system. Its solution is therefore independent of the specific expression for the potential energy of the system as long as it is spherically symmetric. The solution of the radial equation involves $V(r)$ and depends upon the specific expression for it. This solution is obtained for the hydrogen atom with $V(r) = -e^2/r$. We then write the general solution $\psi_{n\ell m}$ which depends upon the three quantum numbers n, ℓ, m while the corresponding energy E_n depends only upon the total quantum number n. The eigenvalues are therefore degenerate, the degree of degeneracy being n^2. The m degeneracy is due to spherical symmetry because as this symmetry is broken, say by applying a magnetic field, the degeneracy is removed. The ℓ degeneracy is due to the Coulomb potential. If the potential deviates from the inverse distance law, even retaining its spherical symmetry, the degeneracy is removed. But the geometrical structure of symmetry causing this degeneracy is still unknown: it is therefore called *accidental degeneracy*.

Problem

7.13. Show that for positive energy E, the Schrödinger wave equation yields a *continuous spectrum of energy*.

Additional problems

7.14. Calculate the most probable radius of the hydrogenic atom B^{+4} and compare it with the experimental value of 0.112 Å.

7.15. A hydrogen atom is in an eigenstate ψ_{200}. Find the probability that the distance of the electron from the nucleus of the atom is smaller than the Bohr radius.

7.16. Determine the energy levels of a three-dimensional isotropic harmonic oscillator. Is there any degeneracy in this case? If so, discuss any degenerate case.

7.17. Solve the radial wave equation for hydrogen-like atom given by

$$\frac{d^2R}{dr^2} + \frac{2m}{\hbar^2}\left(E + \frac{Ze^2}{r}\right)R = \frac{\ell(\ell + 1)}{r^2}R.$$

Find the quantised energy values.

7.18. A hydrogenic atom is in the state described by the normalised energy eigenfunction

$$\psi_{200} = \frac{1}{4\sqrt{2\pi}}\left(\frac{Z}{a_0}\right)^{3/2}(2 - \sigma)e^{-\sigma/2}, \qquad \sigma = \frac{Z}{a_0}r.$$

What is the probability of finding the electron at a distance greater than $2a_0/Z$ from the nucleus?

7.19. A hydrogenic atom is in the state described by the normalised energy eigenfunction

$$\psi_{211} = \frac{1}{8\sqrt{\pi}}\left(\frac{Z}{a_0}\right)^{3/2}\sigma e^{-\sigma/2}\sin\theta\, e^{i\phi}, \qquad \sigma = \frac{Z}{a_0}r.$$

Calculate the expectation values of $1/r$ and $1/r^2$.

7.20. Solve the differential equation

$$\frac{\sin\theta}{P(\theta)}\frac{d}{d\theta}\left(\sin\theta\frac{P(\theta)}{d\theta}\right) + C\sin^2\theta = m^2$$

and find the values C can take.

IOP Publishing

Quantum Mechanics

Mohammad Saleem

Chapter 8

Matrix mechanics

The criterion for a correct theory is that it should be self-consistent and yield results which agree with experiment. However, a theory may include quantities which cannot be observed directly. According to Heisenberg, it is the introduction of such quantities—the unobservables—in atomic theories which leads to incorrect results. For instance, the failure of Bohr's theory of the hydrogen spectrum is attributed to the fact that it involves the concept of unobservables such as the orbit of an electron. The introduction of such a concept in a theory must be avoided. A correct theory should exclude all those quantities which are not observables and hence operationally meaningless. On this basis, about six months earlier than Schrödinger's *wave mechanics*, Heisenberg formulated a theory called *matrix mechanics*. In this theory, the observables are represented by matrices and it is for this reason that the mechanics based on these ideas was labelled matrix mechanics. It has been shown by Schrödinger that matrix mechanics and wave mechanics are mathematically equivalent. These now have a common name, *quantum mechanics*. Since for a physical system, Schrödinger's differential equation is easier to operate with as compared to Heisenberg's matrix mechanics, the former has always overshadowed the scene. In fact, as most of the quantum systems have an infinite number of quantum states, the dynamical variables are usually represented by matrices of infinite order. Such matrices are very difficult to handle compared to the corresponding Schrödinger wave equation. On 26 April 1926, Einstein wrote to Schrödinger: 'I am convinced that you have made a decisive advance with your formation of quantum condition, just as I am equally convinced that the Heisenberg–Born route is off the track'. We will not therefore emphasise Heisenberg's method: it is quite difficult to solve problems using it.

It is interesting to note that Heisenberg and Born, the pioneers of matrix mechanics, were not very familiar with the matrix theory. Therefore, they went to the famous German mathematician Hilbert for help. Hilbert told them that if they looked for the differential equation which had these matrices, they would probably be able to do more with that. However, they thought that was a goofy idea and that

doi:10.1088/978-0-7503-1206-6ch8

Hilbert did not know what he was talking about. Actually, it was they who did not know what Hilbert was talking about! Had they followed his advice, and had Heisenberg been successful in obtaining that differential equation before Schrödinger was able to guess it, then, as a physicist who had established the differential equation forming the basis of wave mechanics, introduced matrix mechanics and originated the uncertainty principle Heisenberg would have achieved an unparalleled status in the domain of quantum mechanics.

8.1 Matrix representation of an operator

Although we will not follow Heisenberg's technique for the formulation of mtrix mechanics, we will indicate how Schrödinger's wave theory can be used to formulate matrix mechanics.

Let us first find the matrix representative of any operator A. Let ψ_1, ψ_2, ... be the eigenfunctions of a Hermitian operator so that these form a complete set. These functions are said to form a coordinate system or a *basis of representation*. For simplicity, we assume that the set ψ_1, ψ_2, ... is discrete. Consider a function ψ_m, i.e. any one of the functions of the complete set. Then, we may write

$$A\psi_m = \phi_m, \tag{8.1}$$

where ϕ_m is the function obtained by the action of the operator A on ψ_m. Since ψ_1, ψ_2, ... form a complete set, every function like ϕ_m can be expanded as a linear combination of them:

$$\phi_m = \sum_j a_{jm}\psi_j. \tag{8.2}$$

Therefore, (8.1) may be written as

$$A\psi_m = \phi_m = \sum_j a_{jm}\psi_j. \tag{8.3}$$

Mutiplying both sides of (8.3) from the left by ψ_n^* and integrating throughout the space, we obtain

$$\begin{aligned}\int \psi_n^* A\psi_m \, \mathrm{d}\tau &= \int \psi_n^* \sum_j a_{jm}\psi_j \, \mathrm{d}\tau \\ &= \sum_j a_{jm}\psi_j \int \psi_n^* \psi_j \, \mathrm{d}\tau = \sum_j a_{jm}\delta_{nj} = a_{nm}.\end{aligned}$$

Thus knowing the operator A and the complete set of eigenfunctions ψ_1, ψ_2, ..., we can in principle calculate all the elements a_{nm}. These elements can be arranged in a square array and possess all the characteristics of a matrix denoted by $A = [a_{nm}]$. The numbers a_{nm} are called the *matrix elements* of A with respect to the functions ψ_n and ψ_m. It may be noted that the matrix representative of an operator gives the same information as the operator itself. As is customary in the literature, we have used the same letter for an operator as well as its matrix representative. The structure of the matrix will be determined by the choice of the basis of representation.

Let us next find the matrix representative of an operator with the complete set of orthonormal eigenfunctions of this very operator as the basis of representation. Let a_{ij} denote the eigenvalues of an operator A corresponding to the eigenfunctions ψ_j. Then

$$A\psi_j = a_{ij}\psi_j.$$

Premultiplying by ψ_i^* and integrating over the entire space, we obtain

$$\int \psi_i^* A\psi_j \, \mathrm{d}\tau = a_{ij}\int \psi_i^*\psi_j \, \mathrm{d}\tau = a_{ij}\delta_{ij}$$

$$= a_{ii}, \quad \text{for } j = 1,$$

$$= 0, \quad \text{for } j \neq i.$$

This shows that with the complete set of its own eigenfunctions as the basis of representation, the matrix representative of any operator A is *diagonal*, the diagonal elements being the eigenvalues of the operator A.

Example 8.1.

Problem

Find the matrix representation of the operator x with energy eigenfunctions of a linear harmonic oscillator as the basis of representation.

Solution

The normalised energy eigenfunctions of the Hamiltonian for a linear harmonic oscillator are

$$\psi_n(x) = A\exp\left(-\frac{\alpha x^2}{2}\right)H_n\left(\sqrt{\alpha}x\right), \quad A = \sqrt{\frac{\sqrt{\alpha}}{\sqrt{\pi}\, 2^n\, n!}}.$$

Then the matrix representation of the operator x is given by

$$x_{nm} = \int_{-\infty}^{\infty} \psi_n^*(x)x\psi_m(x)\,\mathrm{d}x.$$

This integral has been evaluated in chapter 3. Substituting the value of this integral in the above equation, we obtain

$$x_{mn} = \frac{A_n^* A_m}{\alpha}\sqrt{\pi}\, 2^n\, n!, \qquad \text{for } m = n - 1,$$

$$= \frac{A_n^* A_m}{\alpha}\sqrt{\pi}\, 2^n\, (n+1)!\,, \qquad \text{for } m = n + 1,$$

$$= 0, \text{ otherwise}, \qquad \text{i.e. for } m \neq (n \pm 1).$$

Problems

8.1. Show that, over the interval $(-\ell, \ell)$, the functions

$$\frac{1}{2\ell}e^{2n\pi x/\ell}, \quad n = 0, \pm 1, \pm 2, \ldots$$

are the normalised eigenfunctions of the operator

$$x + \frac{d}{dx}.$$

8.2. Find the matrix representative of the operator

$$x + \frac{d}{dx}.$$

with its eigenfunctions forming the basis of representation.

8.2 Change of basis and unitary transformation

Let us next see what happens to the matrix representative of an operator when a basis of representation is changed. Consider an operator A. Let an orthonormal complete set of functions ψ_i, $i = 1, 2, \ldots$ be a basis of representation. Then the matrix elements of the operator A with respect to this basis are given by

$$A_{nm} = \int \psi_n^* A \psi_m \, d\tau. \tag{8.4}$$

Let us now change the basis of representation. We choose another orthonormal complete set of functions ϕ_j as the new basis. Then the matrix elements $\hat{A}_{nm}$ of the same operator with respect to the new basis are given by

$$\hat{A}_{nm} = \int \phi_n^* A \phi_m \, d\tau. \tag{8.5}$$

Let us now see how these two bases of representations are connected to each other. Since, ψ_i form a complete set of functions, each one of the functions ϕ_j can be expanded in terms of them:

$$\phi_m = \sum_j a_{im}\psi_i,$$

$$\phi_n = \sum_j a_{jn}\psi_j.$$

Substituting the expressions for ϕ_m^* and ϕ_n in (8.5), we obtain

$$\begin{aligned}\hat{A}_{nm} &= \int \sum_j a_{jn}^* \psi_j^* A \sum_i a_{im} \psi_i \, \mathrm{d}\tau \\ &= \sum_i \sum_j a_{jn}^* a_{im} \int \psi_j^* A \psi_i \, \mathrm{d}\tau \\ &= \sum_i \sum_j a_{nj}^\dagger a_{im} A_{ji} \\ &= \sum_i \sum_j a_{nj}^\dagger A_{ji} a_{im} \\ &= \left(\alpha^\dagger A \alpha\right)_{nm}. \end{aligned} \tag{8.6}$$

Since the choice of nm is arbitrary, the above equation holds for all values of nm. This is possible only if the matrices on the two sides of the above equation are equal:

$$\hat{A} = \alpha^\dagger A \alpha. \tag{8.7}$$

As the transformation matrix is unitary, $\alpha^\dagger \alpha = I$, and therefore $\alpha^\dagger = \alpha^{-1}$, the above equation may be written as

$$\hat{A} = \alpha^{-1} A \alpha. \tag{8.8a}$$

This may be reversed as

$$A = \alpha \hat{A} \alpha^{-1}. \tag{8.8b}$$

Equations (8.8*a*) and (8.8*b*) show that if an orthonormal basis of representation is changed to another orthonormal basis, a matrix A will transform to matrix $\hat{A} = \alpha^{-1} A \alpha$, where α is a unitary matrix connecting two bases of representation. Such a transformation is called a *unitary transformation*. But under a unitary transformation the algebraic equations remain unchanged. Hence, the expectation value of any operator is independent of the choice of the basic set of functions.

Problem

8.3. Show that under unitary transformations, the trace of a matrix remains invariant.

8.3 Coordinate and momentum representations

When the basis of representation is a complete set of eigenfunctions of the position operator x, then the representation is said to be *coordinate representation* or *x-representation*. As analysed earlier, in this representation, the matrix representing the operator x is diagonal. Let us show this by explicit calculations. The eigenvalue equation for the operator x is given by

$$x\psi_{x'} = x'\psi_{x'}.$$

Here x' is the eigenvalue of the operator x corresponding to the eigenfunction $\psi_{x'}$. The above equation can be written as

$$(x - x')\psi_{x'}(x) = 0.$$

For $x \neq x'$, the function $\psi(x)$ must be zero. For $x = x'$, the function $\psi(x)$ can have any value subject to the condition that

$$\int_{-\infty}^{\infty} \psi_{x'}^{*}\psi_{x'}(x)\, \mathrm{d}x = 1.$$

Actually, in order that the integral may exist, ψ must approach infinity. Since this analysis is valid for all *finite* values of x', the values of x' must form a *continuous spectrum*, i.e. the position coordinates can have any finite value. Thus, the eigenfunction $\psi(x)$ of the position operator x has the following properties:

$$\psi(x) = 0, \qquad \text{for } x \neq x',$$

$$\psi(x) = \infty, \qquad \text{for } x = x',$$

such that

$$\int \psi^{*}(x)\psi(x)\, \mathrm{d}x = 1,$$

the region of integration including the point x'. Hence, it behaves as a delta function:

$$\psi(x) = \delta(x - x').$$

We conclude that the eigenfunction of the position operator is a delta function. Hence, the matrix representative of the operator x is given by

$$x_{x'x''} = \int \delta(x - x')x\delta(x - x'')\, \mathrm{d}x = x\delta(x - x'').$$

This gives a diagonal matrix because all the elements of the matrix x for which $x' \neq x''$ are zeros.

For any operator A, the matrix elements of its matrix representative in the coordination representation are given by

$$A_{x'x''} = \int \delta(x - x')A\delta(x - x')\, \mathrm{d}x.$$

The rows and columns of the matrix A are labelled by the position variables x' and x''.

A matrix representation of any operator with the eigenfunctions of the momentum operator p as the basis of representation is called the *momentum representation* or *p-representation*. In this representation, the matrix representing the momentum operator is diagonal. Since the eigenfunctions of the operator p are $e^{ip_0x/\hbar}$, the Omatrix representation of any operator A is given by

$$A_{p'p''} = \int e^{ip'x/\hbar} A e^{ip''x/\hbar}\, dx.$$

8.4 Continuous distribution of eigenvalues

When an operator has continuous eigenvalues, then it can be shown that most of the statements that are valid for a discrete spectrum are also true in this case. For instance, for discrete eigenvalues, the eigenvalue equation is

$$A\psi_i = a_i\psi_i,$$

where the subscript i is used to denote a particular eigenvalue as well as the corresponding eigenfunction. For a continuous distribution, this equaion is replaced by

$$A\psi_a = a\psi_a,$$

where the continuous eigenvalue is also used as a subscript to designate the correspoding eigenfunction. For a discrete spectrum, the eigenvalues are denoted by $a_1, a_2, a_3, \ldots$ while for a continuous spectrum, the eigenvalues are designated as $a, a', a'', \ldots$.

The orthogonality condition for eigenfunctions corresponding to continuous eigenvalues can be written as

$$\int \psi^*_{a'}(x)\psi_a(x)\, dx = 0, \qquad a' \neq a.$$

Since the eigenvalue spectrum is continuous, this integral would be divergent for $a' = a$ because the wave function is unnormalisable. Therefore,

$$\int \psi^*_{a'}(x)\psi_a(x)\, dx = \infty, \qquad a' = a.$$

Hence, we may write

$$\int \psi^*_{a'}(x)\psi_a(x)\, dx = \delta(a - a').$$

The expansion theorem in the case of a continuous distribution of eigenvalues may be written as

$$\psi = \int c_a\psi_a\, da,$$

or, using a slightly diferent notation, as

$$\psi = \int c(a)\psi(a)\, da.$$

This replaces

$$\psi = \sum_i c_i \psi_i.$$

The closure relation for a discrete spectrum, i.e.

$$\sum_i \psi_i^*(x')\psi_i(x)\,\mathrm{d}x = \delta(x - x')$$

is replaced by

$$\int_{-\infty}^{\infty} \psi_a^*(x')\psi_a(x)\,\mathrm{d}x = \delta(x - x').$$

Problems

8.4. Discuss the case when the eigenvalue spectrum of an operator is discrete over a certain range of values and continuous over a separate range of values.

8.5. Show that under a unitary transformation, the expectation values remain unchanged.

8.6. Show that

$$\frac{1}{2\pi}\int_{-\infty}^{\infty} \mathrm{e}^{\mathrm{i}\gamma(x-x')}\,\mathrm{d}\gamma$$

is a representation of the delta function.

Additional problems

8.7. Show that a change of basis is accomplished by means of a unitary matrix.

8.8 Show that the trace of a matrix remains invariant under a similarity transformation.

8.9. Discuss the case when the eigenvalue spectrum of an operator is discrete over a certain range of values and continuous over another range of values.

IOP Publishing

Quantum Mechanics

Mohammad Saleem

Chapter 9

Angular momentum

The concept of angular momentum plays an important role in the classical as well as quantum mechanical analysis of physical systems. The significance of the law of conservation of angular momentum in the entire domain of physics cannot be overemphasised. It is the strong faith in the conservation of angular momentum that encouraged physicists to open the vista for the introduction of intrinsic angular momentum or spin in quantum mechanics. Moreover, unlike linear momentum, the angular momentum is quantised. We shall now systematically develop the theory of angular momentum and apply it to analyse some interesting problems of atomic and nuclear physics.

9.1 Angular momentum operator

Let $\mathbf{r}$ be the position vector of a particle P with respect to a fixed point O and let $\mathbf{p}$ be its linear momentum. Then in classical mechanics the angular momentum $\mathbf{L}$ of this particle about the point O is defined by the equation

$$\mathbf{L} = \mathbf{r} \wedge \mathbf{p} \tag{9.1}$$

and is called *orbital angular momentum*. This is shown in figure 9.1. The symbol ^ stands for the vector product or the cross product of two vectors. $\mathbf{L}$ is perpendicular to the r–p plane. In terms of the components of $\mathbf{L}$, $\mathbf{r}$ and $\mathbf{p}$ in a rectangular Cartesian coordinate system, (9.1) may be written as

$$\left(L_x, L_y, L_z\right) = \begin{vmatrix} x & y & z \\ p_x & p_y & p_z \\ \mathbf{i} & \mathbf{j} & \mathbf{k} \end{vmatrix} \tag{9.2}$$

where L_x, L_y, L_z are the components of the vector $\mathbf{L}$ along x-, y- and z-axes, p_x, p_y, p_z are the components of the vector $\mathbf{p}$ along these axes and $\mathbf{i}$, $\mathbf{j}$, $\mathbf{k}$ are unit vectors

doi:10.1088/978-0-7503-1206-6ch9

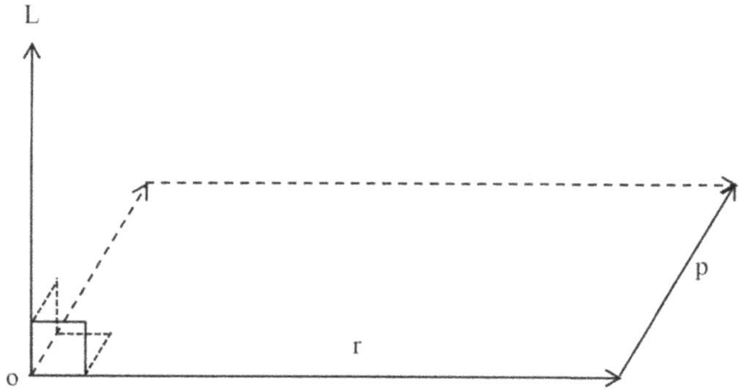

Figure 9.1. Angular momentum.

along them. We have used the notation $(A_x, A_y, A_z) \equiv A_x\mathbf{i} + A_y\mathbf{j} + A_z\mathbf{k}$. Comparing the coefficients of **i**, **j**, **k** on the two sides of equation, we obtain

$$L_x = yp_z - zp_y,$$

$$L_y = zp_x - xp_z,$$

$$L_z = xp_y - yp_x.$$

These equations give the components of angular momentum vector **L** along the three axes.

In the transition from classical to quantum mechanics, we interpret **L**, **r** and **p** as operators such that

$$\mathbf{r} \rightarrow \mathbf{r},$$

$$\mathbf{p} \rightarrow -\mathrm{i}\hbar\nabla$$

and the same letter **L** is used for the angular mementum operator in quantum mechanics. Then, in quantum mechanics, we may write

$$\mathbf{L} = \mathbf{r} \wedge (-i\hbar\nabla),$$

or

$$\left(L_x, L_y, L_z\right) = -\mathrm{i}\hbar \begin{vmatrix} x & y & z \\ \frac{\partial}{\partial x} & \frac{\partial}{\partial y} & \frac{\partial}{\partial z} \\ \mathbf{i} & \mathbf{j} & \mathbf{k} \end{vmatrix}.$$

Comparing the coefficients of **i**, **j**, **k** on the two sides of this equation, we obtain

$$L_x = -\mathrm{i}\hbar\left(y\frac{\partial}{\partial z} - z\frac{\partial}{\partial y}\right), \tag{9.3a}$$

$$L_y = -\mathrm{i}\hbar\left(z\frac{\partial}{\partial x} - x\frac{\partial}{\partial z}\right), \tag{9.3b}$$

$$L_z = -\mathrm{i}\hbar\left(x\frac{\partial}{\partial y} - y\frac{\partial}{\partial x}\right). \tag{9.3c}$$

In quantum mechanics, these equations give the components L_x, L_y, L_z of the angular momentum operator **L** in rectangular Cartesian coordinates.

9.2 Commutators of various components of L

Let us now calculate the commutator of two components, say, L_x and L_y of the angular momentum operator **L**. Let this commutator operate upon an arbitrary function ϕ. Then, we have

$$\left[L_x, L_y\right]\phi = \left[L_xL_y - L_yL_x\right]\phi.$$

By using the expressions for L_x and L_y as given by (9.3a) and (9.3b), we obtain

$$\begin{aligned}
\left[L_x, L_y\right]\phi = -\hbar^2\Bigg[&\left(y\frac{\partial}{\partial z} - z\frac{\partial}{\partial y}\right)\left(z\frac{\partial}{\partial x} - x\frac{\partial}{\partial z}\right) \\
&- \left(z\frac{\partial}{\partial x} - x\frac{\partial}{\partial z}\right)\left(y\frac{\partial}{\partial z} - z\frac{\partial}{\partial y}\right)\Bigg]\phi \\
= -\hbar^2\Bigg[&\left(y\frac{\partial}{\partial z} - z\frac{\partial}{\partial y}\right)\left(z\frac{\partial\phi}{\partial x} - x\frac{\partial\phi}{\partial z}\right) \\
&- \left(z\frac{\partial}{\partial x} - x\frac{\partial}{\partial z}\right)\left(y\frac{\partial\phi}{\partial z} - z\frac{\partial\phi}{\partial y}\right)\Bigg] \\
= -\hbar^2\Bigg[&\left(y\frac{\partial\phi}{\partial x} + yz\frac{\partial^2\phi}{\partial z\partial x} - yx\frac{\partial^2\phi}{\partial z^2} - z^2\frac{\partial^2\phi}{\partial y\partial x} + zx\frac{\partial^2\phi}{\partial y\partial z}\right) \\
&- \left(yz\frac{\partial^2\phi}{\partial x\partial z} - z^2\frac{\partial^2\phi}{\partial x\partial y} - yx\frac{\partial^2\phi}{\partial z^2} + x\frac{\partial\phi}{\partial y} + zx\frac{\partial^2\phi}{\partial z\partial y}\right)\Bigg] \\
= -\hbar^2&\left(y\frac{\partial\phi}{\partial x} - x\frac{\partial\phi}{\partial y}\right) \\
= \hbar^2&\left(x\frac{\partial\phi}{\partial y} - y\frac{\partial\phi}{\partial x}\right) \\
= \mathrm{i}\hbar& L_z\phi, \text{ by virtue of (9.3}c\text{)}.
\end{aligned}$$

As the function ϕ is arbitrary, this equation shows that we must have

$$\left[L_x, L_y\right] = \mathrm{i}\hbar L_z. \tag{9.4a}$$

By cyclic permutation of x, y, z ($x \to y \to z \to x$), we also obtain the following commutation relations:

$$[L_y, L_z] = \mathrm{i}\hbar L_x, \tag{9.4b}$$

$$[L_z, L_x] = \mathrm{i}\hbar L_y. \tag{9.4c}$$

Equations (9.4) show that the components of the angular momentum operator do not mutually commute. Therefore, a simultaneous precise measurement of even any two of them is not possible, unless $L_x = L_y = L_z = 0$. Thus, for a system with non-zero orbital angular momentum, definite values cannot be simultaneously assigned to all its components. In vector form, the above three commutation relations can be put together as

$$\mathbf{L} \wedge \mathbf{L} = \mathrm{i}\hbar\mathbf{L}. \tag{9.5a}$$

The vector product of **L** with itself does not vanish because in quantum mechanics the components of **L** do not commute with each other.

Relations (9.4) can also be written as

$$[L_i, L_j] = \mathrm{i}\hbar\varepsilon_{ijk}L_k, \quad i, j, k = 1, 2, 3 \tag{9.5b}$$

where

$$L_1 \equiv L_x, \quad L_2 \equiv L_y, \quad L_3 \equiv L_z$$

and ε_{ijk} is the Levi–Civita symbol which is equal to +1 if (the distinct indices) i, j, k are in cyclic order, and is otherwise equal to −1.

9.3 Commutator of L^2 and L_z

Let us next find the commutation relation between L^2 and L_z, where L^2 is the square of the magnitude of the total angular momentum operator **L** and L_z is the z-component of the angular momentum operator. It can be mentioned that L_z has been chosen conventionally. We could choose any component of the angular momentum operator; the result would be the same. If we define the square of the magnitude of **L** by the equation

$$L^2 = L_x^2 + L_y^2 + L_z^2,$$

then, we have

$$\begin{aligned}[L^2, L_z] &= [L_x^2 + L_y^2 + L_z^2, L_z] \\ &= [L_x^2, L_z] + [L_y^2, L_z] + [L_z^2, L_z].\end{aligned}$$

Now

$$\begin{aligned}[L_x^2, L_z] &= L_x^2 L_z - L_z L_x^2 \\ &= L_x L_x L_z - L_z L_x L_x,\end{aligned} \tag{9.6}$$

but

$$[L_z, L_x] = i\hbar L_y,$$

or

$$L_z L_x - L_x L_z = i\hbar L_y,$$

or

$$L_x L_z = L_z L_x - i\hbar L_y. \tag{9.7}$$

Substituting the expressions for $L_x L_z$ and $L_z L_x$ from (9.7) in (9.6), we obtain

$$\begin{aligned}\left[L_x^2, L_z\right] &= L_x\left(L_z L_x - i\hbar L_y\right) - \left(L_x L_z + i\hbar L_y\right)L_x \\ &= -i\hbar\left(L_x L_y + L_y L_x\right).\end{aligned}$$

Similarly, we can show that

$$\left[L_y^2, L_z\right] = i\hbar\left(L_y L_x + L_x L_y\right)$$

and

$$\left[L_z^2, L_z\right] = 0.$$

Combining the last three equations, we find that

$$[L^2, L_z] = 0. \tag{9.8}$$

This relation shows that L^2 commutes with L_z. Therefore, it is possible to measure simultaneously the precise values of the square of total angular momentum L^2 and the z-component L_z of **L**. Similarly, it can be shown that L^2 commutes with the other components of angular momentum. Therefore, a simultaneous precise measurement of L^2 *and any one component* of **L** can be made. Hence, we may write

$$[L^2, \mathbf{L}] = 0. \tag{9.9}$$

We note that although L_x, L_y, L_z, the components of the angular momentum operator **L**, do not commute with each other, each one of them commutes with L^2. Mathematically:

$$[L_x, L_y] \neq 0,$$

$$[L_y, L_z] \neq 0,$$

$$[L_z, L_x] \neq 0,$$

$$\left[L^2, L_i\right] = 0, \quad i = x, y, z.$$

Example 9.1.

Problem
Show that the operators x^2 and p_x^2 commute with L_x.

Solution
Let us first calculate the commutator of L_x and x^2. We have

$$\left[L_x, x^2\right] = \left(L_x x^2 - x^2 L_x\right)\phi$$

where ϕ is an arbitrary function. We know that

$$L_x = -\mathrm{i}\hbar\left(y\frac{\partial}{\partial z} - z\frac{\partial}{\partial y}\right). \tag{9.3a'}$$

Substituting this expression for L_x in the equation for the commutator, we obtain

$$\begin{aligned}
\left[L_x, x^2\right]\phi &= \left[-\mathrm{i}\hbar\left(y\frac{\partial}{\partial z} - z\frac{\partial}{\partial y}\right)x^2 - x^2\left\{-\mathrm{i}\hbar\left(y\frac{\partial}{\partial z} - z\frac{\partial}{\partial y}\right)\right\}\right]\phi \\
&= -\mathrm{i}\hbar\left(y\frac{\partial}{\partial z} - z\frac{\partial}{\partial y}\right)\left(x^2\phi\right) - x^2\left\{-\mathrm{i}\hbar\left(y\frac{\partial\phi}{\partial z} - z\frac{\partial\phi}{\partial y}\right)\right\} \\
&= -\mathrm{i}\hbar\left(yx^2\frac{\partial\phi}{\partial z} - zx^2\frac{\partial\phi}{\partial y} - x^2y\frac{\partial\phi}{\partial z} + x^2z\frac{\partial\phi}{\partial y}\right) \\
&= 0.
\end{aligned}$$

This shows that $[L_x, x^2] = 0$, i.e. x^2 commutes with L_x.

We next find the commutator of L_x and p_x^2. We have

$$\begin{aligned}
[L_x, p_x^2] &= [-\mathrm{i}\hbar(y\partial/\partial z - z\partial/\partial y)(-\hbar^2\partial^2\phi/\partial x^2) + \hbar^2\partial^2/\partial x^2(-\mathrm{i}\hbar)(y\partial\phi/\partial z - z\partial\phi/\partial y) \\
&= \mathrm{i}\hbar^3\left(y\frac{\partial^3\phi}{\partial z\partial x^2} - z\frac{\partial^3\phi}{\partial y\partial x^2} - y\frac{\partial^3\phi}{\partial x^2\partial z} + z\frac{\partial^3\phi}{\partial x^2\partial y}\right) \\
&= 0.
\end{aligned}$$

This shows that the operators L_x and p_x^2 also commute.

9.4 Components of the orbital angular momentum operator in spherical polar coordinates

The mathematical discussion of the properties of the angular momentum of a particle becomes convenient if it is expressed in spherical polar coordinates r, θ, ϕ. These coordinates are related to the rectangular Cartesian coordinates x, y, z of a particle by the equations

$$x = r \sin\theta \cos\phi,$$
$$y = r \sin\theta \sin\phi,$$
$$z = r \cos\theta$$

so that

$$r^2 = x^2 + y^2 + z^2,$$
$$\cos\theta = \frac{z}{r},$$
$$\tan\phi = \frac{y}{x}.$$

Let us first express L_z, the z-component of the angular momentum operator, in spherical polar coordinates. We have

$$\begin{aligned}
L_z\psi &= -\mathrm{i}\hbar^2\left(x\frac{\partial\psi}{\partial y} - y\frac{\partial\psi}{\partial x}\right) \\
&= -\mathrm{i}\hbar\left\{x\left(\frac{\partial\psi}{\partial r}\frac{\partial r}{\partial y} + \frac{\partial\psi}{\partial\theta}\frac{\partial\theta}{\partial y} + \frac{\partial\psi}{\partial\phi}\frac{\partial\phi}{\partial y}\right)\right. \\
&\quad \left. - y\left(\frac{\partial\psi}{\partial r}\frac{\partial r}{\partial x} + \frac{\partial\psi}{\partial\theta}\frac{\partial\theta}{\partial x} + \frac{\partial\psi}{\partial\phi}\frac{\partial\phi}{\partial x}\right)\right\} \\
&= -\mathrm{i}\hbar^2\left[r\sin\theta\cos\phi\left(\sin\theta\sin\phi\frac{\partial\psi}{\partial r} + \frac{1}{r}\cos\theta\sin\phi\frac{\partial\psi}{\partial\theta}\right.\right. \\
&\quad \left. + \frac{1}{r}\frac{\cos\phi}{\sin\theta}\frac{\partial\psi}{\partial\phi}\right) \\
&\quad - r\sin\theta\sin\phi\left(\sin\theta\cos\phi\frac{\partial\psi}{\partial r} + \frac{1}{r}\cos\theta\cos\phi\frac{\partial\psi}{\partial\theta}\right. \\
&\quad \left.\left. - \frac{1}{r}\frac{\sin\phi}{\sin\Theta}\frac{\partial\psi}{\partial\phi}\right)\right] \\
&= -\mathrm{i}\hbar\left(\cos^2\phi\frac{\partial\psi}{\partial\phi} + \sin^2\phi\frac{\partial\psi}{\partial\phi}\right) \\
&= -\mathrm{i}\hbar\frac{\partial\psi}{\partial\phi}.
\end{aligned}$$

Thus, in spherical polar coordinates, the z-component of orbital angular momentum is given by

$$L_z = -\mathrm{i}\hbar\frac{\partial}{\partial\phi}. \tag{9.10}$$

Similarly, we can show that

$$L_x = -\mathrm{i}\hbar\left(\sin\phi\frac{\partial}{\partial\theta} + \cot\theta\cos\phi\frac{\partial}{\partial\phi}\right), \tag{9.11}$$

$$L_y = -\mathrm{i}\hbar\left(\cos\phi\frac{\partial}{\partial\theta} - \cot\theta\sin\phi\frac{\partial}{\partial\phi}\right). \tag{9.12}$$

It may be noted that in spherical polar coordinates, the components L_x, L_y, L_z of the angular momentum operator $\mathbf{L}$ involve *only* angular coordinates θ, ϕ and do not depend at all on the radial coordinate r. Moreover, with our choice of orientation axes, the z-component L_z of angular momentum has the simplest form in spherical polar coordinates.

9.5 L^2 in spherical polar coordinates

We have determined the components of orbital angular momentum $\mathbf{L}$ in spherical polar coordinates. Let us next express L^2 in spherical polar coordinates. This operator is defined by the operator equation

$$L^2 = L_x^2 + L_y^2 + L_z^2. \tag{9.13}$$

But

$$\begin{aligned}
L_x^2 &= L_x L_x \\
&= -\hbar^2\left(\sin\phi\frac{\partial}{\partial\theta} + \cot\theta\cos\phi\frac{\partial}{\partial\phi}\right)\left(\sin\phi\frac{\partial}{\partial\theta} + \cot\theta\cos\phi\frac{\partial}{\partial\phi}\right) \\
&= -\hbar^2\left\{\sin^2\phi\frac{\partial^2}{\partial\theta^2} + \sin\phi(-\mathrm{cosec}^2\theta)\cos\phi\frac{\partial}{\partial\phi}\right. \\
&\quad + \sin\phi\cot\theta\cos\phi\frac{\partial^2}{\partial\theta\partial\phi} + \cot\theta\cos\phi\cos\phi\frac{\partial}{\partial\theta} \\
&\quad + \sin\phi\cot\theta\cos\phi\frac{\partial^2}{\partial\theta\partial\phi} + \cot^2\theta\cos\phi(-\sin\phi)\frac{\partial}{\partial\phi} \\
&\quad \left. + \cot^2\theta\cos^2\phi\frac{\partial^2}{\partial\phi^2}\right\}.
\end{aligned}$$

Similarly, we have

$$
\begin{aligned}
L_y^2 &= L_y L_y \\
&= -\hbar^2\left(\cos\phi\frac{\partial}{\partial\theta} - \cot\theta\sin\phi\frac{\partial}{\partial\phi}\right)\left(\cos\phi\frac{\partial}{\partial\theta} - \cot\theta\sin\phi\frac{\partial}{\partial\phi}\right) \\
&= -\hbar^2\left\{\cos^2\phi\frac{\partial^2}{\partial\phi^2} - \cos\phi\left(-\operatorname{cosec}^2\theta\right)\sin\phi\frac{\partial}{\partial\phi}\right. \\
&\quad - \cos\phi\cot\theta\sin\phi\frac{\partial^2}{\partial\theta\partial\phi} - \cot\theta\sin\phi(-\sin\phi)\frac{\partial}{\partial\theta} \\
&\quad - \sin\phi\cot\theta\cos\phi\frac{\partial^2}{\partial\theta\partial\phi} + \cot^2\theta\cos\phi\sin\phi\frac{\partial}{\partial\phi} \\
&\quad \left. + \cot^2\theta\sin^2\phi\frac{\partial^2}{\partial\phi^2}\right\}
\end{aligned}
$$

and

$$L_z^2 = L_z L_z = -\hbar^2\frac{\partial}{\partial\phi}\left(\frac{\partial}{\partial\phi}\right) = -\hbar^2\frac{\partial^2}{\partial\phi^2}.$$

Substituting these expressions in (9.13), we obtain

$$
\begin{aligned}
L^2 &= -\hbar^2\left(\frac{\partial^2}{\partial\theta^2} + \cot\theta\frac{\partial}{\partial\theta} + \cot^2\theta\frac{\partial^2}{\partial\phi^2} + \frac{\partial^2}{\partial\phi^2}\right) \\
&= -\hbar^2\left(\frac{1}{\sin\theta}\frac{\partial}{\partial\theta}\left(\sin\theta\frac{\partial}{\partial\theta}\right) + \frac{1}{\sin^2\theta}\frac{\partial^2}{\partial\phi^2}\right) \\
&= -\hbar^2\Omega, \qquad (9.14a)
\end{aligned}
$$

where

$$\Omega = \frac{1}{\sin\theta}\frac{\partial}{\partial\theta}\left(\sin\theta\frac{\partial}{\partial\theta}\right) + \frac{1}{\sin^2\theta}\frac{\partial^2}{\partial\phi^2}. \qquad (9.14b)$$

The right-hand side of (9.14*a*), with Ω given by (9.14*b*), is the expression for the operator $\mathbf{L}^2$ in spherical polar coordinates.

Applying the operator equation (9.14*a*) on the function $Y(\theta, \phi)$, a function of angular coordinates alone, we obtain

$$
\begin{aligned}
L^2 Y(\theta, \phi) &= -\hbar^2\Omega Y(\theta, \phi) \\
&= -\hbar^2\left[\frac{1}{\sin\theta}\frac{\partial}{\partial\theta}\left(\sin\theta\frac{\partial}{\partial\theta}\right) + \frac{1}{\sin^2\theta}\frac{\partial^2}{\partial\phi^2}\right]Y(\theta, \phi).
\end{aligned}
$$

By virtue of (7.6), this takes the form

$$L^2 Y(\theta, \phi) = -\hbar^2(-CY(\theta, \phi)) = C\hbar^2 Y(\theta, \phi).$$

The parameter C has already been determined in (7.21) as $\ell(\ell + 1)$. Substituting this expression in the above equation, we obtain

$$L^2 Y(\theta, \phi) = \ell(\ell + 1)\hbar^2 Y(\theta, \phi).$$

This shows that $Y(\theta, \phi)$ is the eigenfunction of L^2 corresponding to the eigenvalue $\ell(\ell + 1)\hbar^2$.

Since in spherical polar coordinates, the operators L_x, L_y, L_z are independent of r, the operator $L^2 (\equiv L_x^2 + L_y^2 + L_z^2)$ is also independent of r.

Problem

9.1. Show that in spherical polar coordinates where under the parity operation P, i.e. $\mathbf{r} \to -\mathbf{r}$, i.e. under $r \to r, \theta \to \pi - \theta, \phi \to \pi + \phi$, the operators $\mathbf{L}^2$ and L_z remain unchanged.

Since the operator L^2 remains invariant under the parity operation P, we have

$$P^{-1}L^2P = L^2.$$

Multiplying both sides from the left by P, we obtain

$$L^2P = PL^2,$$

or

$$\left[P, L^2\right] = 0.$$

Similarly, we can show that

$$[P, L_z] = 0.$$

Since these operators commute, they can have simultaneous eigenfunctions; the parity eigenfunctions are also angular momentum eigenfunctions.

9.6 Eigenfunctions and eigenvalues of L_z

Let us now find the eigenfunctions and the corresponding eigenvalues of the third component of angular momentum operator $\mathbf{L}$, i.e. L_z. Let λ be an eigenvalue of the operator L_z corresponding to an eigenfunction ψ. Then, we must have

$$L_z\psi = \lambda\psi. \tag{9.15}$$

In spherical polar coordinates, the operator L_z is given by

$$L_z = -\mathrm{i}\hbar \frac{\partial}{\partial \phi}. \tag{9.10'}$$

Substituting the above expression for L_z in (9.15), we obtain

$$-\mathrm{i}\hbar \frac{\partial \psi}{\partial \phi} = \lambda \psi.$$

The above equation can be written as

$$\frac{1}{\psi} \frac{\partial \psi}{\partial \phi} = \frac{\mathrm{i}}{\hbar} \lambda.$$

Integrating with respect to ϕ, we obtain

$$\ln \psi = \frac{\mathrm{i}}{\hbar} \lambda \phi + \ln A(r, \theta),$$

or

$$\ln \left(\frac{\psi}{A} \right) = \frac{\mathrm{i}}{\hbar} \lambda \phi,$$

or

$$\frac{\psi}{A} = \exp\left(\frac{\mathrm{i}}{\hbar} \lambda \phi \right),$$

or

$$\psi = A \exp\left(\frac{\mathrm{i}}{\hbar} \lambda \phi \right), \tag{9.16}$$

where $A \equiv A(r, \theta)$ is an arbitrary function of r and θ. For $\phi = 0$, the above equation shows that $\psi = A$. For $\phi = 2\pi$, the above equation gives $\psi = A \exp(2\pi \mathrm{i}\lambda/\hbar)$. Now as a physically acceptable wave function must be single-valued, we must have $\psi(2\pi) = \psi(0)$; i.e. ϕ, being a periodic function of period 2π, must remain unaltered as it is changed by 2π. This yields

$$A \exp\left(\frac{2\pi \mathrm{i}\lambda}{\hbar} \right) = A,$$

or

$$A \left[\exp\left(\frac{2\pi \mathrm{i}\lambda}{\hbar} \right) - 1 \right] = 0.$$

Therefore, either A or the expression in square brackets may be zero. The parameter A cannot be zero because otherwise, by virtue of (9.16), the wave function ψ will vanish for all values of ϕ and consequently the probability of finding the particle anywhere will be zero which is impossible. Hence, we obtain

$$\exp\left(\frac{2\pi i\lambda}{\hbar}\right) = 1 = \exp(2im\pi), \quad m = 0, \pm 1, \pm 2, \ldots$$

or

$$\lambda = m\hbar, \quad m = 0, \pm 1, \pm 2, \ldots \tag{9.17}$$

We conclude that if the wave function ψ is to be single-valued, then we must have

$$\lambda = mh, \quad m = 0, \pm 1, \pm 2, \ldots$$

The integer m is known as the *magnetic quantum number* as it plays an important role in the analysis of any physical system placed in a magnetic field. Hence, the eigenvalues λ of the z-component of angular momentum operator are

$$\ldots, -2\hbar, -\hbar, 0, \hbar, 2\hbar, \ldots$$

Equation (9.16) now becomes

$$\psi = A(r, \theta)\exp\left(\frac{i}{\hbar}\lambda\phi\right) = A(r, \theta)\exp(im\phi).$$

These are the eigenfunctions of L_z. To emphasise that an eigenfunction depends upon m, we will write ψ_m for ψ. The above analysis shows that in spherical polar coordinates,

$$\psi_m = A(r, \theta)\exp(im\phi), \quad m = 0, \pm 1, \pm 2, \ldots \tag{9.18a}$$

are the eigenfunctions and $m\hbar$ are the eigenvalues of the operator L_z.

Equation (9.18*a*) shows that under reflection, i.e. under parity operation $\phi \rightarrow \pi + \phi$, we have

$$\Psi_m = Ae^{im\phi} \rightarrow Ae^{im(\pi+\phi)} = Ae^{im\pi}e^{im\phi} = (-1)^m Ae^{im\phi}. \tag{9.18b}$$

Therefore, the parity of the wave function is $(-1)^m$. It is even for even m and is denoted by +; it is odd for odd m and is denoted by −.

Problem

9.2. Show that the operators L_z and L^2 are Hermitian and mutually commute.

9.7 Eigenvalues of L_z and L^2 corresponding to their simultaneous eigenfunctions and ladder operators

The Hermitian operators L_z and L^2 mutually commute. Therefore, there exists a complete set of their simultaneous eigenfunctions. We will try to determine these eigenfunctions and also the eigenvalues of L_z and L^2 corresponding to these functions. Let ψ be a simultaneous eigenfunction of L_z and L^2 corresponding to the eigenvalues a and c, respectively. Since the eigenvalues of L_z can only be $m\hbar$, $m = 0, \pm 1, \pm 2, \ldots$, we write $m\hbar$ in place of a. Then

$$L_z\psi = m\hbar\psi, \quad m = 0, \pm 1, \pm 2, \ldots, \tag{9.19}$$

$$L^2\psi = c\psi. \tag{9.20}$$

Let us denote $L_x + iL_y$ by L_+. Then operating on (9.19) from the left by L_+, we obtain

$$L_+L_z\psi = m\hbar L_+\psi. \tag{9.21a}$$

Now

$$L_+L_z = (L_x + iL_y)L_z = L_xL_z + iL_yL_z, \tag{9.21b}$$

but

$$[L_z, L_x] = L_zL_x - L_xL_z = i\hbar L_y$$

and

$$[L_y, L_z] = L_yL_z - L_zL_y = i\hbar L_x.$$

Using these relations between the components of the angular momentum operator, (9.21b) reduces to

$$\begin{aligned} L_+L_z &= (L_zL_x - i\hbar L_y) + i(L_zL_y + i\hbar L_x) \\ &= L_z(L_x + iL_y) - \hbar(L_x + iL_y) \\ &= L_zL_+ - \hbar L_+. \end{aligned} \tag{9.22}$$

Substituting this expression for L_+L_z in (9.21a), we obtain

$$(L_zL_+ - \hbar L_+)\psi = m\hbar L_+\psi$$

or

$$L_z(L_+\psi) = (m\hbar + \hbar)(L_+\psi) = (m + 1)\hbar(L_+\psi).$$

We note that if $m\hbar$ is an eigenvalue of L_z corresponding to an eigenfunction ψ, then $(L_+\psi)$ is also an eigenfunction of L_z but corresponds to the eigenvalue $(m + 1)\hbar$, provided $L_+\psi \neq 0$. And so on. Thus, if ψ is an eigenfunction of L_z, then a repeated application of the operator L_+ generates new eigenfunctions

$$L_+\psi, L_+^2\psi, L_+^3\psi, \ldots$$

of the operator L_z corresponding to the eigenvalues

$$(m\hbar + \hbar), (m\hbar + 2\hbar), (m\hbar + 3\hbar), \ldots,$$

provided

$$L_+\psi \neq 0, \quad L_+^2\psi \neq 0, \quad L_+^3\psi \neq 0, \ldots.$$

Are these eigenfunctions of L_z also the eigenfunctions of L^2? To find the answer to this question, we operate (9.20) on the left by L_+. This gives

$$L_+L^2\psi = cL_+\psi,$$

or

$$L^2\left(L_+\psi\right) = c\left(L_+\psi\right),$$

because L^2 commutes with L_x as well as L_y, and consequently with their linear combination L_+. This shows that $(L_+\psi)$ is also an eigenfunction of L^2, of course corresponding to the same eigenvalue c as for the eigenfunction ψ. A repeated application of the operator L_+ shows that

$$L_+\psi, \ L_+^2\psi, \ L_+^3\psi, \ \ldots$$

are also eigenfunctions of L^2 but correspond to the *same* eigenvalue c.

To sum up, the functions

$$\psi, \ L_+\psi, \ L_+^2\psi, \ L_+^3\psi, \ \ldots$$

are simultaneous eigenfunctions of the operators L_z and L^2. These correspond to a sequence of eigenvalues

$$m\hbar, \ (m\hbar + \hbar), \ (m\hbar + 2\hbar), \ (m\hbar + 3\hbar), \ \ldots$$

of L_z but to the same eigenvalue c of L^2.

Similarly, a repeated application of the operator L_- $(= L_x - \mathrm{i}L_y)$ shows that

$$L_-\psi, \ L_-^2\psi, \ L_-^3\psi, \ \ldots$$

are simultaneous eigenfunctions of the operators L_z and L^2. These correspond to a sequence of eigenvalues

$$(m\hbar - \hbar), \ (m\hbar - 2\hbar), \ (m\hbar - 3\hbar), \ \ldots$$

of L_z but to the same eigenvalue c of L^2.

Hence, the functions

$$\ldots, \ L_-^3\psi, \ L_-^2\psi, \ L_-\psi, \ \psi, \ L_+\psi, \ L_+^2\psi, \ L_+^3\psi, \ \ldots$$

are *simultaneous eigenfunctions* of L_z and L^2 corresponding to a *series of eigenvalues of* L_z, i.e.

$$\ldots, \ (m\hbar - 3\hbar), \ (m\hbar - 2\hbar), \ (m\hbar - \hbar), \ m\hbar, \ (m\hbar + \hbar), \ (m\hbar + 2\hbar), \ (m\hbar + 3\hbar), \ \ldots \tag{9.23}$$

but to a *fixed eigenvalue* c of L^2. The successive terms of this series differ by $\hbar$. The series extends indefinitely in both directions unless $L_+^n\psi$ or $L_-^n\psi$ vanishes for some n. The operators L_+ and L_- are called *ladder operators.*

Naturally the next question is: will the series terminate? And if so, at one end or both? To find an answer, we note that

$$L^2 = L_x^2 + L_y^2 + L_z^2. \tag{9.13}$$

Taking the average values of the two sides, we obtain

$$\overline{L^2} = \overline{L_x^2 + L_y^2 + L_z^2},$$

or

$$\overline{L^2} = \overline{L_x^2} + \overline{L_y^2} + \overline{L_z^2},$$

or

$$\overline{L^2} - \overline{L_z^2} = \overline{L_x^2} + \overline{L_y^2}. \tag{9.24}$$

Problem

9.3. Show that for the Hermitian operator L_x, we must have

$$\overline{L_x^2} \geqslant 0.$$

As L_x and L_y are Hermitian operators, we must have

$$\overline{L_x^2} \geqslant 0,$$
$$\overline{L_y^2} \geqslant 0.$$

Therefore, by virtue of these relations, (9.24) yields

$$\overline{L^2} - \overline{L_x^2} \geqslant 0,$$

or

$$c - m^2\hbar^2 \geqslant 0,$$

or

$$m^2\hbar^2 \leqslant c.$$

This means that the square of any eigenvalue of L_z should not exceed the eigenvalue of L^2. Thus the series (9.23) for the eigenvalues of L_z *must terminate in each direction* because otherwise it would include terms whose squares are greater than c. Hence, the sequence is bounded above and below; the operator L_z has maximum and minimum eigenvalues. The states are therefore $(2\ell + 1)$-fold degenerate.

We now find the largest and the smallest members of the series. Let ψ_g be a simultaneous eigenfunction of L_z and L^2 corresponding to the largest eigenvalue $g\hbar$ of L_z; $g\hbar$ is the eigenvalue of the top rung. Then

$$L_z\psi_g = g\hbar\psi_g, \tag{9.25}$$

$$L^2\psi_g = c\psi_g. \tag{9.26}$$

Operating (9.25) on the left by L_+, we obtain

$$L_+\left(L_z\psi_g\right) = g\hbar\left(L_+\psi_g\right).$$

By virtue of (9.22), this gives

$$L_z\left(L_+\psi_g\right) = (g\hbar + \hbar)\left(L_+\psi_g\right). \tag{9.27}$$

This equation shows that $(g\hbar + \hbar)$ is also an eigenvalue of L_z unless $L_+\psi_g = 0$. But $(g\hbar + \hbar)$ cannot be an eigenvalue of L_z because its largest eigenvalue is $g\hbar$; no value can be greater than the largest value! Hence we must have

$$L_+\psi_g = 0.$$

Operating this equation on the left by L_-, we obtain

$$L_-L_+\psi_g = 0,$$

or

$$\left(L_x - \mathrm{i}L_y\right)\left(L_x + \mathrm{i}L_y\right)\psi_g = 0,$$

or

$$\left(L_x^2 + \mathrm{i}L_xL_y - \mathrm{i}L_yL_x + L_y^2\right)\psi_g = 0,$$

or

$$\left(L_x^2 + L_y^2 + L_z^2 - L_z^2 + \mathrm{i}\left[L_x, L_y\right]\right)\psi_g = 0,$$

or

$$\left(L_x^2 - L_z^2 + \mathrm{i}\left(\mathrm{i}\hbar L_z\right)\right)\psi_g = 0,$$

or

$$\left(L^2 - L_z^2 - \hbar L_z\right)\psi_g = 0,$$

or

$$\left(c - g^2\hbar^2 - \hbar^2 g\right)\psi_g = 0.$$

Since ψ_g is associated with the eigenvalue $g\hbar$ and is therefore not equal to zero, the above equation yields

$$c - g^2\hbar^2 - \hbar^2 g = 0,$$

or

$$c = g^2\hbar^2 + \hbar^2 g = g(g + 1)\hbar^2. \tag{9.28}$$

Similarly, it can be shown that if $b\hbar$ is the smallest eigenvalue of L_z, then we must have

$$c = b^2\hbar^2 - \hbar^2 b = b(b - 1)\hbar^2. \tag{9.29}$$

Comparing (9.28) and (9.29), we find

$$g(g + 1)\hbar^2 = b(b - 1)\hbar^2,$$

or

$$g^2 - b^2 + (g + b) = 0,$$

or

$$(g + b)(g - b + 1) = 0. \tag{9.30}$$

But $(g - b + 1) \neq 0$ because otherwise b will be equal to $g + 1$ so that $b\hbar$ will be equal to $g\hbar + \hbar$, i.e. the smallest value $b\hbar$ of the eigenvalue series of L_z will be greater than the largest value $g\hbar$ of the series! Hence, by virtue of (9.30), we must have $b = -g$ so that $b\hbar = -g\hbar$. That is, the smallest value of the series is just the negative of its largest value; the maximum and minimum vaues of the L_z eigenvalue series are $g\hbar$ and $-g\hbar$. Hence the eigenvalue spectrum of L_z extending from $-g\hbar$ to $g\hbar$ can be written as

$$-g\hbar,\ (-g\hbar + \hbar),\ (-g\hbar + 2\hbar),\ \ldots,\ (g\hbar - 2\hbar),\ (g\hbar - \hbar),\ g\hbar, \tag{9.31}$$

as, according to (9.23), the successive terms of the series differ by $\hbar$.

What values can $g\hbar$ assume? We know that the eigenvalues of L_z are $m\hbar$, where $m = 0, \pm1, \pm2, \ldots$, i.e. the eigenvalues of L_z are $0, \pm\hbar, \pm2\hbar, \ldots$. As $g\hbar$ is the largest eigenvalue of L_z, it should be either *zero* or a *positive integral multiple* of $\hbar$. Thus, we may write

$$g\hbar = \ell\hbar, \quad \ell = 0, 1, 2, \ldots \tag{9.32}$$

where ℓ is the orbital quantum number. Then the eigenvalue spectrum of L_z becomes

$$-\ell\hbar,\ (-\ell + 1)\hbar,\ (-\ell + 2)\hbar,\ \ldots,\ (\ell - 2)\hbar,\ (\ell - 1)\hbar,\ \ell\hbar. \tag{9.33}$$

That is, the eigenvalues of L_z range from $-\ell\hbar$ to $\ell\hbar$ in integral steps in $\hbar$. This may also be expressed as

$$L_z = m\hbar, \quad m = -\ell, -\ell + 1, \ldots, \ell. \tag{9.34}$$

The magnetic quantum number m takes on all values from $-\ell$ to ℓ in integral steps. The fact that m can assume the values between $-\ell$ and ℓ can be symbolically written as $-\ell \le m \le \ell$. Since the successive terms of the sequence (9.33) differ by $\hbar$, the quantum number m can take $(2\ell + 1)$ values.

The eigenvalues of L^2 are obtained from (9.28) and (9.32):

$$c = g\hbar(g\hbar + \hbar) = \ell(\ell + 1)\hbar^2, \quad \ell = 0, 1, 2, \ldots \tag{9.35}$$

To sum up, we have shown that the eigenvalues of L_z and L^2 corresponding to simultaneous eigenfunctions

$$\ldots, L_-^3\psi, L_-^2\psi, L_-\psi, \psi, L_+\psi, L_+^2\psi, L_+^3\psi, \ldots$$

are given by

$$L_z = m\hbar, \quad m = -\ell, -\ell + 1, \ldots, 1, \tag{9.34'}$$

$$L^2 = \ell(\ell + 1)\hbar^2, \quad \ell = 0, 1, 2, \ldots. \tag{9.35'}$$

Since no two components of the angular momentum operator **L** commute, a precise measurement of angular momentum is not possible. Hence the *angular momentum vector* **L** *is not an observable.* Its magnitude is $\sqrt{\ell(\ell + 1)}\,\hbar$ but its direction is never known. It is significant to note that although the orientation of the vector **L** is not known, it must be so inclined to the z-axis that its projection along this axis gives one of the possible eigenvalues of L_z, i.e., either zero or an integral multiple of $\hbar$. This is illustrated by the vector diagram drawn for $\ell = 2$ as shown in figure 9.2. For $\ell = 2$, we have $|L| = \sqrt{\ell(\ell + 1)}\,\hbar = \sqrt{6}\,\hbar$ and $L_z = -2\hbar, -\hbar, 0, \hbar, 2\hbar$, i.e. there are five

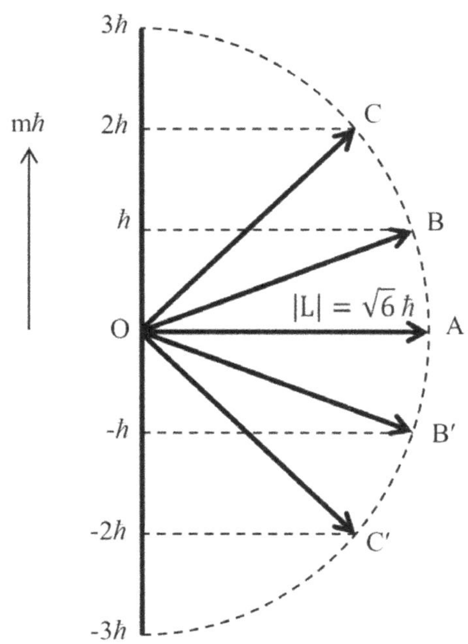

Figure 9.2. Allowed angular momenta for $\ell = 2$.

possible values of L_z. Draw a semi-circle of radius $\sqrt{6}\hbar$ with O as its centre. Take the diameter ZOZ' of the semi-circle as the z-axis. Draw radii OC', OB', OA, OB and OC such that their projections on the z-axis are $-2\hbar$, $-\hbar$, 0, $\hbar$, $2\hbar$, respectively. If **L** is aligned along one of these radii, L_z will have one of these allowed values, i.e. $-2\hbar$, $\hbar$, 0, $\hbar$, $2\hbar$. However, it should *not* be concluded that for a permissible value of L_z, the angular momentum vector **L** must be inclined *only* along one of these directions. For instance, if $L_z = \hbar$, we cannot say that **L** is aligned along OB. On the other hand, **L** may be anywhere on the surface of a cone whose semi-vertical angle is α (where $\cos\alpha = 1/\sqrt{6}$) and whose axis of length $\hbar$ coincides with the z-axis. This is so because the z-component of angular momentum would always be $\hbar$. Thus, even if the eigenvalues of L^2 and L_z are known, the direction of **L** is not determined. In general, for a given value of ℓ, the orbital angular momentum vector **L** will be making a particular angle $\alpha = \cos^{-1}(\ell/\sqrt{\ell(\ell+1)})$ with the z-axis so that it may be anywhere on a cone whose axis is the z-axis and whose semi-vertical angle is α. Since the values of L_z spread from $-\ell\hbar$ to $+\ell\hbar$ with successive values differing by $\hbar$, we will have a series of cones corresponding to appropriate positive and negative values of z, the vector **L** lying on one of these cones as shown for $\ell = 2$ in figure 9.3. It may be emphasised that as the direction of the z-axis was chosen arbitrarily, a component of the orbital angular momentum *in any direction* must be an integral multiple of $\hbar$. This is known as *spatial quantisation*.

In the classical limit, the quantum number ℓ must be very large, i.e. it should tend to infinity. Then $\cos\alpha \to 1$ or $\alpha \to 0$. But when the semi-vertical angle of the cone tends to zero, then the cone itself reduces to the z-axis. Therefore, in agreement with the classical concept, the vector lies along the z-axis and is fixed in space.

Figure 9.3 also shows the angular momentum vectors with $\ell = 2$ precessing about the z-axis in five possible states.

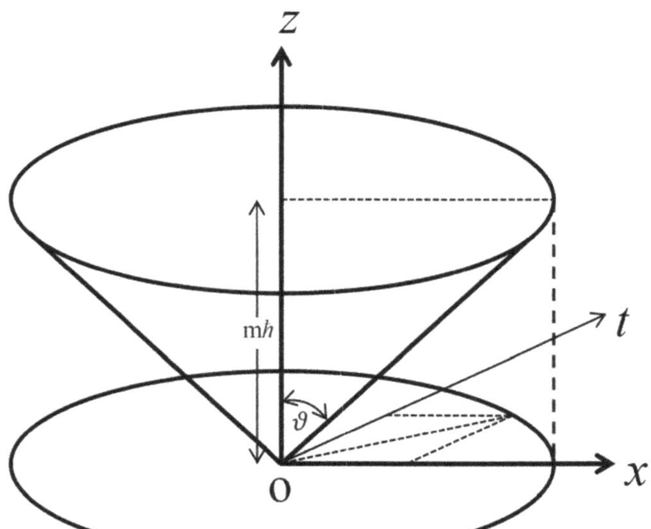

Figure 9.3. Cone for $\ell = 2$.

It is interesting to note that the wave functions $\psi_{n\ell m}$ for the stationary state of a hydrogen atom satisfy the following three differential equations:

$$H\psi_{n\ell m} = E_n\psi_{n\ell m},$$
$$L^2\psi_{n\ell m} = \ell(\ell + 1)\hbar^2\psi_{n\ell m},$$
$$L_z\psi_{n\ell m} = m\hbar\psi_{n\ell m}.$$

9.8 Normal Zeeman effect

It is well-known that the line spectrum of every element is characteristic of that element. It was observed by the Dutch physicist Zeeman (1865–1943) in 1896 that when an element emitting a line spectrum is placed in a magnetic field, then each one of its spectral lines is split into several lines. This is known as the *Zeeman effect*. This is produced by placing a source of light between the poles of a powerful electromagnet. If the field is homogeneous and strong enough, say, of the order of 20 000 gauss, each line when examined perpendicular to the direction of the magnetic field splits into three lines, the two new lines appearing symmetrically on either side of the original line. This is called the *normal Zeeman effect*. This could be explained by classical theory. If the field is weak, a spectral line is split into more than three lines, producing a very complex spectrum. This is called the *anomalous Zeeman effect*. Actually, there is nothing anomalous about it. It was named so because of the complexity of the spectrum *and* because of the fact that it could not be explained by the classical theory. The normal Zeeman effect was obtained by Zeeman and was explained by Lorentz using classical theory. They shared the Nobel Prize in physics in 1902.

Let us see how quantum mechanics explains the normal Zeeman effect. Keeping in view the results which have already been obtained for hydrogenic atoms and angular momentum, it is possible to compute the energy levels of a hydrogenic atom when it is placed in a uniform strong external magnetic field. We first formulate the Hamiltonian of a hydrogenic atom placed in a magnetic field. We know that in the absence of the magnetic field, the Hamiltonian operator H_0 of the *unperturbed hydrogenic atom* is given by

$$H_0 = T + V = -\frac{\hbar^2}{2m}\nabla^2 - \frac{Ze^2}{r}.$$

The Schrödinger wave equation for the unperturbed hydrogenic atom is

$$H_0\psi_{n\ell m} = E_{n_0}\psi_{n\ell m},$$

or

$$\left(-\frac{\hbar^2}{2m}\nabla^2 + V\right)\psi_{n\ell m} = E_{n_0}\psi_{n\ell m}.$$

The corresponding energy eigenvalues are given by

$$E_{n_0} = -\frac{Ze^{f4}\mu}{2n^2\hbar^2}, \quad n = 1, 2, 3, \ldots.$$

For the hydrogen atom, the formula reduces to

$$E_{n0} = -\frac{Ze^{f4}\mu}{2n^2\hbar^2}, \qquad n = 1, 2, 3, \ldots.$$

It is evident from the above equations that while an eigenfunction depends upon three quantum numbers n, ℓ and m, the corresponding energy depends only upon n. For a given value of ℓ, the quantum number m can take $(2\ell + 1)$ values between $-\ell$ and ℓ, the successive values differing by unity. Therefore, for each value of ℓ, the energy eigenvalue is $(2\ell + 1)$-fold degenerate. For instance, for a given n, for $\ell = 1$, the magnetic quantum number m can take three values -1, 0, 1. Therefore, the functions $\psi_{n1,-1}$, ψ_{n10}, ψ_{n11} all correspond to the same energy eigenvalue E_{n0}. This eigenvalue is three-fold degenerate. What happens when a uniform strong magnetic field is applied to the unperturbed hydrogenic atom? There is an interaction of electronic current (produced due to the motion of the charge on the electron) with a uniform strong external magnetic field. This gives rise to the Hamiltonian

$$H' = -\boldsymbol{\mu} \cdot \mathbf{B}$$

where $\boldsymbol{\mu}$ is the *magnetic moment due to the electronic current* and $\mathbf{B}$ is the uniform external magnetic field. This magnetic moment $\boldsymbol{\mu}$ can be calculated *classically* and is given by

$$\boldsymbol{\mu} = -\frac{e}{2\mu c}\,\mathbf{L},$$

where $\mathbf{L}$ is the angular momentum of the circulating electric charge, $-e$ is the charge on the electron and μ is the reduced mass of the electron (This μ has nothing to do with the magnetic moment $\boldsymbol{\mu}$). Eliminating the magnetic moment $\boldsymbol{\mu}$ from the last two equations, we obtain

$$H' = \frac{e}{2\mu c}\mathbf{L} \cdot \mathbf{B}.$$

If the z-axis is taken along the direction of the magnetic field $\mathbf{B}$, we have

$$H' = \frac{e}{2\mu c}L_z B.$$

In going from classical to quantum mechanics, we know that in spherical polar coordinates, the operator L_z corresponding to the z-component of angular momentum is given by

$$L_z \equiv -\,\mathrm{i}\hbar\frac{\partial}{\partial\phi}.$$

Therefore, the equation for H' takes the form

$$H' = \frac{e}{2\mu c}B\left(-\mathrm{i}\hbar\frac{\partial}{\partial\phi}\right).$$

Hence, if the magnetic field is applied in the direction of the z-axis, the Hamiltonian operator of the system becomes

$$H = H_0 + H' = \left[-\frac{\hbar^2}{2m}\nabla^2 + V + \frac{e}{2\mu c}B\left(-i\hbar\frac{\partial}{\partial\phi} \right) \right].$$

Are the eigenfunctions $\psi_{n\ell m}$ of H_0 also the eigenfunctions of the Hamiltonian operator H? We notice that

$$\begin{aligned} H\psi_{n\ell m} &= \left(H_0 + \frac{e}{2\mu c}L_z B \right)\psi_{n\ell m} \\ &= H_0\psi_{n\ell m} + \frac{e}{2\mu c}L_z B\psi_{n\ell m} \\ &= E_{n0}\psi_{n\ell m} + \frac{e}{2\mu c}B\left(-i\hbar\frac{\partial}{\partial\phi} \right)\psi_{n\ell m} \\ &= \left(E_{n0} + \frac{eBm\hbar}{2\mu c} \right)\psi_{n\ell m}. \end{aligned}$$

The above equation shows that $\psi_{n\ell m}$ is also an eigenfunction of the Hamiltonian operator H of the system placed in a uniform strong magnetic field but corresponds to a different eigenvalue $(E_{n0} + eBm\hbar/2\mu c)$. Thus, the energy levels E of the Hamiltonian operator H are given by

$$E = E_{n0} + \frac{eBm\hbar}{2\mu c}.$$

Since m can take $(2\ell + 1)$ integral values from $-\ell$ to ℓ, with a difference of unity between adjacent values, the above equation shows that for a given ℓ, the magnetic field splits each energy level into $2\ell + 1$ levels. The difference between adjacent energy levels is $E' = eB\hbar/2\mu c$.

For $n = 1, \ell = 0$, i.e. for an 1s state, the quantum number m is also zero. Therefore, there is no splitting of the energy level due to the orbital magnetic moment.

For $n = 2, \ell = 1$, i.e. for a 2p state, the quantum number m can take the values -1, 0, 1. Therefore, corresponding to the eigenfunctions $\psi_{21,-1}$, ψ_{210}, ψ_{211}, the energy eigenvalues are

$$E_{21,-1} = E_{20} - \frac{eB\hbar}{2\mu c},$$

$$E_{210} = E_{20},$$

$$E_{211} = E_{20} + \frac{eB\hbar}{2\mu c}.$$

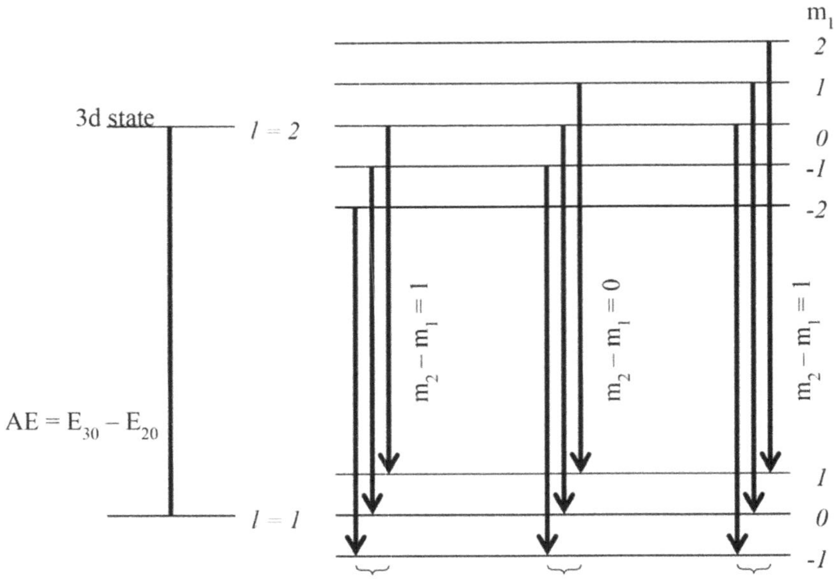

Figure 9.4. Schemetic splitting of spectral lines by a strong uniform magnetic field.

Thus, in this case, each energy level is split up into three. The new energy levels are equally spaced above and below the original energy level. This is shown in figure 9.4 for $n = 2$, $\ell = 1$, $m = -1, 0, 1$.

We note that under the influence of a uniform strong external magnetic field, quantum mechanics determines which energy levels are produced and consequently which spectral lines are emitted as electrons move from one energy shell to another in their orbits of atomic nuclei.

For $n = 3$, $\ell = 2$, i.e. for a 3d state, the quantum number m can take the values -2, -1, 0, 1, 2. Therefore, corresponding to the eigenfunctions $\psi_{32,-2}$, $\psi_{32,-1}$, ψ_{320}, ψ_{321}, ψ_{322}, the energy eigenvalues are

$$E_{32,-2} = E_{30} - \frac{eB\hbar}{\mu c},$$

$$E_{32,-1} = E_{30} - \frac{eB\hbar}{2\mu c},$$

$$E_{320} = E_{30},$$

$$E_{321} = E_{30} + \frac{eB\hbar}{2\mu c},$$

$$E_{322} = E_{30} + \frac{eB\hbar}{\mu c}.$$

Thus, each energy level is split up into five components. Hence, except for $\ell = 0$, the application of the magnetic field removes the degeneracy. The energy levels corresponding to $n = 2$, $\ell = 1$ and $n = 3$, $\ell = 2$ are shown in figure 9.4. The energy difference $\Delta E = E_{30} - E_{20}$. The adjacent energy levels differ by $e\hbar B/2\mu c$. The transitions take place according to a selection rule which states that the spectral lines are emitted *only* when $\Delta m = -1, 0, 1$. Hence, nine transitions occur from the 3d to the 2p state in the form of three lines. This is in agreement with experiments. It may be emphasised that it is owing to the significant role that the quantum number m has played in the theory of the normal Zeeman effect that it is called the magnetic quantum number.

Thus, we see that the normal Zeeman effect can be explained in quantum mechanics in a very simple and logical manner.

It may also be mentioned that in space science, the Zeeman effect is used to measure the magnetic field of the sun and of other stars. This is also used to determine the magnetic field in laboratory plasmas.

9.9 General theory of angular momentum

We have seen that the orbital quantum number ℓ can be either zero or have positive integral values. However, many experimental results can be explained only if angular momentum quantum numbers can also have positive half odd-integral values. The above analysis which gives only non-negative integral values for the angular momentum quantum number should therefore be only a special case of the general theory of angular momentum. We will now show that this is so. We started this chapter with the classical definition of angular momentum, representing the observables by Hermitian operators and determining the various components of the angular momentum operator as well as the commutation relations between these components. We also determined the eigenvalues of the operator L_z by using the expression for L_z *and* the constraint that the eigenfunctions of this operator should be single-valued. To generalise the theory of angular momentum, we do not start with the classical definition of angular momentum. Therefore, we cannot find an expression for its z-component and cannot directly calculate its eigenvalues. We start here with a *definition* of angular momentum operator **J** as that whose components J_x, J_y, J_z satisfy the commuatation relations

$$\left[J_x, J_y\right] = i\hbar J_z, \tag{9.36a}$$

$$\left[J_y, J_z\right] = i\hbar J_x, \tag{9.36b}$$

$$\left[J_z, J_x\right] = i\hbar J_y, \tag{9.36c}$$

We proceed further, remembering all the time that we *cannot directly calculate the eigenvalues of* J_z as *we do not have any expression for this operator*. The removal of this constraint on the eigenvalues of the z-component of angular momentum gives additional values to the angular momentum quantum number.

Problem

9.4. Show that the operators J_z and J^2 $(= J_x^2 + J_y^2 + J_z^2)$ mutually commute.

Since the operators J_z and J^2 mutually commute, there must exist a set of their simultaneous eigenfunctions. We will determine these eigenfunctions and also the eigenvalues of J_z and J^2 corresponding to these eigenfunctions. Let ψ be a simultaneous eigenfunction of J_z and J^2 corresponding to the eigenvalues $a\hbar$ and c, respectively. Then

$$J_z\psi = a\hbar\psi, \tag{9.37}$$

$$J^2\psi = c\psi. \tag{9.38}$$

Let us denote $J_x + iJ_y$ by J_+. Then operating on (9.37) from the left by J_+, we obtain

$$J_+J_z\psi = a\hbar J_+\psi. \tag{9.39a}$$

Now

$$J_+J_z = \left(J_x + iJ_y\right)J_z = J_xJ_z + iJ_yJ_z. \tag{9.39b}$$

But from (9.36*c*), we have

$$\left[J_z, J_x\right] = i\hbar J_y,$$

or

$$J_zJ_x - J_xJ_z = i\hbar J_y,$$

or

$$J_xJ_z = J_zJ_x - i\hbar J_y. \tag{9.39c}$$

Similarly, from (9.36*b*), we have

$$J_yJ_z = J_zJ_y + i\hbar J_x. \tag{9.39d}$$

Substituting these expressions for J_xJ_z and J_yJ_z in (9.39*c*), we obtain

$$\begin{aligned} J_+J_z &= (J_zJ_x - i\hbar J_y) + i(J_zJ_y + i\hbar J_x) \\ &= J_z(J_x + iJ_y) - \hbar(J_x + iJ_y) \\ &= J_zJ_+ - \hbar J_+. \end{aligned} \tag{9.39e}$$

Substituting the expression for J_+J_z from (9.39*e*) in (9.39*a*), we obtain

$$\left(J_zJ_+ - \hbar J_+\right)\psi = a\hbar J_+\psi$$

or

$$J_z\left(J_+\psi\right) = (a\hbar + \hbar)\left(J_+\psi\right) = (a + 1)\hbar\left(J_+\psi\right).$$

This shows that $(J_+\psi)$ is also an eigenfunction of J_z but corresponds to the eigenvalue $(a\hbar + \hbar)$, provided $J_+\psi \neq 0$. And so on. Thus, if ψ is an eigenfunction of J_z, then a repeated application of the operator J_+ will generate new eigenfunctions

$$J_+\psi,\ J_+^2\psi,\ J_+^3\psi,\ \ldots$$

of the operator J_z corresponding to the eigenvalues

$$(a\hbar + \hbar),\ (a\hbar + 2\hbar),\ (a\hbar + 3\hbar),\ \ldots.$$

Are these eigenfunctions of J_z also the eigenfunctions of J^2? To find the answer to this question, we operate (9.38) on the left by J_+. This gives

$$J_+J^2\psi = cJ_+\psi,$$

or

$$J^2\left(J_+\psi\right) = c\left(J_+\psi\right),$$

because J^2 commutes with J_x as well as J_y, and consequently with their linear combination J_+. This shows that $(J_+\psi)$ is also an eigenfunction of J^2, of course corresponding to the same eigenvalue c as for the eigenfunction ψ. A repeated application of the operator J_+ shows that

$$J_+\psi,\ J_+^2\psi,\ J_+^3\psi,\ \ldots$$

are also eigenfunctions of J^2 but correspond to the *same* eigenvalue c.

To sum up, the functions

$$\psi,\ J_+\psi,\ J_+^2\psi,\ J_+^3\psi,\ \ldots$$

are simultaneous eigenfunctions of the operators J_z and J^2. These correspond to a sequence of eigenvalues

$$a\hbar,\ (a\hbar + \hbar),\ (a\hbar + 2\hbar),\ (a\hbar + 3\hbar),\ \ldots$$

of J_z but to the same eigenvalue c of J^2.

Similarly, a repeated application of the operator $J_-\,(= J_x - iJ_y)$ shows that

$$J_-\psi,\ J_-^2\psi,\ J_-^3\psi,\ \ldots$$

are simultaneous eigenfunctions of the operators J_z and J^2. These correspond to a sequence of eigenvalues

$$(a\hbar - \hbar),\ (a\hbar - 2\hbar),\ (a\hbar - 3\hbar),\ \ldots$$

of J_z but to the same eigenvalue c of J^2.

Hence the functions

$$\ldots, J_-^3\psi, J_-^2\psi, J_-\psi, \psi, J_+\psi, J_+^2\psi, J_+^3\psi, \ldots$$

are simultaneous eigenfunctions of J_z and J^2 corresponding to a *series of eigenvalues of* J_z, i.e.

$$\ldots, (a\hbar - 2\hbar), (a\hbar - \hbar), a\hbar, (a\hbar + \hbar), (a\hbar + 2\hbar), \ldots \tag{9.40}$$

but to a *fixed eigenvalue c of* J^2. The successive terms of this series differ by $\hbar$. The series extends indefinitely in both directions unless $J_+^n\psi$ or $J_-^n\psi$ vanishes for some n.

Naturally the next question is: will the series terminate? And if so, at one or both ends? To find an answer, we note that

$$J^2 = J_x^2 + J_y^2 + J_z^2.$$

Taking the average values of the two sides, we obtain

$$\overline{J^2} = \overline{J_x^2 + J_y^2 + J_z^2},$$

or

$$\overline{J^2} = \overline{J_x^2} + \overline{J_y^2} + \overline{J_z^2},$$

or

$$\overline{J^2} - \overline{J_z^2} = \overline{J_x^2} + \overline{J_y^2}. \tag{9.41}$$

Problem

9.5. Show that for the Hermitian operator J_x, we must have

$$\overline{J_x^2} \geqslant 0.$$

But as J_x and J_y are Hermitian, we must have

$$\overline{J_x^2} \geqslant 0,$$

$$\overline{J_y^2} \geqslant 0.$$

Therefore, by virtue of these relations, (9.41) yields

$$\overline{J^2} - \overline{J_z^2} \geqslant 0,$$

or

$$c - a^2\hbar^2 \geqslant 0,$$

or

$$a^2\hbar^2 \leqslant c.$$

This means that the square of any eigenvalue of J_z should not exceed the eigenvalue of J^2. Thus the series (9.40) for the eigenvalues of J_z *must terminate in each direction* because otherwise it would include terms whose squares are greater than c. Hence, the sequence is bounded above and below; the operator J_z has maximum and minimum eigenvalues.

We now find the largest and the smallest members of the series. Let ψ_g be a simultaneous eigenfunction of J_z and J^2 corresponding to the largest eigenvalue $g\hbar$ of J_z; $g\hbar$ is the eigenvalue of the top rung. Then

$$J_z\psi_g = g\hbar\psi_g, \tag{9.42}$$

$$J^2\psi_g = c\psi_g. \tag{9.43}$$

Operating (9.42) on the left by J_+, we obtain

$$J_+J_z\psi_g = g\hbar J_+\psi_g.$$

By virtue of (9.39*c*), this gives

$$\left(J_zJ_+ - \hbar J_+\right)\psi_g = g\hbar J_+\psi_g,$$

or

$$J_z\left(J_+\psi_g\right) = (g\hbar + \hbar)\left(J_+\psi_g\right). \tag{9.44}$$

This equation shows that $(g\hbar + \hbar)$ is also an eigenvalue of J_z unless $J_+\psi_g = 0$. But $(g\hbar + \hbar)$ cannot be an eigenvalue of J_z because its largest eigenvalue is $g\hbar$; no value can be greater than the largest value! Hence, we must have

$$J_+\psi_g = 0.$$

Operating this equation on the left by J_-, we obtain

$$J_-J_+\psi_g = 0,$$

or

$$\left(J_x - \mathrm{i}J_y\right)\left(J_x + \mathrm{i}J_y\right)\psi_g = 0,$$

or

$$\left(J^2 - J_z^2 - \hbar J_z\right)\psi_g = 0,$$

or

$$\left(c - g^2\hbar^2 - \hbar^2 g\right)\psi_g = 0.$$

Since ψ_g is associated with the eigenvalue $g\hbar$ and is therefore not equal to zero, the above equation yields

$$c - g^2\hbar^2 - \hbar^2 g = 0,$$

or

$$c = g^2\hbar^2 + \hbar^2 g = g(g+1)\hbar^2. \tag{9.45}$$

Similarly, it can be shown that if $b\hbar$ is the smallest eigenvalue of J_z, then we must have

$$c = b^2\hbar^2 - \hbar^2 b = b(b-1)\hbar^2. \tag{9.46}$$

Comparing (9.45) and (9.46), we obtain

$$g(g+1)\hbar^2 = b(b-1)\hbar^2,$$

or

$$g^2 - b^2 + (g+b) = 0, \quad \text{because } \hbar \neq 0,$$

or

$$(g+b)(g-b+1) = 0. \tag{9.47}$$

But $(g - b + 1) \neq 0$ because otherwise b will be equal to $g + 1$ so that $b\hbar$ will be equal to $g\hbar + \hbar$, i.e. the smallest value $b\hbar$ of the eigenvalue series of J_z will be greater than the largest value $g\hbar$ of the series! Hence, by virtue of (9.47), we must have $b = -g$ so that $b\hbar = -g\hbar$. That is, the smallest value of the series is just the negative of its largest value; the maximum and minimum vaues of the J_z eigenvalue series are $g\hbar$ and $-g\hbar$. Hence, the eigenvalue spectrum of J_z extending from $-g\hbar$ to $g\hbar$ can be written as

$$-g\hbar, (-g\hbar + \hbar), (-g\hbar + 2\hbar), \ldots, (g\hbar - 2\hbar), (g\hbar - \hbar), g\hbar, \tag{9.48}$$

as, according to (9.40), the successive terms of the series differ by $\hbar$. Consequently, the difference between the highest and lowest eigenvalue must be either zero or a positive integer. That is:

$$g\hbar - (-g\hbar) = 2j\hbar, \quad j = 0, \frac{1}{2}, 1, \frac{3}{2}, 2, \ldots,$$

or

$$g\hbar = j\hbar, \quad j = 0, \frac{1}{2}, 1, \frac{3}{2}, 2, \ldots. \tag{9.49}$$

j is called the *angular momentum quantum number* and can take non-negative integral values or half odd-integral positive values. The lowest and highest eigenvalues of

J_z are $-j\hbar$ and $j\hbar$, where $j = 0, 1/2, 1, 3/2, 2,\ldots$. As successive terms differ by $\hbar$, we have two series of eigenvalues:

$$-\ell\hbar, -(\ell + 1)\hbar, -(\ell + 2)\hbar, \ldots, (\ell - 2)\hbar, (\ell - 1)\hbar, \ell\hbar, \tag{9.50a}$$

where $\ell = 0, 1, 2,\ldots$ and

$$-s\hbar, -(s + 1)\hbar, -(s + 2)\hbar, \ldots, (s - 2)\hbar, (s - 1)\hbar, s\hbar, \tag{9.50b}$$

where $s = 1/2, 3/2,\ldots$. Note that for non-negative integral values of the quantum number, we have written ℓ in place of j. Similarly for positive half odd-integral values, we have written s in place of j.

Thus, we may write

$$J_z\psi(jm) = m\hbar\psi(jm), \quad j = 0, 1/2, 1, 3/2, 2, \ldots. \tag{9.51}$$

The eigenvalues of J^2 are obtained from (9.45):

$$c = g\hbar(g\hbar + \hbar),$$

or

$$c = j(j + 1)\hbar^2, \quad j = 0, \frac{1}{2}, 1, \frac{3}{2}, 2, \ldots, \tag{9.52}$$

where by virtue of (9.49), we replaced g by j. Hence, the eigenvalue of J is $\sqrt{j(j + 1)}\,\hbar$. Thus positive half odd-integral values of angular momentum quantum number emerge as a natural consequence of the general theory of angular momentum.

To sum up: if, instead of defining angular momentum in quantum mechanics by using the classical formula $\mathbf{L} = \mathbf{r} \wedge \mathbf{p}$ and then interpreting $\mathbf{L}, \mathbf{r}, \mathbf{p}$ as certain operators, we use the following commutation relations

$$\left[J_x, J_y\right] = i\hbar J_z, \tag{9.36a'}$$

$$\left[J_y, J_z\right] = i\hbar J_x, \tag{9.36b'}$$

$$\left[J_z, J_x\right] = i\hbar J_y, \tag{9.36c'}$$

between the components J_x, J_y, J_z of the angular momentum operator $\mathbf{J}$ as its definition, then we have seen that the angular momentum j can take not only zero and even integral values but also odd half-integral values. This is a very important result and is used to explain significant experimental data.

The eigenvalue spectra of J_z and J^2 corresponding to simultaneous eigenfunctions are

$$-j\hbar, (j + 1)\hbar, -(j + 2)\hbar, \ldots, (j - 2)\hbar, (j - 1)\hbar, j\hbar, \tag{9.53a}$$

and

$$j(j+1)h^2, \tag{9.53b}$$

$$\text{with } j = 0, \frac{1}{2}, \frac{3}{2}, 2, \ldots$$

One might obtain the idea that for a *given* system, j can have the values 0, 1/2, 1, 3/2, ... This is *not* true. A given system can have *either* integral *or* half odd-integral j.

9.10 Characteristics of ladder operators

In section 9.7, we have seen that in the case of orbital angular momentum, the operators L_+ and L_- defined by

$$L_+ = L_x + iL_y,$$

$$L_- = L_x - iL_y,$$

where L_x and L_y are the x and y components of the orbital angular momentum operator **L**, are called ladder operators. In the general theory of angular momentum, J_+ and J_- defined by

$$J_+ = J_x + iJ_y,$$

$$J_- = J_x - iJ_y,$$

are called *ladder operators.*

We will now determine some characteristics of ladder operators J_+ and J_-. We have shown that if $\psi(jm)$ is a simultaneous eigenfunction of J^2 and J_z with eigenvalues $j(j+1)\hbar^2$ and $m\hbar$, $m = -j, -j+1, \ldots, j-1, j$, then $J_+\psi(jm)$ is also a simultaneous eigenfunction of these operators but with eigenvalues $j(j+1)\hbar^2$ and $(m+1)\hbar$, respectively. That is, if

$$J_z\psi(jm) = m\hbar\psi(jm),$$

$$J^2\psi(jm) = j(j+1)\hbar^2\psi(jm),$$

then

$$J_z\left[J_+\psi(jm)\right] = (m+1)\hbar\left[J_+\psi(jm)\right],$$

$$J^2\left[J_+\psi(jm)\right] = j(j+1)\hbar^2\left[J_+\psi(jm)\right].$$

As J_+ is a ladder operator, we may write

$$J_+\psi(j, m) = c_+\psi(j, m+1) \tag{9.54a}$$

where c_+ is a complex number.

Similarly, with c_- as a complex number, we have

$$J_-\psi(j, m) = c_-\psi(j, m-1) \tag{9.54b}$$

To determine the value of c_+, we proceed as follows. We note that

$$
\begin{aligned}
J_-J_+ &= \left(J_x - \mathrm{i}J_y\right)\left(J_x + \mathrm{i}J_y\right) \\
&= J_x^2 + \mathrm{i}J_xJ_y - \mathrm{i}J_yJ_x + J_y^2 \\
&= J^2 - J_z^2 + \mathrm{i}\left[J_x, J_y\right] \\
&= J^2 - J_z^2 + \mathrm{i}.\ \mathrm{i}\hbar J_z \\
&= J^2 - J_z^2 - \hbar J_z. \qquad (9.55)
\end{aligned}
$$

Now, by virtue of (9.54*a*), the integral

$$
\begin{aligned}
\int \psi^*(jm) J_-\left\{J_+\psi(jm)\right\} \mathrm{d}\tau &= \int \psi^*(jm) J_-\left\{c_+\psi(j, m+1)\right\} \mathrm{d}\tau \\
&= c_+ \int \psi^*\left(jm\right) J_+^\dagger \psi(j, m+1) \mathrm{d}\tau,
\end{aligned}
$$

as J_- is the Hermitian conjugate of J_+. By using the definition of the Hermitian conjugate of an operator for the integral on the right-hand side of the above equation, we obtain

$$
\begin{aligned}
\int \psi^*(jm) J_-\left\{J_+\psi(jm)\right\} \mathrm{d}\tau &= c_+ \int \psi(j, m+1) J_+^* \psi^*(jm) \mathrm{d}\tau \\
&= c_+ \int \psi(j, m+1) c_+^* \psi^*(j, m+1) \mathrm{d}\tau \\
&= c_+^* c_+ \int \psi(j, m+1) \psi^*(j, m+1) \mathrm{d}\tau \\
&= c_+^* c_+ = \left|c_+\right|^2. \qquad (9.56)
\end{aligned}
$$

Also

$$
\begin{aligned}
\int \psi^*(jm) J_-\{J_+\psi(jm)\} \mathrm{d}\tau &= \int \psi^*(jm)\left(J^2 - J_z^2 - \hbar J_z\right)\psi(jm) \mathrm{d}\tau \\
&= \int \psi^*(jm)\left[j(j+1)\hbar^2 - m^2\hbar^2 - m\hbar^2\right]\psi(jm) \mathrm{d}\tau \\
&= [j(j+1) - m(m+1)]\hbar^2. \qquad (9.57)
\end{aligned}
$$

Comparing (9.56) and (9.57), we obtain

$$
\left|c_+\right|^2 = [j(j+1) - m(m+1)]\hbar^2.
$$

If we take the physically unimportant phase as zero and choose c_+ as positive, then we obtain

$$
c_+ = \sqrt{(j(j+1) - m(m+1)}\,\hbar \qquad (9.58)
$$

Equation (9.54*a*) therefore yields:

$$J_{+}\psi(j, m) = \sqrt{(j(j+1) - m(m+1)}\,\hbar\psi(j, m+1). \tag{9.59}$$

Similarly, it can be shown that

$$J_{-}\psi(j, m) = \sqrt{(j(j+1) - m(m-1)}\,\hbar\psi(j, m-1). \tag{9.60}$$

Equations (9.59) and (9.60) show that the ladder operators J_{+}, and J_{-}, respectively, raise and lower the magnetic quantum number in the wave function ψ by unity and multiply it by $\sqrt{(j(j+1) - m(m+1)}\,\hbar$ and $\sqrt{(j(j+1) - m(m-1)}\,\hbar$.

9.11 Electron spin

It has been stated that Niels Bohr was one of the great men of physics who contributed immensely to changing our mode of thought in the twentieth century. His understanding of the subject and its philosophical aspects, his passion for critical analysis of problems and at the same time his affectionate approach at a personal level made him a source of inspiration for all visitors to his Institute at Copenhagen. Heisenberg remarked that 'Bohr's influence on physics and physicists of our century was stronger than that of anyone else, even that of Albert Einstein'. In 1913, he made the most significant contribution of his scientific career. He explained the formation of various spectral series of the hydrogen atom with his quantum theory which was based on the following postulates.

(1) Within the atom, an electron can revolve in certain discrete circular orbits and during its motion in any one of these orbits, it does not radiate energy. These orbits are called *stationary orbits*. Classically, the electron, a charged particle, must radiate energy while being accelerated due to the fact that it is revolving in an orbit. Instead of continuing its motion in that orbit, it should loose energy, precess and ultimately fall into the nucleus. The bold assumption of Bohr forbids the electron from exhibiting its classical characteristics.

(2) The electron can move from one stationary orbit to another by a quantum jump only, losing or gaining energy equal to $h\nu$, where h is Planck's constant and ν is the frequency of radiation emitted or absorbed. Thus if the electron makes a transition from an orbit of higher energy E_i to another orbit of lower energy E_f, then the energy emitted during the process is $E_i - E_f = h\nu$.

(3) The angular momentum of the electron is an integral multiple of $\hbar$, where $\hbar = h/2\pi$ and h is Planck's constant. Therefore, if m is the mass, v is the velocity and r the distance of the electron from the nucleus, then we may write $mvr = nh$ where n is an integer.

By virtue of these postulates, he was able to explain the experimental data but could not provide any justification for violation of classical laws. Ehrenfest, for whom Einstein had great regard and respect, remarked about Bohr's approach: 'If this is the way to reach the goal, I must give up doing physics'. An interesting anecdote related by Weisskopf may also be recalled. It was about 1914 when, during a walk in the neighbourhood of Zurich, two distinguished physicists, Laue and Stern, made a vow that they would give up physics if and when those newfangled ideas by Bohr about the hydrogen atom should turn out to be correct. Fortunately, they did not keep their vow! Laue won the Nobel Prize in physics in 1914 while Stern won the same prize in 1943. But, in general, the physics community appreciated Bohr's theory and he was awarded the Nobel Prize for it in 1922.

It was found with improved spectrographs that in some of the spectral series the so-called 'single lines' had a complex structure and were actually multiplets. To explain this phenomenon, known as *fine structure*, Sommerfeld generalised Bohr's ideas, noting that the orbit of an electron revolving around the nucleus under the inverse square law is an ellipse rather than a circle. But calculations by Sommerfeld showed that the elliptic orbits give the same formula for the energy of the system and consequently the same energy levels and hence were not able to explain the complex nature of the spectrum. Sommerfeld then attributed the complex structure of the hydrogen spectrum to relativistic effects and was able to obtain a formula which gives the correct description of the fine structure of the 'lines'. The spectra of alkali metals, which have a single valence electron, could also be explained by the Sommerfeld formula. For the atoms of alkali metals, the orbits of the outermost electrons are nearly circular and quite far from the nucleus. These electrons are, therefore, moving at relatively low speeds and the variation of their speeds during their motion is small. Thus, on physical grounds, the relativistic effects are negligible and the Sommerfeld formula *should not* have explained the alkali spectra. The fact that it did explain the spectra indicated that the true mechanism of the fine structure phenomenon was not really understood. In addition, it was found that the experimental data required half-integral quantum numbers for their explanation.

To explain this fine structure, in 1926 two young Dutchmen, Uhlenbeck, who had a master's degree in physics, and Goudsmit, a graduate student, proposed that an electron pictured classically as a small sphere not only revolves around the nucleus but also rotates about an intrinsic axis similar to the Earth–Sun system. Therefore, in addition to orbital angular momentum, it possesses angular momentum due to its rotation about that axis. The latter quantity was called *intrinsic angular momentum*, *spin angular momentum* or simply *spin*. It is interesting to note that they wrote up their results for their adviser Ehrenfest. He asked them to seek the advice of Lorentz who, when informed, told these young men that a classical model for the spin of the electron required that it must be spinning with a speed on the surface about ten times the speed of light. When Uhlenbeck and Goudsmit informed their adviser Ehrenfest about the remarks of that great mathematician, he told them that he had already submitted their paper for publication. He, however, asked them not to worry as they were young enough to afford such stupidity! In any case, with this hypothesis, they were able to interpret one-electron spectra correctly. But by this time Schrödinger had formulated his differential equation. The solution of this differential equation for the hydrogen atom gives exactly the same energy levels and the same series of lines as those obtained by Bohr's theory, although in a more systematic and logical way. Pauli, thereupon, suggested that the idea of spin should be introduced in quantum mechanics as an ad hoc hypothesis. Since the classical picture of spin did not help, in order to obtain agreement with experimental results he introduced spin as an intrinsic property of an electron, defined by the quantum mechanical operators S_x, S_y, S_z satisfying the equations

$$\left[S_x, S_y\right] = \mathrm{i}\hbar S_z, \tag{9.61a}$$

$$\left[S_y, S_z\right] = \mathrm{i}\hbar S_x, \tag{9.61b}$$

$$\left[S_z, S_x\right] = \mathrm{i}\hbar S_y, \tag{9.61c}$$

such that the component of spin vector **S** in any direction can have one of the two values: $+\hbar/2$ and $-\hbar/2$. This hypothesis has proved to be correct. Spin is now regarded as one of the intrinsic properties of an elementary particle. He assumed further that the magnetic moment $\boldsymbol{\mu}$ of the electron due to spin was given by

$$\boldsymbol{\mu} = -\frac{e}{mc}\mathbf{S}, \tag{9.62}$$

where m is the mass of the electron, $-e$ is the charge on it and c is the velocity of light in a vacuum. Therefore, the component of $\boldsymbol{\mu}$ along any axis can only have two values: $e\hbar/2mc$ and $-e\hbar/2mc$. The quantity $e\hbar/2mc$ is called a *Bohr magnetron*. Since the component of spin vector **S** along *any* direction is $\pm\hbar/2$, we may write

$$S_x = \pm\frac{1}{2}\hbar, \quad S_y = \pm\frac{1}{2}\hbar, \quad S_z = \pm\frac{1}{2}\hbar. \tag{9.63}$$

The measurement of S_x, S_y or S_z should yield one of the two values $\pm\hbar/2$. As S_x, S_y and S_z do not commute, a simultaneous precise measurement of the eigenvalues of any two of these operators is impossible. Conventionally, we consider the z-component of spin: it is equal to $\hbar/2$ or $-\hbar/2$. When it is $\hbar/2$, the particle is said to be in a *spin up* position. For $S_z = -\hbar/2$, it is said to be in a *spin down* position. It is important to remember that in a spin up position, the spin (angular momentum) vector is actually inclined at an angle of $\cos^{-1}(1/\sqrt{3}) \approx 55°$ to the z-axis. From (9.63), we have

$$S_x^2 = S_y^2 = S_z^2 = \frac{1}{4}\hbar^2. \tag{9.64}$$

Therefore the eigenvalue of S^2, the square of the spin operator, is given by

$$S^2 = S_x^2 + S_y^2 + S_z^2 = \frac{3}{4}\hbar^2. \tag{9.65}$$

Hence the eigenvalue of S is

$$S = \sqrt{\frac{3}{4}}\hbar = \sqrt{\frac{1}{2}\left(\frac{1}{2}+1\right)}\hbar. \tag{9.66}$$

For $s = 1/2$, this is consistent with the formula

$$S = \sqrt{s(s+1)}\,\hbar.$$

s is the *spin quantum number*, generally called *spin*. Thus we say that the spin of the electron is 1/2. In terms of s, we may write

$$S_z = m\hbar, \qquad m = -s, \qquad -s+1, \ldots, s \tag{9.67a}$$

and

$$S = \sqrt{s(s+1)}\,\hbar. \tag{9.67b}$$

It is significant that when we talk about the spin of a particle, say spin 1/2 of the electron, we are actually referring to the spin quantum number for that particle and *not* its magnitude. Thus, the 'spin of the electron is $\frac{1}{2}$' means that its spin quantum number is 1/2 while the magnitude of its spin is

$$\sqrt{\frac{1}{2}\left(\frac{1}{2}+1\right)}\,\hbar \;\; \frac{\sqrt{3}}{2}\hbar.$$

Similar statements are valid for particles with spin 1, spin 3/2, etc.

The beauty of quantum mechanics is that in its domain, the idea of spin is naturally introduced as one of the intrinsic properties of elementary particles.

9.12 Matrix representations of S_x, S_y, S_z

We will now determine the matrix representations of the operators S_x, S_y, S_z for electron spin. It is significant to record that as spin is an intrinsic property of a particle, it does *not* depend on the space coordinates. Consequently, its eigenfunctions *cannot* be regarded as functions of space coordinates. We will therefore use the notation $\psi(s, m)$ instead of $\Psi(x, y, z, t)$. Therefore, the elements of a matrix representing an operator S are

$$S_{ss'mm'} = \int \psi^*(s, m)\, S\psi(s', m')\,\mathrm{d}\tau, \tag{9.68}$$

and according to (9.59), (9.60) and (9.51), which hold for every angular momentum operator, we must have

$$S_+\psi(s, m) = \sqrt{s(s+1) - m(m+1)}\,\hbar\psi(s, m+1), \tag{9.69}$$

$$S_-\psi(s, m) = \sqrt{s(s+1) - m(m-1)}\,\hbar\psi(s, m-1), \tag{9.70}$$

$$S_z\psi(s, m) = m\hbar\psi(s, m), \qquad m = -s, -s+1, \ldots, s-1, s. \tag{9.71}$$

The matrix elements of the operator S_+ are therefore given by

$$S_{+ss'mm'} = \int \psi^*(s', m')\, S_+\psi(s, m)\,\mathrm{d}\tau.$$

By virtue of (9.69), this yields

$$\begin{aligned} S_{+ss'mm'} &= \int \psi^*(s', m')\sqrt{s(s+1) - m(m+1)}\,\hbar\psi(s', m'+1)\,d\tau \\ &= \sqrt{s(s+1) - m(m+1)}\,\hbar\delta_{s's}\delta_{m',m+1}. \end{aligned} \tag{9.72}$$

For an electron, $s' = s = 1/2$. Substituting this value in (9.72) and, for the sake of convenience, writing $S_{+mm'}$ for $S_{+ss'mm'}$, we obtain

$$S_{+mm'} = \sqrt{\left(\frac{3}{4}\right) - m(m+1)}\,\hbar\delta_{m',m+1}. \tag{9.73}$$

We can obtain all the matrix elements by forming different combinations of the two quantum numbers, m and m'. Since each of these quantum numbers can only take the values $\pm 1/2$, we will have, in all, four matrix elements:

$$S_+\left(m = \frac{1}{2}, m' = \frac{1}{2}\right),\ S_+\left(\frac{1}{2}, -\frac{1}{2}\right),\ S_+\left(-\frac{1}{2}, \frac{1}{2}\right),\ S_+\left(-\frac{1}{2}, -\frac{1}{2}\right).$$

These can be arranged to form a 2×2 matrix. Their values are given by (9.73) in which the symbol $\delta_{m,m'+1}$ shows that the only non-zero element is that for which $m = m' + 1$. We arrange these matrix elements such that

(1) Elements with the same m' are in the same row.
(2) Elements with the same m are in the same column.
(3) The value of m' decreases as we go from the first to the second row.
(4) The value of m decreases as we go from the first to second column.

Thus, the matrix S_+ is given by

$$S_+ = \begin{bmatrix} 0 & \hbar \\ 0 & 0 \end{bmatrix} = \hbar \begin{bmatrix} 0 & 1 \\ 0 & 0 \end{bmatrix}. \tag{9.74}$$

Since $S_+^\dagger = S_-$, we have

$$S_- = \begin{bmatrix} 0 & 0 \\ \hbar & 0 \end{bmatrix} = \hbar \begin{bmatrix} 0 & 0 \\ 1 & 0 \end{bmatrix}, \tag{9.75}$$

but $S_\pm = S_x \pm \mathrm{i}S_y$. Therefore, adding (9.74) to (9.75) and simplifying, we obtain

$$S_x = \frac{\hbar}{2}\begin{bmatrix} 0 & 1 \\ 1 & 0 \end{bmatrix}. \tag{9.76}$$

Subtracting (9.75) from (9.74) and simplifying, we obtain

$$S_y = \frac{\hbar}{2}\begin{bmatrix} 0 & -\mathrm{i} \\ \mathrm{i} & 0 \end{bmatrix}. \tag{9.77}$$

To find the matrix corresponding to the operator S_z, we note that its matrix elements are given by

$$S_{zss'mm'} = \int \psi^*(s', m') S_z \psi(s, m) \mathrm{d}\tau$$
$$= m'\hbar \delta_{s's} \delta_{m'm}. \tag{9.78}$$

For an electron, $s' = s = 1/2$. Therefore, $\delta_{s's} = 1$. Substituting this value in the above equation and writing $S_{zm'm}$ for $S_{zs'sm'm}$, we obtain

$$S_{zmm'} = m'\hbar \delta_{m'm}. \tag{9.79}$$

Non-zero elements are those for which $m' = m$ and these are obtained from (9.79) by putting $m' = m = 1/2$ and $m' = m = -1/2$. Thus, the matrix S_z is given by

$$S_z = \frac{\hbar}{2}\begin{bmatrix} 1 & 0 \\ 0 & -1 \end{bmatrix}. \tag{9.80}$$

The eigenvectors of the matrices S_x, S_y, S_z are called *spin vectors.*

The matrices

$$\begin{bmatrix} 0 & 1 \\ 1 & 0 \end{bmatrix}, \begin{bmatrix} 0 & -\mathrm{i} \\ \mathrm{i} & 0 \end{bmatrix}, \begin{bmatrix} 1 & 0 \\ 0 & -1 \end{bmatrix}$$

associated with the components of a spin-half particle, were first discovered by Pauli, and are known as *Pauli's spin matrices.* We will denote them by σ_x, σ_y, σ_z, respectively:

$$\sigma_x = \begin{bmatrix} 0 & 1 \\ 1 & 0 \end{bmatrix}, \tag{9.81a}$$

$$\sigma_y = \begin{bmatrix} 0 & -\mathrm{i} \\ \mathrm{i} & 0 \end{bmatrix}, \tag{9.81b}$$

$$\sigma_z = \begin{bmatrix} 1 & 0 \\ 0 & -1 \end{bmatrix}. \tag{9.81c}$$

Therefore, in terms of Pauli's spin matrices, we may write

$$\mathbf{S} = \frac{\hbar}{2}\boldsymbol{\sigma}.$$

Problem

9.6. Define non-singular, Hermitian and unitary matrices. Show that the Pauli spin matrices possess all these characteristics.

Example 9.2.

Problem

Show, by direct calculations, that

(i) $\sigma_x^2 = \sigma_y^2 = \sigma_z^2 = I$,

(ii) $\sigma^2 = \sigma_x^2 + \sigma_y^2 + \sigma_z^2 = 3I$,

(iii) $[\sigma_i, \sigma_j] = 2\mathrm{i}\sigma_k$ (i, j, k a cyclic permutation of x, y, z),

(iv) $\sigma_i\sigma_j + \sigma_j\sigma_i = 2\delta_{ij}$.

Solution

(i)

$$\sigma_x^2 = \sigma_x\,\sigma_x = \begin{bmatrix} 0 & 1 \\ 1 & 0 \end{bmatrix}\begin{bmatrix} 0 & 1 \\ 1 & 0 \end{bmatrix} = \begin{bmatrix} 1 & 0 \\ 0 & 1 \end{bmatrix} = I,$$

$$\sigma_y^2 = \sigma_y\,\sigma_y = \begin{bmatrix} 0 & -\mathrm{i} \\ \mathrm{i} & 0 \end{bmatrix}\begin{bmatrix} 0 & -\mathrm{i} \\ \mathrm{i} & 0 \end{bmatrix} = \begin{bmatrix} 1 & 0 \\ 0 & 1 \end{bmatrix} = I,$$

$$\sigma_z^2 = \sigma_z\,\sigma_z = \begin{bmatrix} 1 & 0 \\ 0 & -1 \end{bmatrix}\begin{bmatrix} 1 & 0 \\ 0 & -1 \end{bmatrix} = \begin{bmatrix} 1 & 0 \\ 0 & 1 \end{bmatrix} = I.$$

(ii) $\sigma^2 = \sigma_x^2 + \sigma_y^2 + \sigma_z^2 = 3I$.

(iii) We will now prove that $[\sigma_x, \sigma_y] = 2\mathrm{i}\sigma_z$. We have

$$\begin{aligned}
[\sigma_x, \sigma_y] &= \sigma_x\sigma_y - \sigma_y\sigma_x \\
&= \begin{bmatrix} 0 & 1 \\ 1 & 0 \end{bmatrix}\begin{bmatrix} 0 & -\mathrm{i} \\ \mathrm{i} & 0 \end{bmatrix} - \begin{bmatrix} 0 & -\mathrm{i} \\ \mathrm{i} & 0 \end{bmatrix}\begin{bmatrix} 0 & 1 \\ 1 & 0 \end{bmatrix} \\
&= \begin{bmatrix} \mathrm{i} & 0 \\ 0 & -\mathrm{i} \end{bmatrix} - \begin{bmatrix} -\mathrm{i} & 0 \\ 0 & \mathrm{i} \end{bmatrix} \\
&= \begin{bmatrix} 2\mathrm{i} & 0 \\ 0 & -2\mathrm{i} \end{bmatrix} \\
&= 2\mathrm{i}\begin{bmatrix} 1 & 0 \\ 0 & -1 \end{bmatrix} \\
&= 2\mathrm{i}\sigma_z.
\end{aligned}$$

(iv) For $j = i$, we have to prove that

$$\sigma_i\sigma_i + \sigma_i\sigma_i = 2 \quad \text{(no summation over dummy index)},$$

or

$$\sigma_i^2 = 1, \qquad i = x, y, z.$$

This has been proved above.

For $j \neq i$, we have to show that

$$\sigma_i \sigma_j + \sigma_j \sigma_i = 0,$$

i.e. σ_i and σ_j anticommute.

We will prove it for $i = x, j = z$. We have

$$\begin{aligned}\sigma_x \sigma_z + \sigma_z \sigma_x &= \begin{bmatrix} 0 & 1 \\ 1 & 0 \end{bmatrix}\begin{bmatrix} 1 & 0 \\ 0 & -1 \end{bmatrix} + \begin{bmatrix} 1 & 0 \\ 0 & -1 \end{bmatrix}\begin{bmatrix} 0 & 1 \\ 1 & 0 \end{bmatrix} \\ &= \begin{bmatrix} 0 & -1 \\ 1 & 0 \end{bmatrix} + \begin{bmatrix} 0 & 1 \\ -1 & 0 \end{bmatrix} \\ &= \begin{bmatrix} 0 & 0 \\ 0 & 0 \end{bmatrix} \\ &= 0.\end{aligned}$$

9.13 Eigenvectors of S_z

Let us now determine the eigenvectors of the matrix

$$S_z = \frac{\hbar}{2}\begin{bmatrix} 1 & 0 \\ 0 & -1 \end{bmatrix}. \tag{9.80}$$

Since S_z is a 2×2 matrix, its eigenvector should have only two rows. If we denote it by

$$\begin{bmatrix} a \\ b \end{bmatrix}, \tag{9.81}$$

then as the eigenvalues of S_z are $\pm\hbar/2$, we must have

$$\frac{\hbar}{2}\begin{bmatrix} 1 & 0 \\ 0 & -1 \end{bmatrix}\begin{bmatrix} a \\ b \end{bmatrix} = \pm\frac{\hbar}{2}\begin{bmatrix} a \\ -b \end{bmatrix},$$

or

$$\begin{bmatrix} a \\ -b \end{bmatrix} = \pm\begin{bmatrix} a \\ b \end{bmatrix}. \tag{9.82}$$

First, considering the vector with the positive sign and comparing the corresponding coefficients on the two sides of this equation, we obtain $b = 0$. Similarly, for the vector with the negative sign, we obtain $a = 0$. Hence, there are two linearly independent eigenvectors (two vectors u_1 and u_2 are said to be linearly independent if it is impossible to make their linear combination $a_1 u_1 + a_2 u_2$ equal to zero except for $a_1 = a_2 = 0$):

$$\begin{bmatrix} a \\ 0 \end{bmatrix} \quad \text{and} \quad \begin{bmatrix} 0 \\ b \end{bmatrix}. \tag{9.83}$$

In order that the spin vector

$$\begin{bmatrix} a \\ 0 \end{bmatrix}$$

may be normalised, we must have

$$\begin{bmatrix} a \\ 0 \end{bmatrix}^{\dagger}\begin{bmatrix} a \\ 0 \end{bmatrix} = \begin{bmatrix} a^* & 0 \end{bmatrix}\begin{bmatrix} a \\ 0 \end{bmatrix} = a^*a = 1.$$

Taking the physically unimportant phase factor as unity, we may write

$$a^2 = 1,$$

or

$$a = 1.$$

Thus, the normalised eigenvector, say α_1, corresponding to the eigenvalue $\hbar/2$, is given by

$$\alpha_1 = \begin{bmatrix} 1 \\ 0 \end{bmatrix}.$$

Similarly, the other independent normalised vector, say α_2, corresponding to the eigenvalue $-\hbar/2$ is given by

$$\alpha_2 = \begin{bmatrix} 0 \\ 1 \end{bmatrix}.$$

Problem

9.7. Show that the spin vectors are orthogonal to each other.

The conditions of orthogonality and normalisation for the two vectors α_1 and α_2 may be expressed as

$$\alpha_i^{\dagger}\alpha_j = \delta_{ij}, \quad i = 1, 2. \tag{9.84}$$

Since the Hermitian matrix S_z has only two linearly independent eigenvectors α_1 and α_2 corresponding to eigenvalues $+\hbar/2$ and $-\hbar/2$, they form a *complete set of vectors in the 'spin space'*. Every spin vector (usually called a *spinor*) can be expanded in terms of them. Thus, the most general spin vector α is a linear combination of α_1 and α_2:

$$\alpha = c_1\alpha_1 + c_2\alpha_2 = c_1\begin{bmatrix} 1 \\ 0 \end{bmatrix} + c_2\begin{bmatrix} 0 \\ 1 \end{bmatrix} = \begin{bmatrix} c_1 \\ c_2 \end{bmatrix}.$$

In order that α may be normalised, we must have

$$\alpha^\dagger \alpha = 1,$$

but

$$\alpha^\dagger = \begin{bmatrix} c_1^* c_2^* \end{bmatrix}.$$

Therefore, $\alpha^\dagger \alpha = 1$ yields

$$\begin{bmatrix} c_1^* c_2^* \end{bmatrix} \begin{bmatrix} c_1 \\ c_2 \end{bmatrix} = 1,$$

or

$$c_1^* c_1 + c_2^* c_2 = 1. \tag{9.85}$$

This suggests that $c_1^* c_1$ and $c_2^* c_2$ may be interpreted as the probabilities of finding the electron with a z-component of spin as $\hbar/2$ and $-\hbar/2$, respectively. The electron is said to be in the *spin-up* or *spin-down* state according to whether the z-component of its spin is $\hbar/2$ or $-\hbar/2$.

Problems

9.8. Find the normalised eigenvectors of the components S_x, S_y of the electron spin operator **S**.

9.9. Prove the following relations:

$$\sigma_x \alpha_1 = \alpha_2,$$

$$\sigma_y \alpha_1 = \mathrm{i}\alpha_2,$$

$$\sigma_z \alpha_1 = \alpha_1,$$

and

$$\sigma_x \alpha_2 = \alpha_1,$$

$$\sigma_y \alpha_2 = -\mathrm{i}\alpha_1,$$

$$\sigma_z \alpha_2 = -\alpha_2.$$

9.10. Show that for a spin-half particle, the operator S^2 $(= S_x^2 + S_y^2 + S_z^2)$ is given by

$$S^2 = \frac{3\hbar^2}{4} \begin{bmatrix} 1 & 0 \\ 0 & 1 \end{bmatrix}.$$

9.14 The wave function for the electron

We have seen that the z-component S_z of electron spin can take two values: $\hbar/2$ and $-\hbar/2$. That is

$$S_z = \pm\frac{\hbar}{2} = m_s\hbar, \quad m_s = \pm\frac{1}{2}.$$

Thus, m_s may be treated as a variable which can assume only two values. The wave function representing an electron will, therefore, depend upon m_s in addition to x, y, z, t:

$$\psi = \psi\left(x, y, z, t, m_s\right).$$

Since m_s can take only two values, ψ may be considered as having two components: $\psi_+(x, y, z, t)$ and $\psi_-(x, y, z, t)$. The former corresponds to the $\hbar/2$ spin state while the latter refers to the $-\hbar/2$ state. These two spin states can be taken as elements of a column matrix:

$$\Psi = \begin{bmatrix} \psi_+(x, y, z, t) \\ \psi_-(x, y, z, t) \end{bmatrix}. \tag{9.86}$$

Hence, the wave function for an electron may be written as a column vector with two elements. The function $\psi_+^*(x, y, z, t)\psi_+(x, y, z, t)$ may be interpreted as the probability that at time t the electron has the coordinates x, y, z with the spin $+\hbar/2$ in the z-direction. Similarly, $\psi_-^*(x, y, z, t)\,\psi_-(x, y, z, t)$ may be interpreted as the probability that at time t the electron has the coordinates x, y, z with spin $-\hbar/2$ in the z-direction. It may be emphasised that in general these two functions are quite different functions of x, y, z.

Example 9.3.

Problem

Determine the spin matrices S_x, S_y, S_z for a particle of spin 1.

Solution

To construct matrices for particles of spin 1, i.e. for particles with $s = 1$, we proceed as follows.

From (9.69)–(9.71), for $s = 1$, we have

$$S_+\psi(s, m) = \sqrt{2 - m(m+1)}\,\hbar\psi(s, m+1), \tag{9.87a}$$

$$S_-\psi(s, m) = \sqrt{2 - m(m-1)}\,\hbar\psi(s, m-1), \tag{9.87b}$$

$$S_z\psi(s, m) = m\hbar\psi(s, m), \tag{9.87c}$$

where $m = -1, 0, 1$. Therefore:

$$S_{+s'sm'm} = \int \psi^*(s', m') S_+ \psi(s, m) \mathrm{d}\tau$$
$$= \sqrt{2 - m(m+1)}\,\hbar \int \psi^*(s', m') \psi(s, m+1) \mathrm{d}\tau$$
$$= \sqrt{2 - m(m+1)}\,\hbar \delta_{m',m+1},$$

or

$$S_{+mm'} = \sqrt{2 - m(m+1)}\,\hbar \delta_{m',m+1}. \tag{9.88}$$

From this equation, we can write down the matrix elements of the operator S_+. These are

$$S_+ = \begin{bmatrix} 0 & \sqrt{2}\hbar & 0 \\ 0 & 0 & \sqrt{2}\hbar \\ 0 & 0 & 0 \end{bmatrix}. \tag{9.89}$$

Moreover as $S_- = S_+^\dagger$, we have

$$S_- = \begin{bmatrix} 0 & 0 & 0 \\ \sqrt{2}\hbar & 0 & 0 \\ 0 & \sqrt{2}\hbar & 0 \end{bmatrix} \tag{9.90}$$

and $S_{zmm'} = m'\hbar\delta_{mm'}$ yields

$$S_z = \begin{bmatrix} \hbar & 0 & 0 \\ 0 & 0 & 0 \\ 0 & 0 & -\hbar \end{bmatrix}. \tag{9.91}$$

Adding (9.89) to (9.90), we obtain

$$S_+ + S_- = 2S_x = \begin{bmatrix} 0 & \sqrt{2}\hbar & 0 \\ \sqrt{2}\hbar & 0 & \sqrt{2}\hbar \\ 0 & \sqrt{2}\hbar & 0 \end{bmatrix},$$

or

$$S_x = \frac{\hbar}{\sqrt{2}} \begin{bmatrix} 0 & 1 & 0 \\ 1 & 0 & 1 \\ 0 & 1 & 0 \end{bmatrix}. \tag{9.92}$$

Subtracting (9.90) from (9.89) and simplifying, we obtain

$$S_y = -\mathrm{i}\frac{\hbar}{\sqrt{2}}\begin{bmatrix} 0 & 1 & 0 \\ -1 & 0 & 1 \\ 0 & -1 & 0 \end{bmatrix}. \tag{9.93}$$

Finally, from (9.91), we have

$$S_z = \hbar\begin{bmatrix} 1 & 0 & 0 \\ 0 & 0 & 0 \\ 0 & 0 & -1 \end{bmatrix}. \tag{9.94}$$

These are the required matrices for a particle of spin 1.

Problems

9.11. Find the normalised eigenvectors of spin matrices for $s = 1$.
9.12. Determine the spin matrices for a particle of spin 2.

9.15 Spins of elementary particles

Spin is not a characteristic of electrons only. It is now regarded as one of the fundamental properties of a particle. For instance, nucleons have spin 1/2; π-mesons have spin zero; photons, treated as particles, have spin 1. Since for each value of s, the z-component of spin can have all the values between $-s\hbar, \ldots, s\hbar$, successive terms in the series differing by $\hbar$, the total number of values is $2s + 1$. It is important to note that as *s is fixed for every particle*, we have

$$\lim \sqrt{s(s+1)}\,\hbar = 0,$$
$$\hbar \to 0.$$

This shows that *spin has no classical analogue*.

9.16 The average value of spin

The average value of spin represented by a matrix A for a particle in a state α is *defined* by the equation

$$\langle A \rangle = \alpha^{\dagger} A \alpha.$$

Example 9.4

Problem

Find the average values of S_z, S_x and S_y when the electron is in the state

$$\begin{bmatrix} 1 \\ 0 \end{bmatrix}.$$

Solution

The average value of the z-component of the spin is given by

$$\begin{aligned}\langle S_z \rangle &= \alpha^\dagger S_z \alpha \\ &= \begin{bmatrix} 1 & 0 \end{bmatrix} \frac{\hbar}{2} \begin{bmatrix} 1 & 0 \\ 0 & 1 \end{bmatrix} \begin{bmatrix} 1 \\ 0 \end{bmatrix} \\ &= \frac{\hbar}{2} \begin{bmatrix} 1 & 0 \end{bmatrix} \begin{bmatrix} 1 \\ 0 \end{bmatrix} = \frac{\hbar}{2},\end{aligned}$$

and

$$\begin{aligned}\langle S_x \rangle &= \alpha^\dagger S_x \alpha \\ &= \begin{bmatrix} 1 & 0 \end{bmatrix} \frac{\hbar}{2} \begin{bmatrix} 0 & 1 \\ 1 & 0 \end{bmatrix} \begin{bmatrix} 1 \\ 0 \end{bmatrix} \\ &= \frac{\hbar}{2} \begin{bmatrix} 1 & 0 \end{bmatrix} \begin{bmatrix} 0 \\ 1 \end{bmatrix} = 0,\end{aligned}$$

and

$$\begin{aligned}\langle S_y \rangle &= \alpha^\dagger S_y \alpha \\ &= \begin{bmatrix} 1 & 0 \end{bmatrix} \frac{\hbar}{2} \begin{bmatrix} 0 & -\mathrm{i} \\ \mathrm{i} & 0 \end{bmatrix} \begin{bmatrix} 1 \\ 0 \end{bmatrix} \\ &= \frac{\hbar}{2} \begin{bmatrix} 1 & 0 \end{bmatrix} \begin{bmatrix} 0 \\ \mathrm{i} \end{bmatrix} = 0.\end{aligned}$$

Example 9.5

Problem

Find the expectation value of S_z when the electron is in the state

$$\alpha = \begin{bmatrix} c_1 \\ c_2 \end{bmatrix}.$$

Solution

The average value of the z-component of the spin is given by

$$\begin{aligned}\langle S_z \rangle &= \alpha^{\dagger} S_z \alpha \\ &= \begin{bmatrix} c_1^* & c_2^* \end{bmatrix} \frac{\hbar}{2} \begin{bmatrix} 1 & 0 \\ 0 & -1 \end{bmatrix} \begin{bmatrix} c_1 \\ c_2 \end{bmatrix} \\ &= \frac{\hbar}{2} \begin{bmatrix} c_1^* & c_2^* \end{bmatrix} \begin{bmatrix} c_1 \\ -c_2 \end{bmatrix} \\ &= \frac{\hbar}{2} \left(c_1^* c_1 - c_2^* c_2 \right) \\ &= \frac{\hbar}{2} \left(|c_1|^2 - |c_2|^2 \right).\end{aligned}$$

9.17 Spin and statistics

It has been found that various phenomena can be explained only by assuming that some elementary particles possess half odd-integral spins, i.e. 1/2, 3/2, 5/2, ... These particles are called *fermions*. These elementary particles with spins 1/2, 3/2, 5/2, ... are found to obey Pauli's exclusion principle: two identical fermions cannot occupy the same state. They are therefore described by antisymmetric wave functions. It is observed that they obey Fermi–Dirac statistics. On the other hand, elementary particles with intrinsic spins 0, 1, 2, ... obey Bose–Einstein statistics, are called *bosons* and are described by symmetric wave functions. Thus, electrons, neutrons and protons with spin 1/2 are fermions while photons (spin 1) and π mesons (spin 0) are bosons.

Let us consider a system of two composite particles each containing n fermions. Suppose that the interactions between these composite particles are sufficiently weak so that the internal structure of any individual composite particle is not disturbed. The interchange of any pair of fermions of two composite particles would merely change the sign of the wave function describing the system. In order to interchange the composite particles, we have to interchange n fermions of one with n fermions of the other. The sign of the wave function for the system would therefore change n times. If n is even, the wave function for the composite particle will remain unchanged; it is a symmetrical wave function and the composite particles are bosons. If n is odd, the composite particles are fermions.

9.18 Addition of angular momenta

So far, we have considered the angular momentum properties of a single particle. Sometimes, we have to combine the angular momenta of two particles so that we may express the angular momentum of the composite system in terms of the angular momenta of the individual particles. Let us consider the addition of angular momenta of two *non-interacting particles* with angular momenta $\mathbf{J}_1$ and $\mathbf{J}_2$, the operators J_{1x}, J_{1y}, J_{1z} and J_{2x}, J_{2y}, J_{2z} denoting their components. Since the particles are non-interacting, the angular momenta $\mathbf{J}_1$ and $\mathbf{J}_2$ are independent and, therefore, commute with each other, i.e. each component of $\mathbf{J}_1$ commutes with every component of $\mathbf{J}_2$:

$$\left[J_{1x}, J_{2x}\right] = \left[J_{1x}, J_{2y}\right] = \cdots = \left[J_{1z}, J_{2z}\right] = 0. \tag{9.95}$$

Problem

9.13. By using the set of equation (9.95), show that the operators J_1^2, J_2^2, J_{1z}, J_{2z} mutually commute.

Let us denote the vector sum of non-interacting angular momenta $\mathbf{J}_1$ and $\mathbf{J}_2$ by $\mathbf{J}$ so that

$$\mathbf{J} = \mathbf{J}_1 + \mathbf{J}_2. \tag{9.96}$$

We will now show that $\mathbf{J}$ itself is an angular momentum operator. Since $\mathbf{J}_1$ and $\mathbf{J}_2$ are angular momentum operators, we must have

$$\mathbf{J}_1 \wedge \mathbf{J}_1 = i\hbar\mathbf{J}_1 \tag{9.97}$$

and

$$\mathbf{J}_2 \wedge \mathbf{J}_2 = i\hbar\mathbf{J}_2. \tag{9.98}$$

We know that

$$\mathbf{J} = \mathbf{J}_1 + \mathbf{J}_2. \tag{9.96'}$$

Therefore,

$$\begin{aligned}\mathbf{J} \wedge \mathbf{J} &= \left(\mathbf{J}_1 + \mathbf{J}_2\right) \wedge \left(\mathbf{J}_1 + \mathbf{J}_2\right) \\ &= \mathbf{J}_1 \wedge \mathbf{J}_1 + \mathbf{J}_1 \wedge \mathbf{J}_2 + \mathbf{J}_2 \wedge \mathbf{J}_1 + \mathbf{J}_2 \wedge \mathbf{J}_2,\end{aligned} \tag{9.99}$$

but

$$\mathbf{J}_1 \wedge \mathbf{J}_2 + \mathbf{J}_2 \wedge \mathbf{J}_1 = \mathbf{J}_1 \wedge \mathbf{J}_2 - \mathbf{J}_1 \wedge \mathbf{J}_2 = 0$$

and

$$\mathbf{J}_1 \wedge \mathbf{J}_1 + \mathbf{J}_2 \wedge \mathbf{J}_2 = i\hbar\mathbf{J}_1 + i\hbar\mathbf{J}_2 = i\hbar\left(\mathbf{J}_1 + \mathbf{J}_2\right) = i\hbar\mathbf{J}.$$

Consequently, (9.99) reduces to

$$\mathbf{J} \wedge \mathbf{J} = i\hbar\mathbf{J}. \tag{9.100}$$

This equation shows that **J**, the vector sum of angular momenta $\mathbf{J}_1$ and $\mathbf{J}_2$, is itself an angular momentum operator. Its components, therefore, should satisfy the relevant commutation relations.

Problem

9.14. If J_z is the z-component of the angular momentum operator $\mathbf{J} = \mathbf{J}_1 + \mathbf{J}_2$, show that
(i) $[J^2, J_z] = 0$,
(ii) $[J_1^2, J^2] = [J_2^2, J^2] = 0$,
(iii) $[J_1^2, J_z] = [J_2^2, J_z] = 0$.

We know that the operators J_1^2, J_2^2, J_{1z}, J_{2z} mutually commute. Let Γ, J_1^2, J_2^2, J_{1z}, J_{2z} constitute a complete set of commuting operators, the independent operators Γ forming with others a complete set for the system. Since all the operators of the above set commute, there exists a complete set of their simultaneous orthonormal eigenfunctions. We denote these eigenfunctions by

$$\psi\left(\gamma j_1 j_2 m_1 m_2\right)$$

where $\gamma, j_1, j_2, m_1, m_2$ are the quantum numbers labelling the eigenvalues of Γ, J_1^2, J_2^2, J_{1z}, J_{2z}, respectively. Now the operators Γ commute with all the components of angular momenta $\mathbf{J}_1$ and $\mathbf{J}_2$, and hence with all the operators J_1^2, J_2^2, $\mathbf{J}^2$, J_z. Therefore the operators Γ, J_1^2, J_2^2, J^2, J_z form another complete commuting set for the system because this set contains as many operators as the first one. The simultaneous orthonormal eigenfunctions corresponding to this set may be denoted by $\phi(\gamma j_1\, j_2\, j\, m)$, where j, m are the quantum numbers labelling the eigenvalues of J^2 and J_z. Thus, there are at least two complete sets of mutually commuting operators. For the sake of brevity, we will drop the symbols Γ and γ in further analysis and conventionally call the set of remaining four operators J_1^2, J_2^2, J_{1z}, J_{2z} or J_1^2, J_2^2, $\mathbf{J}^2$, J_z a complete set.

We note that $\mathbf{J}_1$ and $\mathbf{J}_2$ are given as angular momenta, with J_{1z} and J_{2z} as their z-components. Therefore, if j_1 and j_2 are the quantum numbers associated with the angular momentum, then we must have

$$J_1^2\psi(j_1 m_1) = j_1(j_1 + 1)\hbar^2\psi(j_1 m_1),$$

$$J_{1z}\psi(j_1 m_1) = m_1\hbar\psi(j_1 m_1), \quad m_1 = -j_1, -j_1 + 1, \ldots, j_1,$$

and

$$J_2^2, \psi(j_2 m_2) = j_2(j_2 + 1)\hbar^2\psi(j_2 m_2),$$

$$J_{2z}, \psi(j_2 m_2) = m_2\hbar\psi(j_2 m_2), \quad m_2 = -j_2, -j_2 + 1, \ldots, j_2.$$

We have shown that in this situation, $\mathbf{J}$ (= $\mathbf{J}_1 + \mathbf{J}_2$) is also an angular momentum operator and J_z describes its z-component. Then if the corresponding quantum number is denoted by j, we have

$$J^2\psi(jm) = j(j+1)\hbar^2\psi(jm),$$

$$J_z\psi(jm) = m\hbar\psi(jm), \quad m = -j, -j+1, \ldots, j.$$

Our addition problem is to determine the eigenvalues and simultaneous eigenfunctions of J^2 and J_z in terms of those of the separate independent angular momentum operators and their z-components.

Remark

If $\psi(j_1m_1)$ is a simultaneous eigenfunction of J_1^2 and J_{1z}, and $\psi(j_2\, m_2)$ is a simultaneous eigenfunction of J_2^2 and J_{2z}, and $\psi(j_1\, j_2\, m_1\, m_2)$ is a simultaneous eigenfunction of the set of operators J_1^2, J_2^2, $\mathbf{J}^2$, J_z, then

$$\psi\left(j_1j_2m_1m_2\right) = \psi\left(j_1m_1\right)\psi\left(j_2m_2\right).$$

Thus, in quantum mechanics, the problem of the addition of angular momenta reduces to: given the eigenvalues of the operators, J_1, J_2, J_{1z}, J_{2z} and the simultaneous eigenfunctions $\psi(j_1j_2m_1m_2)$ of the system corresponding to the operators J_1^2, J_2^2, J_{1z}, J_{2z}, to determine the eigenvalues of J^2 and J_z and the simultaneous eigenfunctions $\phi(j_1j_2\, jm)$ in terms of those of the set J_1^2, J_2^2, J^2, J_z.

Let us try to solve this problem. Since $\mathbf{J} = \mathbf{J}_1 + \mathbf{J}_2$, we have

$$J_z = J_{1z} + J_{2z}, \tag{9.101}$$

therefore

$$\begin{aligned} J_z\psi(j_1j_2m_1m_2) &= (J_{1z} + J_{2z})\psi(j_1j_2m_1m_2) \\ &= (m_1\hbar + m_2\hbar)\psi(j_1, j_2m_1m_2),\ m_1 = -j_1, -j_1+1, \ldots, j_1, \\ &= m_2 = -j_2, -j_2+1, \ldots, j_2, \\ &= m\hbar\, \psi(j_1j_2m_1m_2). \end{aligned} \tag{9.102}$$

This shows that the eigenvalues of J_z are obtained by merely adding every eigenvalue of J_{1z} to each of the eigenvalues of J_{2z}.

Now the eigenvalues of J_{1z} and J_{2z} corresponding to quantum numbers j_1, j_2 associated with J_1^2, J_2^2 are

$$-j_1\hbar, \left(-j_1+1\right)\hbar, \left(-j_1+2\right)\hbar, \ldots, \left(j_1-2\right)\hbar, \left(j_1-1\right)\hbar, j_1\hbar \tag{9.103}$$

and

$$-j_2\hbar,\ \left(-j_2+1\right)\hbar,\ \left(-j_2+2\right)\hbar,\ \ldots,\ \left(j_2-2\right)\hbar,\ \left(j_2-1\right)\hbar,\ j_2\hbar, \tag{9.104}$$

respectively. Therefore, the eigenvalues of J_z are given by

$$\begin{array}{ll} -(j_1+j_2)\hbar, & \text{which occurs once,} \\ -(j_1+j_2-1)\hbar, & \text{which occurs twice,} \\ -(j_1+j_2-2)\hbar, & \text{which occurs three times,} \\ & \ldots \\ (j_1+j_2-2)\hbar, & \text{which occurs three times,} \\ (j_1+j_2-1)\hbar, & \text{which occurs twice,} \\ (j_1+j_2)\hbar, & \text{which occurs once.} \end{array}$$

These eigenvalues of J_z can be rearranged in different series such that no eigenvalue occurs more than once in any series, the successive terms in each series differing by $\hbar$, and the lowest value in each series is just the negative of the highest value. Then the series are:

$$\begin{aligned} &-(j_1+j_2)\hbar,\ -(j_1+j_2-1)\hbar,\ \ldots (j_1+j_2-1)\hbar,\ (j_1+j_2)\hbar, \\ &-(j_1+j_2-1)\hbar,\ -(j_1+j_2-2)\hbar,\ \ldots,\ (j_1+j_2-1)\hbar,\ (j_1+j_2-1)\hbar, \\ &-(j_1+j_2-2)\hbar,\ \ldots,\ (j_1+j_2-2)\hbar, \end{aligned} \tag{9.105}$$

etc.

Let us consider the *first of these series*, i.e.

$$-\left(j_1+j_2\right)\hbar,\ -\left(j_1+j_2-1\right)\hbar,\ \ldots,\ \left(j_1+j_2-1\right)\hbar,\ \left(j_1+j_2\right)\hbar.$$

Since in this series the eigenvalue spectrum of J_z extends from $-(j_1+j_2)\hbar$ to $(j_1+j_2)\hbar$, the quantum number specifying the eigenvalue of J^2 corresponding to simultaneous eigenfunctions is j_1+j_2. Therefore, the eigenvalue of J^2 is $(j_1+j_2)(j_1+j_2+1)\hbar^2$ and the simultaneous eigenfunctions of $J^2$and J_z are $\phi(j_1\,j_2\,j_1+j_2\,m)$, where (j_1+j_2) is fixed for the series while m varies from $-(j_1+j_2)$ to (j_1+j_2). Thus, the simultaneous eigenfunctions of J^2 and J_z for this series are non-degenerate.

For the *second series*, the quantum number specifying the eigenvalue of J^2 is j_1+j_2-1 and the simultaneous eigenfunctions are $\phi(j_1\,j_2\,j_1+\mathrm{j}_2-1, m')$, where m' varies from $-(j_1+j_2-1)$ to (j_1+j_2-1). These eigenfunctions are also non-degenerate, and so on.

Thus, all the simultaneous eigenfunctions of J^2 and J_z are non-degenerate and the quantum numbers specifying the eigenvalues of J^2 for different series are

$$\left(j_1+j_2\right),\ \left(j_1+j_2-1\right),\ \left(j_1+j_2-2\right),\ \ldots \tag{9.106}$$

Where does this series terminate? By virtue of the set of equations (9.61), the *last series* of the eigenvalues of J_z may be written as

$$-(j_1 + j_2 - k)\hbar, \ldots, (j_1 + j_2 - k)\hbar$$

where k is yet to be determined and $j_1 + j_2 - k \geqslant 0$ because in writing down each series we start with the smallest member which is just the negative of its largest member. Suppose that $j_2 \geqslant j_1$. Then from the previous discussion, the first member of the last series is obtained by adding last member of the J_{1z} spectrum to the first member of the J_{2z} spectrum, i.e.

$$-(j_1 + j_2 - k)\hbar = (j_1 - j_2)\hbar,$$

or

$$k = j_1.$$

The *last series*, therefore, becomes

$$-(j_2 - j_1)\hbar, \ldots, (j_2 - j_1)\hbar,$$

the quantum number specifying the eigenvalues of J^2 is $j_2 - j_1$. If $j_1 \geqslant j_2$, the *last series* would be

$$-(j_1 - j_2)\hbar, \ldots, (j_1 - j_2)\hbar.$$

Hence, for the last series of J_z eigenvalues, as $j_1 > j_2$, the quantum number specifying the eigenvalue of J^2 is $j_1 - j_2$. For $j_2 = j_1$, it is evidently the same, i.e. zero in both cases. Putting this all together, for the last series, we may write the J^2 quantum number as $|j_1 - j_2|$. This gives the *last member of the series* (9.106) which, now, becomes

$$j_1 + j_2, \quad j_1 + j_2 - 1, \ldots, |j_1 - j_2|. \tag{9.107}$$

To sum up: if j_1 and j_2 are the quantum numbers specifying the eigenvalues of the squares of two independent angular momenta $\mathbf{J}_1$ and $\mathbf{J}_2$ so that

$$J_1^2 = j_1(j_1 + 1)\hbar^2, \qquad J_2^2 = j_2(j_2 + 1)\hbar^2,$$

$$J_{1z} = -j_1, \hbar, \ldots, j_1 h, \qquad J_{2z} = -j_2\hbar, \ldots, j_2\hbar,$$

then the eigenvalues of the z-component of the resultant angular momentum $\mathbf{J}$ are

$$\begin{gathered}
-(j_1 + j_2)\hbar, -(j_1 + j_2 - 1)\hbar, \ldots, (j_1 + j_2 - 1)\hbar, (j_1 + j_2)\hbar \\
-(j_1 + j_2 - 1)\hbar, \ldots, (j_1 + j_2 - 1)\hbar \\
\cdots \\
-|j_1 - j_2|\,\hbar, \ldots, |j_1 - j_2|\hbar
\end{gathered} \tag{9.108}$$

while the eigenvalues of J^2 are given by

$$J^2 = j(j+1)\hbar^2,$$

where the allowed values of j are

$$(j_1 + j_2),\ (j_1 + j_2 - 1),\ \ldots,\ |j_1 - j_2|,$$
$$\text{i.e. } |j_1 - j_2| \leqslant j \leqslant j_1 + j_2.$$

This is known as the *triangular condition.* We shall illustrate it using an example.

Let us consider two independent angular momenta with $j_1 = 2$ and $j_2 = 3/2$. The eigenvalue spectrum of J^2 corresponds to quantum numbers

$$(j_1 + j_2),\ (j_1 + j_2 - 1),\ \ldots,\ |j_1 - j_2|.$$

Therefore in this case, we have the quantum numbers of J^2 forming the eigenvalue spectra of J_z as

$$-7/2\hbar,\ -5/2\hbar,\ -3/2\hbar,\ -1/2\hbar,\ 1/2\hbar,\ 3/2\hbar,\ 5/2\hbar,\ 7/2\hbar,$$
$$-5/2\hbar,\ -3/2\hbar,\ -1/2\hbar,\ 1/2\hbar,\ 3/2\hbar,\ 5/2\hbar,$$
$$-3/2\hbar,\ -1/2\hbar,\ 1/2\hbar,\ 3/2\hbar,$$
$$-1/2\hbar,\ 1/2\hbar.$$

This shows that the eigenvalues of J_z occur as

$$\begin{matrix} -7/2\hbar, & -5/2\hbar, & -3/2\hbar, & -1/2\hbar, & 1/2\hbar, & 3/2\hbar, & 5/2\hbar, & 7/2\hbar, \\ (1) & (2) & (3) & (4) & (4) & (3) & (2) & (1). \end{matrix}$$

The numbers in brackets indicate how many times each eigenvalue occurs.

If we have to find the resultant of three independent angular momenta, we first add two angular momenta and then add the third one to the resultant of the first two. We can find the resultant of any finite number of independent angular momenta by repeating this process.

The solution is not yet complete. We have not yet obtained the eigenfunctions ϕ of the second set in terms of the functions ψ of the first set. To achieve this, we note that as all the operators of the complete set J_1^2, J_2^2, J_{1z}, J_{2z} represent observables, each one of them is Hermitian. Therefore, the orthonormal functions $\psi(j_1 j_2 m_1 m_2)$, which are the simultaneous eigenfunctions of these Hermitian operators, form a complete set. Since for the given values of j_1 and j_2, the quantum numbers m_1 and m_2 take $(2j_1 + 1)$ and $(2j_2 + 1)$ different values, respectively, the number of allowed states of the system for fixed values of j_1 and j_2 is $(2j_1 + 1)(2j_2 + 1)$.

Example 9.6.

Problem
Show that the total number of non-degenerate states $\phi(j_1\, j_2, j\, m)$ for all j is also $(2j_1 + 1)(2j_2 + 1)$.

Solution
For each j, there are $(2j + 1)$ non-degenerate states of the system. Therefore, the total number of non-degenerate states $\phi(j_1\, j_2, jm)$ for all j is

$$\sum_{|j_1-j_2|}^{j_1+j_2} (2j + 1). \tag{9.109}$$

For convenience, writing a and b for $|j_1 - j_2|$ and $j_1 + j_2$, respectively, the above sum yields

$$\sum_a^b (2j + 1) = \left(2\sum_a^b j\right) + (b - a + 1)$$

where $(b - a + 1) = n$, say, denotes the number of terms in the series. Then, we have

$$\begin{aligned}
\sum_a^b j &= a + (a + 1) + (a + 2) + \cdots + b \\
&= \frac{n}{2}[2a + (n - 1) \cdot 1] \\
&= \frac{1}{2}(b - a + 1)[2a + (b - a + 1) - 1] \\
&= \frac{1}{2}(b - a + 1)(a + b).
\end{aligned}$$

Substituting this value of $\sum_a^b j$ in (9.109), we obtain

$$\begin{aligned}
\sum_a^b (2j + 1) &= (b - a + 1)(a + b) + (b - a + 1) \\
&= (b + 1 - a)(b + 1 + a) = (b + 1)^2 - a^2 \\
&= b^2 + 2b + 1 - a^2 \\
&= \left(j_1 + j_2\right)^2 + 2\left(j_1 + j_2\right) + 1 - \left(j_1 - j_2\right)^2 \\
&= 4j_1 j_2 + 2j_1 + 2j_2 + 1 \\
&= \left(2j_1 + 1\right)\left(2j_2 + 1\right).
\end{aligned}$$

This is the same as the number of independent ψ.

9.19 Clebsch–Gordan coefficients

Now consider the functions $\phi(j_1\, j_2\, j\, m)$ corresponding to the second complete set of operators Γ, J_1^2, J_2^2, J^2, J_z. Since the functions $\psi(j_1\, j_2\, m_1\, m_2)$ form a complete set, the functions $\phi(j_1\, j_2\, jm)$ can be expressed in terms of them:

$$\phi\left(j_1 j_2 jm\right) = \sum_{m_1 m_2} C\left(j_1 j_2\, m_1 m_2;\, jm\right)\psi\left(j_1 j_2\, m_1 m_2\right). \tag{9.110}$$

The summation is carried over m_1 and m_2 only as j_1 and j_2 are assumed fixed. The coefficients $C(j_1\ j_2\ m_1\ m_2;\ jm)$ are called the *Clebsch–Gordan coefficients* or the *Wigner coefficients* or the *vector addition coefficients*. These coefficients are also denoted by $C(j_1\ j_2\ j\ m_1\ m_2\ m)$. A knowledge of C determines ϕ in terms of ψ. As quantum numbers j_1 and j_2 are fixed, we may omit them in the case of ϕ. Hence, (9.110) may be written as

$$\phi(jm) = \sum_{m_1 m_2} C\left(j_1 j_2\, m_1 m_2;\, jm\right)\psi\left(j_1 j_2\, m_1 m_2\right). \tag{9.111}$$

To find the values of C, we proceed as follows. Applying J_z $(= J_{1z} + J_{1z})$ to (9.111), we obtain

$$J_z\phi(jm) = \sum_{m_1 m_2} C\left(j_1 j_2\, m_1 m_2;\, jm\right)\left(J_{1z} + J_{2z}\right)\psi\left(j_1 j_2\, m_1 m_2\right)$$

or

$$m\hbar\phi(jm) = \sum_{m_1 m_2} C\left(j_1 j_2\, m_1 m_2;\, jm\right)\left(m_1 + m_2\right)\hbar\psi\left(j_1 j_2\, m_1 m_2\right).$$

Substituting the expression for $\phi(jm)$ from (9.111) in (9.112), we obtain

$$m\hbar\sum_{m_1 m_2} C\left(j_1 j_2\, m_1 m_2;\, jm\right)\psi\left(j_1 j_2\, m_1 m_2\right) = \sum_{m_1 m_2} C\left(j_1 j_2\, m_1 m_2;\, jm\right)\left(m_1 + m_2\right)\hbar\psi\left(j_1 j_2\, m_1 m_2\right),$$

or

$$\sum_{m_1 m_2} C(m - m_1 - m_2)\hbar C\left(j_1 j_2\, m_1 m_2;\, jm\right)\psi\left(j_1 j_2\, m_1 m_2\right) = 0. \tag{9.112}$$

Since all the ψ are linearly independent, the above equation can hold only if the coefficient of each ψ is separately zero:

$$(m - m_1 - m_2)C\left(j_1 j_2\, m_1 m_2;\, jm\right) = 0.$$

Therefore, we must have

$$C\left(j_1 j_2\, m_1 m_2;\, jm\right) = 0, \quad \text{unless } m = m_1 + m_2. \tag{9.113}$$

This shows that, for non-zero Clebsch–Gordan coefficients, the quantum numbers m_1, m_2 and m, are not independent; they are related by the equation $m = m_1 + m_2$. The double sum in (9.111) is actually a single sum:

$$\phi(jm) = \sum_{m_1} C\left(j_1 j_2 m_1,\, m - m_1;\, jm\right)\psi\left(j_1 j_2 m_1 m - m_1\right). \tag{9.114}$$

Now as $m = m_1 + m_2$, we must have $m_{max} = m_{1max} + m_{2max}$. But for given values of j_1 and j_2, the maximum values of m_1 and m_2 are j_1 and j_2, respectively. That is,

$$m_{1max} = j_1,$$

$$m_{2max} = j_2.$$

Therefore, equation $m_{max} = m_{1max} + m_{2max}$ yields

$$m_{max} = j_1 + j_2. \tag{9.115}$$

Moreover, according to the series (9.107), $j_1 + j_2$ is the maximum value of j:

$$j_{max} = j_1 + j_2. \tag{9.116}$$

Hence, the above analysis shows that

$$j_1 + j_2 = j_{max} = m_{max} = m_{1max} + m_{2max}. \tag{9.117}$$

Thus, when j and m have their maximum values, m_1 can take only one value: the maximum ($= j_1$). Equation (9.114) *in this case* reduces to

$$\phi\left(j_1 + j_2 j_1 + j_2\right) = C\left(j_1 j_2 j_1 j_2;\, j_1 + j_2 j_1 + j_2\right)\psi\left(j_1 j_2 j_1 j_2\right). \tag{9.118}$$

The normalisation condition yields

$$C^*\left(j_1 j_2 j_1 j_2;\, j_1 + j_2 j_1 + j_2\right)C\left(j_1 j_2 j_1 j_2;\, j_1 + j_2 j_1 + j_2\right) = 1. \tag{9.119}$$

The standard convention is to take the above Clebsch–Gordan coefficient as real and positive. Then (9.119) gives

$$C\left(j_1 j_2 j_1 j_2;\, j_1 + j_2 j_1 + j_2\right) = 1. \tag{9.120}$$

Substituting this value of C in (9.118), we obtain

$$\phi\left(j_1 + j_2 j_1 + j_2\right) = \psi\left(j_1 j_2 j_1 j_2\right). \tag{9.121}$$

Applying $J_- = J_{1-} + J_{2-}$ to (9.121), we obtain

$$J_-\phi\left(j_1 + j_2 j_1 + j_2\right) = \left(J_{1-} + J_{2-}\right)\psi\left(j_1 j_2 j_1 j_2\right). \tag{9.122}$$

Using equation

$$J_- u(j, m) = \sqrt{(j + m)(j - m + 1)}\,\hbar u(j, m - 1), \tag{9.123}$$

we obtain

$$\sqrt{2(j_1+j_2)}\,\hbar\phi(j_1+j_2 j_1+j_2-1) = \sqrt{2j_1}\,\hbar\psi(j_1 j_2 j_1-1 j_2) + \sqrt{2j_2}\,\hbar\psi(j_1 j_2 j_1 j_2-1),$$

or

$$\phi(j_1+j_2 j_1+j_2-1) = \sqrt{\frac{j_1}{j_1+j_2}}\,\psi(j_1 j_2 j_1-1 j_2) + \sqrt{\frac{j_2}{j_1+j_2}}\,\psi(j_1 j_2 j_1 j_2-1). \quad (9.124)$$

A comparison with (9.114) gives us the values of two Clebsch–Gordan coefficients:

$$C(j_1 j_2 j_1-1 j_2; j_1+j_2 j_1+j_2-1) = \sqrt{\frac{j_1}{j_1+j_2}}, \quad (9.125)$$

$$C(j_1 j_2 j_1 j_2-1; j_1+j_2 j_1+j_2-1) = \sqrt{\frac{j_2}{j_1+j_2}}. \quad (9.126)$$

Operating (9.124) by J_-, we can obtain more transformation coefficients. A continuation of this process will give all the coefficients for which $j = j_1 + j_2$. However, this process is very lengthy. To avoid it, wherever possible, we make use of the following theorem.

Theorem 9.1.

If u and v are two orthonormal functions and $w_1 = a_1 u + b_1 v$ and $w_2 = a_2 u + b_2 v$ are two normalised linear combinations of them, then w_2 is orthogonal to w_1 provided that $a_2 = b_1^*$ and $b_2 = -a_1^*$.

Proof

Since w_1 and w_2 are normalised, we must have

$$|a_1|^2 + |b_1|^2 = 1, \quad (9.127)$$

$$|a_2|^2 + |b_2|^2 = 1. \quad (9.128)$$

As w_1 and w_2 are to be orthogonal, we should have

$$a_1^* a_2 + b_1^* b_2 = 0. \quad (9.129)$$

Substituting the value of b_2 from (9.129) in (9.128), we obtain

$$|a_2|^2 + \left|\frac{a_1^* a_2}{b_1^*}\right|^2 = 1,$$

or

$$|a_2|^2 + \frac{|a_1|^2 |a_2|^2}{|b_1|^2} = 1,$$

or

$$|a_2|^2 \, |b_1|^2 + |a_1|^2 |a_2|^2 = |b_1|^2,$$

or

$$|a_2|^2 \left(|a_1|^2 + |b_1|^2 \right) = |b_1|^2.$$

Since by virtue of (9.127), the expression in brackets in the above equation is unity, this equation yields

$$|a_2|^2 = |b_1|^2.$$

The phase factor is undetermined if we choose it so that $a_2 = b_1^*$. Then, from (9.129), we have $b_2 = -a_1^*$. This proves the theorem.

Applying the above theorem to (9.124), we obtain another normalised linear combination of ψ which is orthogonal to $\phi(j_1 + j_2 \, j_1 + j_2 - 1)$. Denoting it by ϕ', we have

$$\phi' = \sqrt{\frac{j_2}{j_1 + j_2}} \psi \left(j_1 j_2 j_1 - 1 j_2 \right) - \sqrt{\frac{j_1}{j_1 + j_2}} \psi \left(j_1 j_2 j_1 j_2 - 1 \right). \tag{9.130}$$

To what does ϕ' correspond? Operating both sides by J_z $(= J_{1z} + J_{2z})$, we obtain

$$J_z J_z \phi' = \left(J_{1z} + J_{2z} \right) \sqrt{\frac{j_2}{j_1 + j_2}} \psi \left(j_1 j_2 j_1 - 1 j_2 \right) - \left(J_{1z} + J_{2z} \right) \sqrt{\frac{j_1}{j_1 + j_2}} \psi \left(j_1 j_2 j_1 j_2 - 1 \right). \tag{9.131}$$

Problem

9.15. Show that the right-hand side of (9.131) is equal to $(j_1 + j_2 - 1)\hbar\phi'$.

Equation (9.131) then reduces to

$$J_z \phi' = \left(j_1 + j_2 - 1 \right) \hbar \phi'. \tag{9.132}$$

This shows that the function ϕ' is an eigenfunction of J_z corresponding to the eigenvalue $(j_1 + j_2 - 1)\hbar$. It must also be an eigenfunction of J^2. But what is the value of the quantum number j corresponding to the eigenvalue of J^2? To answer this question, we argue as follows.

(1) The quantum number j cannot be greater than $j_1 + j_2$ because $j_1 + j_2$ is the maximum value of j.

(2) It cannot be $j_1 + j_2$ because we have an eigenfunction for which the J^2 quantum number is $j_1 + j_2$ and a corresponding eigenvalue of J_z is $(j_1 + j_2 - 1)\hbar$. And the previous discussion has shown that all the simultaneous eigenfunctions of J^2 and J_z are non-degenerate.

(3) It cannot be less than $j_1 + j_2 - 1$ because for a simultaneous eigenfunction, the J^2 quantum number cannot be less than the J_z quantum number.

(4) Therefore, the J^2 quantum number must be $j_1 + j_2 - 1$.

Hence, $\phi' = \phi(j_1 + j_2 - 1\, j_1 + j_2 - 1)$. Equation (9.130) therefore becomes

$$\phi\left(j_1 + j_2 - j_1 + j_2 - 1\right) = \sqrt{\frac{j_2}{j_1 + j_2}}\,\psi\left(j_1 j_2 j_1 - 1 j_2\right) - \sqrt{\frac{j_1}{j_1 + j_2}}\,\psi\left(j_1 j_2 j_1 j_2 - 1\right). \tag{9.133}$$

This enables us to read two more Clebsch–Gordan coefficients.

Proceeding along these lines, making use of both techniques, we can determine all the C.

Problem

9.16. Add two non-interacting angular momenta $j_1 = 1$ and $j_2 = 1/2$, and find the values of the Clebsch–Gordan coefficients.

Additional problems

9.17. Show that the absolute value of the total angular momentum of an electron in a spherically symmetric field is a constant of the motion.

9.18. Find the normalised eigenvectors of the component of the electron spin operator along the direction of the z'-axis, making an angle Θ with the z-axis. If the spin component of an electron along the z'-axis has been established to be $+\hbar/2$, show that the probability that the measurement of the spin component along the z-axis will yield $+\hbar/2$ is $\cos^2 \Theta/2$.

9.19. Find the expectation values of L^2 for a hydrogen atom in a state characterised by $n = 2$, $\ell = 0$, $m = 0$. The corresponding wave function ψ_{200} is given by

$$\psi_{200} = \frac{1}{4\sqrt{2\pi}}\left(\frac{1}{a_0}\right)^{3/2}\left(2 - \frac{r}{a_0}\right)\exp\left(\frac{-r}{2a_0}\right).$$

9.20. Show that the eigenvalues of the three components of the spin operator **S** are $\pm\hbar/2$ and find their normalised eigenvectors.
9.21. Find out whether the operator $L_yL_z + L_zL_y$ is Hermitian or not.
9.22. Determine the normalised Legendre polynomials of degree 4.
9.23. Find the product of the Pauli spin matrices.
9.24. In the basis of the eigenvectors of L_z and L^2, determine the matrix representations of L_z and L^2 for $\ell = 1$.
9.25. Show that the only matrix which commutes with the Pauli spin matrices is a scalar matrix.
9.26. Calculate the Clebsch–Gordan coefficients for
(a) $j_1 = \frac{1}{2}$ and $j_2 = \frac{1}{2}$,
(b) $j_1 = 1$ and $j_2 = \frac{1}{2}$.
9.27. What are the possible total spin quantum numbers of a system of three particles each of spin $\frac{1}{2}$. Also, find the corresponding spin function.
9.28. Find the eigenvalues and eigenvectors of Pauli's spin matrices.
9.29. Why do we obtain different characteristics of angular momentum when we start with classical and quantum mechanical definitions of angular momentum?

IOP Publishing

Quantum Mechanics

Mohammad Saleem

Chapter 10

Perturbation theory

10.1 Introduction

We have so far considered only those physical systems for which the Schrödinger amplitude equation could be exactly solved. But for most of the physical systems encountered in nature, the Schrödinger amplitude equation cannot be solved exactly as appropriate mathematical tools are not available, and we have to resort to approximation. The most important method for approximate solutions is the *perturbation method.* This method is used to obtain approximate solutions of the Schrödinger amplitude equation for a given system provided that the Hamiltonian of this system can be expressed as a sum of two parts, one of which is large and characterises some system, called the *unperturbed system*, for which the Schrödinger amplitude equation is *exactly solvable*, while the other part is very small compared to the first one and can be treated as causing perturbation. The given system is said to be a *perturbed system*. For instance, the Hamiltonian of the electron of a hydrogen atom in the Coulomb field of the proton (the nucleus of the hydrogen atom) could be treated as a Hamiltonian of the unperturbed system. The Schrödinger amplitude equation for this Hamiltonian can be solved exactly. The perturbation could be, say, due to the application of a small electric field. We will first show how this method developed for the steady state by Schrödinger himself can be used in practice for solving the time-independent Schrödinger wave equation. There are two cases: (1) the energy spectrum of the unperturbed system is non-degenerate and (2) the energy spectrum of the unperturbed system has some degeneracy. We will consider these two cases individually. We will then consider the application of perturbation theory to the time-dependent Schrödinger wave equation.

10.2 Time-independent perturbation theory for nondegenerate states

Consider a physical system. Let H be its Hamiltonian operator which does not depend upon time explicitly. Therefore, it is sufficient to solve the Schrödinger amplitude equation. Suppose that the amplitude equation

$$H\psi_n = E_n\psi_n \tag{10.1}$$

doi:10.1088/978-0-7503-1206-6ch10

for the system cannot be solved exactly. Then the exact expression for ψ_n and the exact value of E_n cannot be obtained. Suppose further that the Hamiltonian H can be expressed as a sum of two parts,

$$H = H_0 + \lambda H' \tag{10.2}$$

where H_0 is the Hamiltonian operator for an unperturbed system for which the corresponding Schrödinger wave equation

$$H_0\psi_n^{(0)} = E_n^{(0)}\psi_n^{(0)} \tag{10.3}$$

is exactly solvable. The part H' of the Hamiltonian of the given system is very small compared to H_0 and is called the *perturbation Hamiltonian*. The arbitrary parameter λ has been introduced only for mathematical facility to keep a track of the order of the various terms and is finally made equal to unity. The function $\psi_n^{(0)}$ and the energy $E_n^{(0)}$ denote the eigenfunction and the eigenvalue of the unperturbed system in the nth state. The energy spectrum of H_0 is *assumed* to be discrete and non-degenerate. However, the theory developed here can be easily extended to the case when the energy spectrum of H_0 is partly continuous provided that $E_n^{(0)}$ is one of the discrete sets of its eigenvalues. The basic assumption in the perturbation technique is that the unperturbed Hamiltonian H_0 evolves continuously into the perturbed Hamiltonian H as the parameter λ goes from zero to unity.

Since H' is very small compared to H_0, it is reasonable to assume that the perturbation will cause a very small change in the eigenfunctions and eigenvalues of the unperturbed system. We may therefore express ψ_n and E_n as:

$$\psi_n = \psi_n^{(0)} + \lambda\psi_n^{(1)} + \lambda^2\psi_n^{(2)} + \cdots, \tag{10.4}$$

$$E_n = E_n^{(0)} + \lambda E_n^{(1)} + \lambda^2 E_n^{(2)} + \cdots, \tag{10.5}$$

where the functions $\psi_n^{(0)}$, $\psi_n^{(1)}$,... and energies $E_n^{(0)}$, $E_n^{(1)}$,... are independent of λ. As the perturbation H' is very small, the above two series are expected and therefore assumed to converge rapidly. The functions $\psi_n^{(1)}$, $\psi_n^{(2)}$,... are the first order, second order,... corrections to the eigenfunction $\psi_n^{(0)}$of the unperturbed system while the energies $E_n^{(1)}$, $E_n^{(2)}$,... are the first-order, second-order... corrections to the eigenvalue $E_n^{(0)}$ of the unperturbed system. That is, the power of λ in any term reflects the order of smallness of the corresponding ψ_n or E_n. As it facilitates dealing with orthonormal wave functions, it will be assumed in the following discussion that this has been done in the case of $\psi_i^{(0)}$.

To find an approximate solution of the Schrödinger wave equation for the given system, i.e.

$$H\psi_n = E_n\psi_n \tag{10.1'}$$

we insert in it the expressions for H, ψ_n and E_n as given, respectively, by (10.2), (10.4) and (10.5). This yields

$$(H_0 + \lambda H')\left(\psi_n^{(0)} + \lambda\psi_n^{(1)} + \lambda^2\psi_n^{(2)} + \cdots\right)$$
$$= \left(E_n^{(0)} + \lambda E_n^{(1)} + \lambda^2 E_n^{(2)} + \cdots\right)\left(\psi_n^{(0)} + \lambda\psi_n^{(1)} + \lambda^2\psi_n^{(2)} + \cdots\right),$$

or

$$H_0\psi_n^{(0)} + \lambda\left(H_0\psi_n^{(1)} + H'\psi_n^{(0)}\right) + \lambda^2\left(H_0\psi_n^{(2)} + H'\psi_n^{(1)}\right) + \cdots$$
$$= E_n^{(0)}\psi_n^{(0)} + \lambda\left(E_n^{(0)}\psi_n^{(1)} + E_n^{(1)}\psi_n^{(0)}\right) + \lambda^2\left(E_n^{(0)}\psi_n^{(2)} + E_n^{(1)}\psi_n^{(1)}\right.$$
$$\left. + E_n^{(2)}\psi_n^{(0)}\right) + \cdots \tag{10.6}$$

The coefficients of various powers of λ on the two sides of this equation must be equal. Equating the terms not containing λ, we obtain

$$H_0\psi_n^{(0)} = E_n^{(0)}\psi_n^{(0)}. \tag{10.3'}$$

This is known as the *zero order equation*. It represents the unperturbed system and as already mentioned it is exactly solvable. The solution gives us $E_n^{(0)}$ and $\psi_n^{(0)}$. The *first-order equation* is obtained by equating the coefficients of λ on the two sides of (10.6). This gives

$$H_0\psi_n^{(1)} + H'\psi_n^{(0)} = E_n^{(0)}\psi_n^{(1)} + E_n^{(1)}\psi_n^{(0)},$$

or

$$H_0\psi_n^{(1)} - E_n^{(0)}\psi_n^{(1)} = E_n^{(1)}\psi_n^{(0)} - H'\psi_n^{(0)}. \tag{10.7}$$

In this equation, $\psi_n^{(1)}$ and $E_n^{(1)}$, the first-order corrections to the unperturbed eigenfunction and eigenvalue corresponding to the nth state, are the only unknown quantities. The two equations, i.e. (10.7) and the equation obtained by using the normalisation condition, are sufficient to determine these two unknown quantities. Let us see how this can be achieved.

To obtain the value of $E_n^{(1)}$ and the expression for $\psi_n^{(1)}$, we proceed as follows. Recalling that a Hamiltonian operator is always Hermitian and therefore its eigenfunctions form a complete set, we express $\psi_n^{(1)}$ occuring in (10.7) in terms of eigenfunctions $\psi_j^{(0)}$ of H_0 as

$$\psi_n^{(1)} = \sum_j a_{nj}\psi_j^{(0)}. \tag{10.8}$$

The first index of a_{nj} in the above expansion corresponds to the state of the perturbed system while the second index refers to the unperturbed state. The first-order correction $\psi_n^{(1)}$ of the eigenfunction $\psi_n^{(0)}$ is completely determined if we can evaluate all a. Substituting the expression for $\psi_n^{(1)}$ from (10.8) in (10.7), we obtain

$$H_0 \sum_j a_{nj}\psi_j^{(0)} - E_n^{(0)} \sum_j a_{nj}\psi_j^{(0)} = \left(E_n^{(1)} - H'\right)\psi_n^{(0)},$$

or

$$\sum_j a_{nj} H_0 \psi_j^{(0)} - \sum_j a_{nj} E_n^{(0)} \psi_j^{(0)} = \left(E_n^{(1)} - H'\right)\psi_n^{(0)},$$

or

$$\sum_j a_{nj}\left(E_j^{(0)} - E_n^{(0)}\right)\psi_j^{(0)} = \left(E_n^{(1)} - H'\right)\psi_n^{(0)},$$

where we have used (10.3) with $n = j$. Multiplying the last equation from the left by $\psi_m^{(0)*}$ (m may or may not be equal to n) and integrating *over all space*, we obtain

$$\int \sum_j a_{nj}(E_j^0 - E_n^{(0)})\psi_m^{(0)*}\psi_j^{(0)}\, \mathrm{d}\tau = \int \psi_m^{(0)*}(E_n^{(1)} - H')\psi_n^{(0)}\, \mathrm{d}\tau,$$

or

$$\sum_j a_{nj}(E_j^{(0)} - E_n^{(0)}) \int \psi_m^{(0)*}\psi_j^{(0)}\, \mathrm{d}\tau = E_n^{(1)} \int \psi_m^{(0)*}\psi_n^{(0)}\, \mathrm{d}\tau - \int \psi_m^{(0)*} H' \psi_n^{(0)}\, \mathrm{d}\tau. \quad (10.9)$$

Since the eigenfunctions of the Hermitian operator H_0 of the unperturbed system with non-degenerate eigenvalues form an orthonormal set, we have

$$\int \psi_m^{(0)*}\psi_j^{(0)}\, \mathrm{d}\tau = \delta_{mj}.$$

Equation (10.9) therefore becomes

$$\sum_j a_{nj}\left(E_j^{(0)} - E_n^{(0)}\right)\delta_{mj} = E_n^{(1)}\delta_{mn} - H'_{mn,} \quad (10.10)$$

where

$$H'_{mn} = \int \psi_m^{(0)*} H' \psi_n^{(0)}\, \mathrm{d}\tau \quad (10.11)$$

is called the *matrix element of* H' referred to $\psi_m^{(0)}$ and $\psi_n^{(0)}$. Summing over j on the left-hand side of (10.10), we find that all those terms for which $j \neq m$ vanish because then $\delta_{mj} = 0$ and we are left with only one term, i.e. that for which $j = m$. Equation (10.10) therefore reduces to

$$a_{nm}(E_m^{(0)} - E_n^{(0)}) = E_n^{(1)}\delta_{mn} - H'_{mn}. \quad (10.12)$$

10.3 First-order correction to energy

To find the first-order correction to energy, i.e. $E_n^{(1)}$, we take $m = n$ in (10.12). The left-hand side then becomes zero, the term containing a_{nn} disappears and, since $\delta_{nn} = 1$, this equation yields

$$E_n^{(1)} = H'_{nn} = \int \psi_n^{(0)*} H' \psi_n^{(0)} \, d\tau. \tag{10.13}$$

Equation (10.13) determines the first-order correction $E_n^{(1)}$ to energy. This correction can now be evaluated because all the quantities in the integral in (10.13) are known. We observe that to obtain this correction, we need not have any knowledge of the first-order corrections to the unperturbed eigenfunctions. This equation also tells us that the first-order correction to energy, i.e. $E_n^{(1)}$, is just the expectation value of the perturbation Hamiltonian H' calculated using the unperturbed eigenfunction $\psi_n^{(0)}$. Thus the energy of the perturbed system, correct to the first order, is

$$E_n = E_n^{(0)} + E_n^{(1)} = E_n^{(0)} + H'_{nn}. \tag{10.14}$$

To sum up: the first-order correction to energy was obtained by expressing the unknown first-order correction $\psi_n^{(1)}$ to the eigenfunction as a linear combination of the complete set of eigenfunctions of the unperturbed Hamiltonion H_0:

$$\psi_n^{(1)} = \sum_j a_{nj} \psi_j^{(0)}. \tag{10.8'}$$

Thus, by computing the a, we could find an expression for the first-order correction $\psi_n^{(1)}$ to the eigenfunction. However, during this process it was noticed that all the terms containing a, except one, disappeared. For $m = n$, the remaining term containing a_{nn} also becomes zero and we obtain a value for the first-order correction $E_n^{(1)}$ to energy.

Example 10.1.

Problem

A linear harmonic oscillator is perturbed by an additional potential energy bx^4. Find the first-order correction to energy in the ground state of the linear oscillator.

Solution

The eigenfunction for the ground state of an unpertubed linear harmonic oscillator is given by

$$\psi_0(x) = \left(\frac{\alpha}{\pi}\right)^{1/4} \exp\left(-\frac{\alpha x^2}{2}\right).$$

As shown above, the first-order correction to energy is

$$E_0^{(1)} = \int \psi_0^{(0)*} H' \psi_0^{(0)} \, d\tau.$$

Since the perturbation Hamiltonian is $H' = bx^4$, the above equation takes the form

$$E_0^{(1)} = b\left(\frac{\alpha}{\pi}\right)^{1/2} \int x^4 e^{-\alpha x^2}\, dx = b\sqrt{\frac{\alpha}{\pi}}\sqrt{\frac{9\pi}{16\alpha^5}} = \frac{3b}{4a^2}.$$

This gives the first-order correction to the energy of a linear harmonic oscillator in its ground state under a perturbation Hamiltonian equivalent to bx^4.

Example 10.2.

Problem

A linear harmonic oscillator is perturbed by a potential energy bx^3. Calculate the first-order correction to energy in the nth state.

Solution

According to the perturbation theory, the first-order correction $E_n^{(1)}$ to the energy of the system in the nth state is given by

$$E_n^{(1)} = \int_{-\infty}^{\infty} \psi_n^{(0)*} H' \psi_n^{(0)}\, dx$$

$$= \int_{-\infty}^{\infty} \psi_n^{(0)*} bx^3 \psi_n^{(0)}\, dx.$$

Since the wave function is even or odd according to whether n is even or odd, the product $\psi_n^{(0)*}\psi_n^{(0)}$ will always be even. The product of this even function with x^3, an odd function of x, gives an odd function of x. Hence, between the limits from $-\infty$ to $+\infty$, the integral will be zero. We conclude that with such a perturbation, the first-order correction to energy is always zero.

10.4 The anomalous Zeeman effect

While explaining the normal Zeeman effect, we stated that in a *strong* magnetic field with a strength of more than 20 000 gauss, interaction of the field with magnetic moment due to the electronic current dominates. However, when the hydrogen atom is placed in a *weak* magnetic field, it is found that the magnetic moment due to the spin of the electron and the spin–orbit coupling play equally significant roles. Let us see how this helps us in explaining the anomalous Zeeman effect. We know from classical electromagnetic theory that the interaction energy between the magnetic moment $\boldsymbol{\mu}$ and the field $\mathbf{B}$ is given by

$$E = -\boldsymbol{\mu} \cdot \mathbf{B}.$$

The magnetic moment arises due to the orbital angular momentum, the spin angular momentum and the coupling between them. The expressions for $\boldsymbol{\mu}_L$ and $\boldsymbol{\mu}_S$ are given by

$$\boldsymbol{\mu}_L = -\frac{e}{2mc}\mathbf{L}, \tag{10.15a}$$

$$\boldsymbol{\mu}_S = -\frac{e}{2mc}2\mathbf{S} \tag{10.15b}$$

If there was no spin–orbit coupling or it was weak enough to be ignored, then the energy operator H' would be given by

$$H' = -(\boldsymbol{\mu}_L + \boldsymbol{\mu}_S)\cdot\mathbf{B} = \frac{e}{2mc}\left(L + \frac{e}{2mc}2\mathbf{S}\right)\cdot\mathbf{B} \tag{10.16}$$

However, if *LS*-coupling is taken into account, the scenario changes. The magnetic moment $\boldsymbol{\mu}_L$ is antiparallel to **L** and the magnetic moment $\boldsymbol{\mu}_S$ is antiparallel to **S**, but the magnetic moment $\boldsymbol{\mu}_J$ is *not* antiparallel to the resultant angular momentum $\mathbf{J} = \mathbf{L} + \mathbf{S}$. This is due to the reason that $\boldsymbol{\mu}_J$ is proportional to $\mathbf{L} + 2\mathbf{S}$ and *not* to $\mathbf{L} + \mathbf{S}$. Therefore, as shown in figure 10.1, although $\boldsymbol{\mu}_L$ and **L** are in the same line, and $\boldsymbol{\mu}_S$ and **S** are in the same line, the vectors $\boldsymbol{\mu}_J$ and **J** are *not* in the same line. In fact, if we consider the components of the magnetic moment vector $\boldsymbol{\mu}_J$ along and perpendicular to the vector **J**, the component perpendicular to the vector **J** vanishes on the average as the vector $\boldsymbol{\mu}_J$ precesses about **J**. Consequently, we have to take into consideration the component of $\boldsymbol{\mu}_J$ along **J**. This is given by

$$\boldsymbol{\mu}_J \cdot \hat{\mathbf{J}}$$

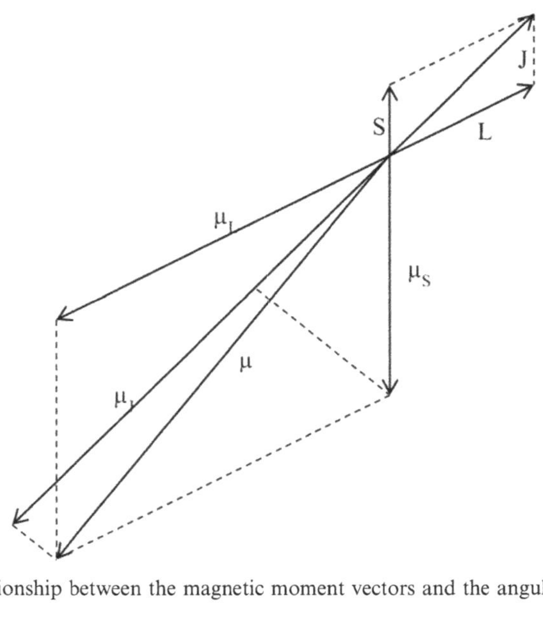

Figure 10.1. The relationship between the magnetic moment vectors and the angular momentum vectors.

where $\hat{\mathbf{J}}$ is a unit vector in the direction of $\mathbf{J}$. Taking the direction into account, we can write it as

$$\boldsymbol{\mu}_J \cdot \hat{\mathbf{J}}\hat{\mathbf{J}}.$$

Therefore, using (10.15) and (10.16), we obtain

$$\begin{aligned}\boldsymbol{\mu}_J \cdot \hat{\mathbf{J}}\hat{\mathbf{J}} &= \left(\boldsymbol{\mu}_L + \boldsymbol{\mu}_S\right) \cdot \hat{\mathbf{J}}\hat{\mathbf{J}} \\ &= -\frac{e}{2mc}(\mathbf{L} + 2\mathbf{S}) \cdot \hat{\mathbf{J}}\hat{\mathbf{J}} \\ &= -\frac{e}{2mc}(\mathbf{J}\,1+\mathbf{S}) \cdot \hat{\mathbf{J}}\hat{\mathbf{J}} \\ &= -\frac{e}{2mc}\frac{(\mathbf{J} + \mathbf{S}) \cdot \mathbf{J}\mathbf{J}}{\mathbf{J}^2},\end{aligned} \tag{10.17}$$

where we have made use of the fact that $\mathbf{J} = |\mathbf{J}|\,\hat{\mathbf{J}}$. Hence, the perturbation Hamiltonian is

$$\begin{aligned}H' = -\boldsymbol{\mu} \cdot \hat{\mathbf{J}}\,\hat{\mathbf{J}} \cdot \mathbf{B} &= \frac{e}{2mc}\frac{(\mathbf{J} + \mathbf{S}) \cdot \mathbf{J}}{\mathbf{J}^2}(\mathbf{J} \cdot \mathbf{B}) \\ &= \frac{e}{2mc}\frac{\left(\mathbf{J}^2 + \mathbf{S} \cdot \mathbf{J}\right)}{\mathbf{J}^2}(\mathbf{J} \cdot \mathbf{B}).\end{aligned} \tag{10.18}$$

Now

$$\mathbf{J} = \mathbf{L} + \mathbf{S},$$

or

$$\mathbf{J} - \mathbf{S} = \mathbf{L}.$$

Squaring both sides of the above equation, we obtain

$$\mathbf{J}^2 + \mathbf{S}^2 - 2\mathbf{J} \cdot \mathbf{S} = \mathbf{L}^2,$$

or

$$2\mathbf{J} \cdot \mathbf{S} = \mathbf{J}^2 + \mathbf{S}^2 - \mathbf{L}^2.$$

Substituting this expression for $\mathbf{J} \cdot \mathbf{S}$ in (10.18) and taking the magnetic field $\mathbf{B}$ along the z-axis, we obtain

$$H' = \frac{e}{2mc}\frac{\left[\mathbf{J}^2 + \frac{1}{2}\left(\mathbf{J}^2 + \mathbf{S}^2 - \mathbf{L}^2\right)\right]}{\mathbf{J}^2}J_z B. \tag{10.19}$$

Operating it upon the function $\psi_{n\ell m}$, we have

$$H'\psi_{n\ell m} = \left[\frac{e}{2mc}\frac{\left[\mathbf{J}^2 + \frac{1}{2}\left(\mathbf{J}^2 + \mathbf{S}^2 - \mathbf{L}^2\right)J_z B\right]}{\mathbf{J}^2}\right]\psi_{n\ell m}$$

$$= \left[\frac{eB}{2mc}\frac{\left\{\mathbf{J}^2 + \frac{1}{2}\left(\mathbf{J}^2 + \mathbf{S}^2 - \mathbf{L}^2\right)\right\}}{\mathbf{J}^2}\right]m_j\hbar\psi_{n\ell m}, \quad \text{because } J_z\psi_{n\ell m} = m_j\hbar\psi_{n\ell m},$$

$$= \frac{e\hbar B}{2mc}m_j\frac{\left[\mathbf{J}^2 + \frac{1}{2}\left(\mathbf{J}^2 + \mathbf{S}^2 - \mathbf{L}^2\right)\right]}{\mathbf{J}^2}\psi_{n\ell m}. \tag{10.20}$$

Noting that $\mathbf{J}^2\psi_{n\ell m} = j(j+1)\hbar^2$, etc, we obtain

$$H'\psi_{n\ell m} = \frac{e\hbar B}{2mc}m_j\left[1 + \frac{1}{2}\frac{j(j+1) + (s(s+1) - \ell(\ell+1)}{j(j+1)}\right]\psi_{n\ell m}. \tag{10.21}$$

The expression in square brackets is called Lande's g-factor and is denoted by g:

$$g = \left[1 + \frac{1}{2}\frac{j(j+1) + (s(s+1) - \ell(\ell+1)}{j(j+1)}\right], \tag{10.22}$$

so that (10.21) takes the form

$$H'\psi_{n\ell m} = \frac{e\hbar B}{2mc}m_j g\psi_{n\ell m}. \tag{10.23}$$

Multiplying from the left by $\psi^*_{n\ell m}$ and integrating, we obtain

$$\int\psi^*_{n\ell m}H'\psi_{n\ell m}\,\mathrm{d}\tau = \frac{e\hbar B}{2mc}m_j g\int\psi^*_{n\ell m}\psi_{n\ell m}\,\mathrm{d}\tau. \tag{10.23a}$$

Now the first-order correction in energy is given by

$$\Delta E = \int\psi^*_{n\ell m}H'\psi_{n\ell m}\,\mathrm{d}\tau.$$

Substituting this expression in (10.23a) and recalling that the functions $\psi_{n\ell m}$ are normalised, we obtain

$$\Delta E = \frac{e\hbar B}{2mc}m_j g. \tag{10.24}$$

This shows that in the presence of a weak magnetic field, the energy levels are split according to (10.24).

If $s = 0$, then $j = \ell + s = \ell$ so that $g = 1$ and (10.24) yields

$$\Delta E = \frac{e\hbar B}{2mc} m_j.$$

Let us next consider the first-order correction to the eigenfunction.

10.5 The first-order correction to the eigenfunction

The first-order correction to energy has been obtained using (10.12) with $m = n$. The same equation can be used to obtain the first-order correction $\psi_n^{(1)}$ to the eigenfunction $\psi_n^{(0)}$ but with $m \neq n$. For $m \neq n$, this equation becomes

$$a_{nm}\left(E_m^{(0)} - E_n^{(0)}\right) = -H'_{mn},$$

or

$$a_{nm} = \frac{H'_{mn}}{E_n^{(0)} - E_m^{(0)}}, \quad m \neq n. \tag{10.25}$$

This equation gives all a except a_{nn} because then the right-hand side of the above equation becomes indeterminate.

To determine the value of a_{nn}, we make use of the normalisation condition for ψ_n. If the wave function ψ_n is normalised to unity, then

$$\int \psi_n^* \psi_n \, \mathrm{d}\tau = 1.$$

Substituting the expression for ψ_n from (10.4) in the above equation, we obtain

$$\int \left(\psi_n^{(0)} + \lambda\psi_n^{(1)} + \lambda^2\psi_n^{(2)} + \cdots\right)^* \left(\psi_n^{(0)} + \lambda\psi_n^{(1)} + \lambda^2\psi_n^{(2)} + \cdots\right) \mathrm{d}\tau = 1,$$

or

$$\int \psi_n^{(0)*}\psi_n^{(0)} \, \mathrm{d}\tau + \lambda \int \left(\psi_n^{(0)*}\psi_n^{(1)} + \psi_n^{(1)*}\psi_n^{(0)}\right) \mathrm{d}\tau$$
$$+ \lambda^2 \int \left(\psi_n^{(0)*}\psi_n^{(2)} + \psi_n^{(1)*}\psi_n^{(1)} + \psi_n^{(2)*}\psi_n^{(0)}\right) \mathrm{d}\tau + \cdots = 1.$$

Since $\psi_n^{(0)}$ are normalised to unity, the first integral on the left-hand side is equal to unity, and the above equation reduces to

$$\lambda \int \left(\psi_n^{(0)*}\psi_n^{(1)} + \psi_n^{(1)*}\psi_n^{(0)}\right) \mathrm{d}\tau + \lambda^2 \int \left(\psi_n^{(0)*}\psi_n^{(2)} + \psi_n^{(1)*}\psi_n^{(1)} + \psi_n^{(2)*}\psi_n^{(0)}\right) \mathrm{d}\tau + \cdots = 0.$$

Since λ is an arbitrary parameter, the coefficient of each power of λ should be separately zero:

$$\int \left(\psi_n^{(0)*}\psi_n^{(1)} + \psi_n^{(1)*}\psi_n^{(0)}\right) \mathrm{d}\tau = 0, \tag{10.26a}$$

$$\int \left(\psi_n^{(0)*}\psi_n^{(2)} + \psi_n^{(1)*} + \psi_n^{(1)} + \psi_n^{(2)*}\psi_n^{(0)}\right) \mathrm{d}\tau = 0, \tag{10.26b}$$

and so on. Substituting the expression for $\psi_n^{(1)}$ from (10.8) in (10.26*a*), we obtain

$$\int\left[\left(\psi_n^{(0)*}\sum_j a_{nj}\psi_j^{(0)} + \sum_j a_{nj}^*\psi_j^{(0)*}\psi_n^{(0)}\right)\right] d\tau = 0,$$

or

$$\sum_j a_{nj}\int \psi_n^{(0)*}\psi_j^{(0)}\, d\tau + \sum_j a_{nj}^*\int \psi_j^{(0)*}\psi_j^{(0)*}\psi_n^{(0)}\, d\tau = 0,$$

or

$$\sum_j a_{nj}\delta_{nj} + \sum_j a_{nj}^*\delta_{nj} = 0.$$

Summing over *j*, we obtain

$$a_{nn} + a_{nm}^* = 0.$$

Problem

10.1. Show that if the sum of a complex number and its complex conjugate is equal to zero, then the number must be pure imaginary.

The above relation shows that a_{nn} must be pure imaginary. We may write it as

$$a_{nn} = \mathrm{i}\beta,$$

where β is real. We now show that $a_{nn} = \mathrm{i}\beta$ has no physical significance. We proceed as below. In the first approximation, we have

$$\psi_n = \psi_n^{(0)} + \lambda\psi_n^{(1)}.$$

Substituting the expression for $\psi_n^{(1)}$ in terms of $\psi_j^{(0)}$, we obtain

$$\psi_n = \psi_n^{(0)} + \lambda\sum_j a_{nj}\psi_j^{(0)}.$$

By separating the $j = n$ term from the sigma sign, we can write the above equation as

$$\psi_n = \psi_n^{(0)} + \lambda\left(a_{nn}\psi_n^{(0)} + \sum_j{}' a_{nj}\psi_j^{(0)},\right)$$

where Σ' means that the summation is over all the values of *j* except $j = n$. Recalling that a_{nn} is pure imaginary and writing $\mathrm{i}\beta$ for it in the above equation, we obtain

$$\psi_n = \psi_n^{(0)} + \lambda\left(\mathrm{i}\beta\psi_n^{(0)} + \sum_j{}' a_{nj}\psi_j^{(0)}\right),$$

or

$$\psi_n = (1 + i\lambda\beta)\psi_n^{(0)} + \sum_j{}' a_{nj}\psi_j^{(0)}. \tag{10.27a}$$

Now, *to a first approximation*, $(1 + i\lambda\beta)$ is equal to $e^{i\lambda\beta}$. Therefore, the above equation may be written as

$$\psi_n = \psi_n^{(0)} e^{i\lambda\beta} + \sum_j{}' a_{nj}\psi_j^{(0)}. \tag{10.27b}$$

This shows that the constant β merely multiplies the function $\psi_n^{(0)}$ by the phase factor $e^{i\lambda\beta}$. Since all the ψ are mutually orthogonal and have been individually normalised and physical quantities always involve $\psi^*\psi$, the factor $e^{i\lambda\beta}$ is physically unimportant. Hence, we may put β (and consequently a_{nn}) equal to zero:

$$a_{nn} = 0. \tag{10.27c}$$

Equation (10.27*b*) therefore reduces to

$$\psi_n = \psi_n^{(0)} + \sum_j{}' a_{nj}\psi_j^{(0)}. \tag{10.28}$$

This gives the eigenfunction corrected to the first order; all a occurring in this equation can be calculated using (10.25). Equation (10.28) shows that in order to determine the expression for the correction to *only* one of the eigenfunctions, we have to find in principle an infinite number of a. However, as the series converges rapidly, in practice we need only a finite and a small number of a.

10.6 Second-order non-degenerate perturbation

In some physical problems, the first-order correction to energy is either zero or too small to yield a result consistent with experiments. The second-order corrections then become imperative. To determine such corrections to the energy eigenvalue and eigenfunction, we equate the coefficients of λ^2 on the two sides of (10.6). We obtain

$$H_0\psi_n^{(2)} + H'\psi_n^{(1)} = E_n^{(0)}\psi_n^{(2)} + E_n^{(1)}\psi_n^{(1)} + E_n^{(2)}\psi_n^{(0)}. \tag{10.29}$$

With the first-order quantities $E_n^{(1)}$ and $\psi_n^{(1)}$ determined above, the only unknowns involved in this equation are $E_n^{(2)}$ and $\psi_n^{(2)}$. These can be determined in the following manner. We express $\psi_n^{(1)}$ and $\psi_n^{(2)}$ in terms of the complete set of eigenfunctions $\psi_j^{(0)}$. We have already expanded $\psi_n^{(1)}$ as:

$$\psi_n^{(1)} = \sum_j a_{nj}\psi_j^{(0)}. \tag{10.8'}$$

We now expand $\psi_n^{(2)}$ in terms of the complete set of unperturbed eigenfunctions $\psi_j^{(0)}$:

$$\psi_n^{(2)} = \sum_j b_{nj}\psi_j^{(0)}. \tag{10.30}$$

Substituting for $\psi_n^{(2)}$ from (10.30) and for $\psi_n^{(1)}$ from (10.8′) in (10.29), we obtain

$$H_0 \sum_j b_{nj}\psi_j^{(0)} + H' \sum_j a_{nj}\psi_j^{(0)} = E_n^{(0)} \sum_j b_{nj}\psi_j^{(0)} + E_n^{(1)} \sum_j a_{nj}\psi_j^{(0)} + E_n^{(2)}\psi_n^{(0)},$$

or

$$\sum_j b_{nj} H_0 \psi_j^{(0)} + \sum_j a_{nj} H' \psi_j^{(0)} = \sum_j b_{nj} E_n^{(0)} \psi_j^{(0)} + \sum_j a_{nj} E_n^{(1)} \psi_j^{(0)} + E_n^{(2)}\psi_n^{(0)}.$$

Since $H_0\psi_j^{(0)} = E_j^{(0)}\psi_j^{(0)}$, the above equation gives

$$\sum_j b_{nj}\left(E_j^{(0)} - E_n^{(0)}\right)\psi_j^{(0)} + \sum_j a_{nj} H' \psi_j^{(0)} = \sum_j a_{nj} E_n^{(1)} \psi_j^{(0)} + E_n^{(2)}\psi_n^{(0)}.$$

Multiplying the above equation from the left by $\psi_m^{(0)*}$ and integrating over all space, we obtain

$$\sum_j b_{nj}\left(E_j^{(0)} - E_n^{(0)}\right) \int \psi_m^{(0)*}\psi_j^{(0)}\, d\tau + \sum_j a_{nj} \int \psi_m(0)^* H' \psi_j^{(0)}\, d\tau$$
$$= \sum_j a_{nj} E_n^{(1)} \int \psi_m^{(0)*}\psi_j^{(0)}\, d\tau + E_n^{(2)} \int \psi_m^{(0)*}\psi_n^{(0)}\, d\tau. \qquad (10.31)$$

We now make use of the orthonormalisation condition for the unperturbed eigenfunctions $\psi_j^{(0)}$, which, being the eigenfunctions of a Hermitian operator corresponding to non-degenerate eigenvalues, are also mutually orthogonal. That is, these unperturbed eigenfunctions form an orthonormal set. Therefore, we may write

$$\int \psi_m^{(0)*}\psi_j^{(0)}\, d\tau = \delta_{mj}$$

and, consequently, (10.31) can be written as

$$\sum_j b_{nj}\left(E_j^{(0)} - E_n^{(0)}\right)\delta_{mj} + \sum_j a_{nj} \int \psi_m^{(0)*} H' \psi_j^{(0)}\, d\tau = \sum_j a_{nj} E_n^{(1)} \delta_{mj} + E_n^{(2)}\delta_{mn}.$$

Summing over j in the first two terms on the two sides of the above equation, we obtain

$$b_{nm}\left(E_m^{(0)} - E_n^{(0)}\right) + \sum_j a_{nj} \int \psi_m^{(0)*} H' \psi_j^{(0)}\, d\tau = a_{nm} E_n^{(1)} + E_n^{(2)}\delta_{mn}. \qquad (10.32)$$

This equation can be used to obtain the second-order correction to energy as well as the eigenfunction.

10.7 The second-order correction to energy

To obtain the second-order correction to energy, we proceed as follows.

For $m = n$, (10.32) yields

$$\sum_j a_{nj} \int \psi_n^{(0)*} H' \psi_j^{(0)}\, d\tau = a_{nn} E_n^{(1)} + E_n^{(2)},$$

or

$$\sum_j a_{nj} H'_{nj} = a_{nn} E_n^{(1)} + E_n^{(2)}.$$

But, as shown earlier, a_{nn} can be taken as zero. The above equation therefore reduces to

$$E_n^{(2)} = \sum_j a_{nj} H'_{nj},$$

or

$$E_n^{(2)} = \sum_j{}' a_{nj} H'_{nj} \tag{10.33}$$

where Σ' means that the summation over j excludes $j = n$ as $a_{nn} = 0$. Substituting the expression for a_{nj} from (10.25) in (10.33), we obtain

$$E_n^{(2)} = \sum_j{}' \frac{H'_{jn} H'_{nj}}{E_n^{(0)} - E_j^{(0)}}. \tag{10.34a}$$

This gives the second-order correction to energy. All the quantities on the right-hand side of the above equation being known, $E_n^{(2)}$ can be calculated.

Problem

10.2. Show that if H' is real and is a multiplication operator, then

$$H'_{nj} = H'^{*}_{jn}$$

hence

$$|H'_{nj}|^2 = |H'^{*}_{jn}|^2.$$

Equation (10.34a) can now be written as

$$E_n^{(2)} = \sum_j{}' \frac{H'^{*}_{nj} H'_{nj}}{E_n^{(0)} - E_j^{(0)}}$$

$$= \sum_j{}' \frac{|H'_{nj}|^2}{E_n^{(0)} - E_j^{(0)}}$$

$$= \sum_j{}' \frac{|H'_{jn}|^2}{E_n^{(0)} - E_j^{(0)}} \tag{10.34b}$$

10.8 The second-order correction to the eigenfunction

Let us now calculate the second-order correction to the eigenfunction. We have seen that for $m = n$, (10.32) can be used to obtain the second-order correction to energy. For $m \neq n$, the same equation can be used to obtain the second-order correction to eigenfunction. For $m \neq n$, this equation yields

$$b_{nm}\left(E_m^{(0)} - E_n^{(0)}\right) + \sum_j a_{nj} H'_{mj} = a_{nm} E_n^{(1)},$$

or

$$\begin{aligned} b_{nm} &= \frac{a_{nm} E_n^{(1)}}{E_m^{(0)} - E_n^{(0)}} - \frac{\sum_j a_{nj}\, H'_{mj}}{E_m^{(0)} - E_n^{(0)}} \\ &= \frac{a_{nm} E_n^{(1)}}{E_m^{(0)} - E_n^{(0)}} - \frac{\sum'_j a_{nj} H'_{mj}}{E_m^{(0)} - E_n^{(0)}} \end{aligned} \tag{10.35}$$

because $a_{nn} = 0$. This equation can be used to calculate all b except b_{nn}. To find the value of b_{nn}, we make use of the result obtained from the normalisation condition of ψ_n as given by (10.26b):

$$\int \left(\psi_n^{(0)*}\psi_n^{(2)} + \psi_n^{(1)*}\psi_n^{(1)} + \psi_n^{(2)*}\psi_n^{(0)}\right) \mathrm{d}\tau = 0. \tag{10.26b'}$$

Substituting the expressions for $\psi_n^{(1)}$ and $\psi_n^{(2)}$ from (10.8) and (10.30) in the above equation, we obtain

$$\int \psi_n^{(0)*} \sum'_j b_{nj}\psi_j^{(0)}\, \mathrm{d}\tau + \int \sum_j a_{nj}^* \psi_j^{(0)*} \sum_k a_{nk}\psi_k^{(0)}\, \mathrm{d}\tau + \int \sum_j b_{nj}^* \psi_j^{(0)*}\psi_n^{(0)}\, \mathrm{d}\tau = 0.$$

By making use of the orthonormality of the eigenfunctions of the unperturbed system, this reduces to

$$\sum_j b_{nj}\delta_{nj} + \sum_j a_{nj}^* \sum_k a_{nk}\delta_{jk} + \sum_j b_{nj}^*\delta_{jn} = 0.$$

Summing over j in the first and third terms and over k in the second term, we obtain

$$b_{nn} + \sum_j a_{nj}^* a_{nj} + b_{nn*} = 0.$$

Since, by virtue of (10.27c), $a_{nn} = 0$, the above equation reduces to

$$b_{nn} + b_{nn}^* = -\sum'_j a_{nj}^* a_{nj} = -\sum'_j \left|a_{nj}\right|^2. \tag{10.36}$$

Problem

10.3. Show that the phase of b has no physical significance.

As the phase of b_{nn} has *no physical significance*, this phase may be chosen as zero. Then (10.36) may be written as

$$b_{nn} = -\frac{1}{2}\sum_j{}' a_{nj}^* a_{nj} = -\frac{1}{2}\sum_j{}' |a_{nj}|^2.$$

By virtue of (10.25), this takes the form

$$b_{nn} = -\frac{1}{2}\sum_j{}' \frac{|H'_{jn}|^2}{\left(E_n^{(0)} - E_j^{(0)}\right)^2} \tag{10.37}$$

Hence, (10.30) yields

$$\psi_n^{(2)} = \sum_m b_{nm}\psi_m^{(0)} = \sum_m{}' b_{nm}\psi_m^{(0)} + b_{nn}\psi_n^{(0)},$$

where b_{nm} and b_{nn} are given by (10.35) and (10.37), respectively.
This gives the second-order correction to the eigenfunction.

10.9 First-order perturbation: energy correction in a two-fold degenerate case

So far, the discussion has been confined to non-degenerate cases. Let us now consider a system which in the *absence of perturbation* is two-fold degenerate so that there are two linearly independent eigenfunctions, say $\psi_n^{(0)}$ and $\psi_m^{(0)}$, corresponding to the same eigenvalue $E_n^{(0)}$ (i.e. $E_n^{(0)} = E_m^{(0)}$). Then

$$H_0\psi_n^{(0)} = E_n^{(0)}\psi_n^{(0)} \tag{10.38a}$$

and

$$H_0\psi_m^{(0)} = E_n^{(0)}\psi_m^{(0)}, \tag{10.38b}$$

where H_0 is the unperturbed Hamiltonian operator with a discrete energy spectrum. We have shown that for the first-order non-degenerate perturbation

$$\psi_n^{(1)} = \psi_n^{(0)} + \sum_j{}' a_{nm}\psi_m^{(0)}, \tag{10.28'}$$

where

$$a_{nm} = \frac{H'_{mn}}{E_n^{(0)} - E_m^{(0)}}, \quad m \neq n, \tag{10.25'}$$

and $a_{nn} = 0$. Since, here the difference between the energy levels corresponding to linearly independent eigenfunctions $\psi_n^{(0)}$ and $\psi_m^{(0)}$ is zero, the above equation for the first-order perturbation does not give valid solutions as the right-hand side of (10.25′) diverges. The perturbation theory developed for non-degenerate states breaks down in this case and a new method is needed.

In fact, the perturbation theory for the non-degenerate case was developed on the basis that, as the perturbation Hamiltonian H' is very small compared to the unperturbed Hamiltonian H_0, the perturbed wave function ψ_n differs only slightly from the unperturbed eigenfunction $\psi_n^{(0)}$ while the corresponding energy also differs slightly from the unique energy eigenvalue associated with the corresponding unperturbed eigenfunction. However, *now* there are two linearly independent eigenfunctions of the unperturbed system which correspond to the same eigenvalue. Moreover, if there are two linearly independent eigenfunctions corresponding to a degenerate eigenvalue as is the case here, a linear combination of them is *also* an eigenfuncion corresponding to the *same eigenvalue*. Thus, in addition to two linearly independent eigenfunctions which correspond to the same eigenvalue $E_n^{(0)}$, there are linear combinations of them which *also* correspond to the same eigenvalue $E_n^{(0)}$. It may be remembered that these linear combinations can be chosen so that they form an orthonormal set with the remaining eigenfunctions of the unperturbed system. We shall assume that orthonormalisation has been performed. Now, for a *non-degenerate perturbation theory*, if the perturbation is switched off, the state

$$\psi_n = \psi_n^{(0)} + \psi_n^{(1)}$$

with energy

$$E_n = E_n^{(0)} + E_n^{(1)}$$

goes to

$$\psi_n = \psi_n^{(0)}$$

with energy

$$E_n = E_n^{(0)}.$$

An important question arises here. If the perturbation for a *degenerate* case is switched off so that the energy goes from

$$E_n = E_n^{(0)} + E_n^{(1)}$$

to

$$E_n = E_n^{(0)},$$

then *to which unperturbed wave function will the perturbed wavefunction reduce*? For instance, in the case mentioned above, ψ_n may reduce to $\psi_n^{(0)}$ *or* $\psi_m^{(0)}$ or a linear combination of them because each one of them corresponds to the same unperturbed energy eigenvalue $E_n^{(0)}$. The answer to this question is that in general ψ_n reduces to a linear combination of $\psi_n^{(0)}$ and $\psi_m^{(0)}$ because the last two functions are themselves only special cases of this linear combination. For convenience in further discussion, the eigenfunctions $\psi_n^{(0)}$ and $\psi_m^{(0)}$ corresponding to the degenerate eigenvalue $E_n^{(0)}$ may be written as $\psi_{n_1}^{(0)}$ and $\psi_{n_2}^{(0)}$. Then (10.38*a*) and (10.38*b*) may be expressed as

$$H_0\psi_{n_1}^{(0)} = E_n^{(0)}\psi_{n_1}^{(0)}, \tag{10.39}$$

$$H_0\psi_{n_2}^{(0)} = E_n^{(0)}\psi_{n_2}^{(0)}. \tag{10.40}$$

Next, proceeding on the same lines as for a non-degenerate case, we may write

$$H = H_0 + \lambda H',$$
$$\psi_n = a_1\psi_{n_1}^{(0)} + a_2\psi_{n_2}^{(0)} + \lambda\psi_n^{(1)} + \lambda^2\psi_n^{(2)} + \cdots,$$
$$E_n = E_n^{(0)} + \lambda E_n^{(1)} + \lambda^2 E_n^{(2)} + \cdots.$$

Substituting these expressions in

$$H\psi_n = E_n\psi_n,$$

we obtain

$$(H_0 + \lambda H')\left(a_1\psi_{n_1}^{(0)} + a_2\psi_{n_2}^{(0)} + \lambda\psi_n^{(1)} + \lambda^2\psi_n^{(2)} + \cdots\right)$$
$$= \left(E_n^{(0)} + \lambda E_n^{(1)} + \lambda^2 E_n^{(2)} + \cdots\right)\left(a_1\psi_{n_1}^{(0)} + a_2\psi_{n_2}^{(0)} + \lambda\psi_n^{(1)} + \lambda^2\psi_n^{(2)} + \cdots\right).$$

Comparing the coefficients of various powers of λ on the two sides of this equation, we obtain

$$H_0\left(a_1\psi_{n_1}^{(0)} + a_2\psi_{n_2}^{(0)}\right) = E_n^{(0)}\left(a_1\psi_{n_1}^{(0)} + a_2\psi_{n_2}^{(0)}\right), \qquad (10.41)$$

$$H_0\psi_n^{(1)} + H'\left(a_1\psi_{n_1}^{(0)} + a_2\psi_{n_2}^{(0)}\right) = E_n^{(0)}\psi_n^{(1)} + E_n^{(1)}\left(a_1\psi_{n_1}^{(0)} + a_2\psi_{n_2}^{(0)}\right). \qquad (10.42)$$

Multiplying from the left by $\psi_{n_1}^{(0)*}$ and integrating over all space, we obtain

$$\int\psi_{n_1}^{(0)*}H_0\psi_n^{(1)}\,\mathrm{d}\tau + \int\psi_{n_1}^{(0)*}H'\left(a_1\psi_{n_1}^{(0)} + a_2\psi_{n_2}^{(0)}\right)\mathrm{d}\tau$$
$$= E_n^{(0)}\int\psi_{n_1}^{(0)*}\psi_n^{(1)}\,\mathrm{d}\tau + \int\psi_{n_1}^{(0)*}E_n^{(1)}\left(a_1\psi_{n_1}^{(0)} + a_2\psi_{n_2}^{(0)}\right)\mathrm{d}\tau. \qquad (10.43)$$

Problem

10.4. Show that

$$\int\psi_{n_1}^{(0)*}H_0\psi_n^{(1)}\,\mathrm{d}\tau = E_n^{(0)}\int\psi_{n_1}^{(0)*}\psi_n^{(1)}\,\mathrm{d}\tau.$$

In view of the above result, (10.43) becomes

$$a_1\int\psi_{n_1}^{(0)*}H'\psi_{n_1}^{(0)}\,\mathrm{d}\tau + a_2\int\psi_{n_1}^{(0)*}H'\psi_{n_2}^{(0)}\,\mathrm{d}\tau$$
$$= a_1E_n^{(1)}\int\psi_{n_1}^{(0)*}\psi_{n_1}^{(0)}\mathrm{d}\tau + a_2E_n^{(1)}\int\psi_{n_1}^{(0)*}\psi_{n_2}^{(0)}\,\mathrm{d}\tau.$$

Since the functions $\psi_{n_1}^{(0)}$ and $\psi_{n_2}^{(0)}$ are individually normalised and mutually orthogonal, the first integral on the right-hand side of the above equation is unity while the second integral on the same side of the above equation is zero. The equation therefore reduces to

$$a_1 \int \psi_{n_1}^{(0)*} H' \psi_{n_1}^{(0)} \, d\tau + a_2 \int \psi_{n_1}^{(0)*} H' \psi_{n_2}^{(0)} \, d\tau = a_1 E_n^{(1)}. \tag{10.44}$$

If we use the notation

$$\int \psi_{nj}^{(0)*} H' \psi_{ni}^{(0)} \, d\tau = H'_{ji}, \tag{10.45}$$

then (10.44) reads

$$a_1 H'_{11} + a_2 H'_{12} = a_1 E_n^{(1)}. \tag{10.46}$$

Similarly, multiplying (10.43) from the left by $\psi_{n_2}^{(0)}$ and proceeding in exactly the same way as above, we obtain

$$a_1 H'_{21} + a_2 H'_{22} = a_2 E_n^{(1)}. \tag{10.47}$$

Remember that

$$|a_1|^2 + |a_2|^2 = 1. \tag{10.48}$$

Let us now compute $E_n^{(1)}$. Equations (10.46) and (10.47) can be written as

$$\left(H'_{11} - E_n^{(1)}\right)a_1 + H'_{12}a_2 = 0 \tag{10.49}$$

and

$$H'_{21}a_1 + \left(H'_{22} - E_n^{(1)}\right)a_2 = 0 \tag{10.50}$$

These two equations in a_1 and a_2 can have non-trivial solutions only if the determinant of their coefficients vanishes, i.e. if

$$\begin{vmatrix} H'_{11} - E_n^{(1)} & H'_{12} \\ H'_{21} & H'_{22} - E_n^{(1)} \end{vmatrix} = 0. \tag{10.51}$$

This is called the *secular equation*. This is a quadratic equation in energy $E_n^{(1)}$ and gives two values which we denote by $(E_n^{(1)})_1$ and $(E_n^{(1)})_2$. If, as is usually the case, these values are distinct, then the energy level splits into a doublet with energy values as $E_n^{(0)} + (E_n^{(1)})_1$ and $E_n^{(0)} + (E_n^{(1)})_2$ and consequently the *perturbation removes the degeneracy*. However, if the two values, i.e. the two roots of (10.51) are equal, $(E_n^{(1)})_1 = (E_n^{(1)})_2$, then the degeneracy is not removed to the first order. The perturbation has displaced the energy level but it does not split it. It may be emphasised that as the degeneracy in an unperturbed system is frequently removed by perturbation, this is an important and powerful technique for understanding the structure of such systems or situations. Substituting these values of $E_n^{(1)}$ in (10.49) and (10.50) and using the condition that the eigenfunctions must be normalised, two sets of values for the coefficients a_1 and a_2 are obtained. The two linear combinations with these values of a, therefore, form the appropriate eigenfunctions for the problem and can be used with the unperturbed eigenfunctions for these other states.

10.10 The application of perturbation theory to the Stark effect

It is found experimentally that when an electric field is applied to the hydrogenic atom for which the potential energy, $-Ze^2/r$, is spherically symmetric, its single lines are split into multiplets. This is called the *Stark effect*. It was discovered by Stark, a German physicist, who was awarded the Nobel Prize in 1919. He was such a strong supporter of 'German physics' that he dismissed quantum theory and the theory of relativity, two of the most brilliant theories of the twentieth century, as products of Jewish thinking! To explain the Stark effect on perturbation theory, we proceed as follows.

Let us consider a one-electron atom in a uniform electric field of charge Ze on the atomic nucleus. The unperturbed Hamiltonian of the hydrogenic atom is

$$H_0 = \frac{p^2}{2\mu} - \frac{Ze^2}{r}$$

where μ is the reduced mass of the hydrogenic atom. As already stated, the potential energy is spherically symmetric. For such a potential, the energy levels and energy eigenfunctions are given by

$$E_n = -|E_n| = -\frac{Z^2 e^4 \mu}{2n^2\hbar^2},$$

$$\psi_{n\ell m}(r, \theta, \phi) = R_{n\ell}(r) Q_{\ell m}(\theta) F_m(\phi)$$

$$= R_{n\ell}(r)(-1)^{-\frac{1}{2}(m+|m|)} Y_{\ell m}(\theta, \phi).$$

Suppose that a uniform external electric field $\mathcal{E}$ is applied to the hydrogenic atom. If we choose the z-axis in the direction of the electric field, then

$$e\mathcal{E} \cdot \mathbf{r} = e\mathcal{E}z$$

is the perturbation potential. The vector $\mathbf{r}$ is the position vector of the electron with respect to the proton. Since the radius vector $\mathbf{r}$ points from the proton to the electron, the electric dipole moment of the hydrogen atom is $-e\mathbf{r}$ ($\equiv \boldsymbol{\mu}_e$) as it points from the negative to the positive charge.

Let us first consider the hydrogen atom emitting a line spectrum in its ground state for which $n = 1$. We have seen that while the solution $\psi_{n\ell m}$ of the Schrödinger amplitude equation depends upon the total, orbital and magnetic quantum numbers n, ℓ, m, the energy E_n depends only upon the total quantum number n. The quantum number ℓ can take the values $0, 1, 2, \ldots, n-1$ while m can take the values $-\ell, -\ell+1, \ldots, \ell$. Therefore, for a given n, there exists a degenerate energy value E_n corresponding to n^2 linearly independent functions for different values of ℓ and m. Evidently, for

$n = 1$, since ℓ can have only one value zero and consequently m can also only be zero, the energy value E_1 is non-degenerate. We can use the non-degenerate perturbation theory for calculating the first-order correction to energy of the hydrogen atom in the ground state. This unperturbed state is described by the eigenfunction

$$\psi_{100} = \frac{1}{\sqrt{\pi a^3}} e^{-2r/a_0}.$$

When a uniform external electric field is applied, the system is perturbed. Since the unperturbed ground state of the hydrogen atom is non-degenerate, we can use the non-degenerate perturbation theory to calculate the perturbation effects. The first-order correction to energy is given by

$$E^{(1)}_{(100)} = \int \left(\psi^{(0)}_{100}\right)^* H' \psi^{(0)}_{100}\, \mathrm{d}\tau = \int \left(\psi^{(0)}_{100}\right)^* e\mathcal{E}z\psi^{(0)}_{100}\, \mathrm{d}\tau$$
$$= e\mathcal{E} \int Z \left|\psi_{100}\right|^2 \mathrm{d}\tau.$$

Since $|\psi_{100}|^2$ is an even function of parity and the perturbing potential $e\mathcal{E}z$ is an odd function under reflection, the product corresponding to the ground state of the hydrogen atom is an odd function of parity. Hence, the integral on the right-hand side of the above equation vanishes and consequently the first-order correction to the ground state energy of the hydrogen atom when it is placed in an electric field is zero:

$$E^{(1)}_{(100)} = 0.$$

This shows that to the first order, the energy level of the hydrogen atom in the ground state is not changed by the application of a uniform external electric field.

Problem

10.5. Show that the second-order correction $E^{(2)}_{n=1}$ to the ground state energy of the hydrogen atom is

$$E^{(2)}_{n=1} = -\frac{9}{4} e^2 \mathcal{E}^2.$$

Let us next compute the first-order correction to the energy in the *first excited state* of the hydrogen atom emitting a line spectrum when it is placed in a uniform electric field. For this state, the total quantum number $n = 2$. Therefore, the orbital quantum number ℓ can take the values 0 and 1 and consequently m can be 0 and −1, 0, 1. The energy eigenvalue is therefore four-fold degenerate; there are four linearly independent eigenfunctions corresponding to the same energy eigenvalue E_2. This degeneracy also holds for hydrogenic atoms for which the potential is $-Ze^2/r$. (However, for non-hydrogenic atoms, due to the screening effects of other electrons, the potential energy is not spherically symmetric, and therefore the situation is different.) This will be an interesting phenomenon for the application of the *degenerate perturbation theory*. Hence, there are four linearly independent eigenfunctions corresponding to the same energy eigenvalue which depend only on the total quantum number n:

$$\begin{aligned} &\ell = 0, \quad m = 0,\\ &n = 2,\\ &\ell = 1, \quad m = -1, 0, 1. \end{aligned}$$

The four solutions are:

$$\psi_{200} = \frac{1}{4\sqrt{2\pi}}\left(\frac{1}{a_0}\right)^{3/2}(2-\sigma)\mathrm{e}^{-\sigma/2}, \quad \sigma = \frac{r}{a_0},$$

$$\psi_{21,-1} = \frac{1}{8\sqrt{\pi}}\left(\frac{1}{a_0}\right)^{3/2}\sigma\mathrm{e}^{-\sigma/2}\sin\theta\,\mathrm{e}^{-\mathrm{i}\phi},$$

$$\psi_{210} = \frac{1}{4\sqrt{2\pi}}\left(\frac{1}{a_0}\right)^{3/2}\sigma\mathrm{e}^{-\sigma/2}\cos\theta,$$

$$\psi_{211} = \frac{1}{8\sqrt{\pi}}\left(\frac{1}{a_0}\right)^{3/2}\sigma\mathrm{e}^{-\sigma/2}\sin\theta\,\mathrm{e}^{\mathrm{i}\phi}.$$

Since, the first excited state of the hydrogen atom, for which $n=2$, is four-fold degenerate, by extending the earlier analysis for the two-fold degenerate case, we can write the secular equation for this case as

$$\begin{vmatrix} H'_{11} - E_n^{(1)} & H'_{12} & H'_{13} & H'_{14} \\ H'_{21} & H'_{22} - E_n^{(1)} & H'_{23} & H'_{24} \\ H'_{31} & H'_{32} & H'_{33} - E_n^{(1)} & H'_{34} \\ H'_{41} & H'_{42} & H'_{43} & H'_{44} - E_n^{(1)} \end{vmatrix} = 0, \tag{10.52}$$

where

$$H'_{ij} = \int \psi_{ni}{}^{(0)*} H' \psi_{nj}{}^{(0)}\,\mathrm{d}\tau. \tag{10.45'}$$

This is a quartic in $E_n^{(1)}$ and should give four values. In order to obtain these values, we have to evaluate all H'_{ij}, $i, j = 1, 2, 3, 4$. This evaluation is very much simplified by noting the following points.

1. The perturbing potential $e\mathcal{E}z$ ($= e\mathcal{E}r\cos\theta$) commutes with L_z ($\equiv \mathrm{i}\hbar\partial/\partial\phi$). Thus, we may write

$$L_z z = z L_z.$$

Multiplying from the left and from the right by $\psi^*_{n\ell m'}$ and $\psi_{n\ell m}$, respectively, and integrating throughout the space, we obtain

$$\int \psi^*_{n\ell' m'} L_z z \psi_{n\ell m}\, d\tau = \int \psi^*_{n\ell' m'} z L_z \psi_{n\ell m}\, d\tau. \tag{10.53}$$

By making use of the definition of the Hermitian conjugate of an operator, the integral on the left-hand side of the above equation gives

$$\begin{aligned}\int \psi^*_{n\ell' m'} L_z(z\psi_{n\ell m})\, d\tau &= \int (z\psi_{n\ell m}) L_z^\dagger \psi^*_{n\ell' m'}\, d\tau \\ &= \int (z\psi_{n\ell m}) L_z \psi^*_{n\ell' m'}\, d\tau,\end{aligned}$$

since L_z is Hermitian. By making use of the eigenvalue equation for L_z, i.e. $L_z\psi = m\hbar\psi$, we obtain

$$\begin{aligned}\int \psi^*_{n\ell' m'} L_z z \psi_{n\ell m}\, d\tau &= \int (z\psi_{n\ell m}) m'\hbar \psi^*_{n\ell' m'}\, d\tau \\ &= m'\hbar \int (z\psi_{n\ell m}) \psi^*_{n\ell' m'}\, d\tau \\ &= m'\hbar \int \psi^*_{n\ell' m'} z \psi_{n\ell m}\, d\tau. \end{aligned} \tag{10.54}$$

The right-hand side of (10.53) gives:

$$\begin{aligned}\int \psi^*_{n\ell' m'} z L_z \psi_{n\ell m} d\tau &= \int \psi^*_{n\ell, m,} z m'\hbar \psi_{n\ell m}\, d\tau \\ &= m\hbar \int \psi^*_{n\ell' m'} z \psi_{n\ell m}\, d\tau. \end{aligned} \tag{10.55}$$

Substituting the expressions for the integrals from (10.54) and (10.55) in (10.53), we obtain

$$(m - m')\hbar \int \psi^*_{n\ell' m'} z \psi_{n\ell m}\, d\tau = 0. \tag{10.56}$$

This means that only those matrix elements of z can be different from zero for which $m' = m$; all others for which $m' \neq m$ will disappear.

2. We also notice that

$$\int \psi^*_{n\ell' m'} z \psi_{n\ell m}\, d\tau$$

can be non-zero only if $\ell' \neq \ell$. This is so because for $\ell' = \ell$, the function $\psi^*_{n\ell m}\psi_{n\ell m}$ being of even parity, the integrand $\psi^*_{n\ell m} z \psi_{n\ell m}$ becomes an odd function and the above integral disappears. Thus, only those matrix elements can be non-zero for which $\ell' \neq \ell$ and $m' = m$.

Let us now use a different notation for the matrix elements; this notation will clearly demonstrate their dependence on quantum numbers. We write

$$H'_{ij}(\ell m, \ell' m') = \int \psi_{ni}^{(0)*}(\ell m) H' \psi_{nj}^{(0)}(\ell' m')\, \mathrm{d}\tau, \tag{10.57}$$

where $\ell = 0, 1$ and $m = -1, 0, 1$. Then the secular equation may be written as

$$\begin{vmatrix} H'_{11}(00, 00) - E_n^{(1)} & H'_{12}(00, 10) & H'_{13}(00, 11) & H'_{14}(00, 1-1) \\ H'_{21}(10, 00) & H'_{22}(10, 10) - E_n^{(1)} & H'_{23}(10, 11) & H'_{24}(10, 1-1) \\ H'_{31}(11, 00) & H'_{32}(11, 10) & H'_{33}(11, 11) - E_n^{(1)} & H'_{34}(11, 1-1) \\ H'_{41}(1-1, 1-1) & H'_{42}(1-1, 10) & H'_{43}(1-1, 10) & H'_{44}(1-1, 1-1) - E_n^{(1)} \end{vmatrix} = 0. \tag{10.58}$$

By virtue of the constraints that all elements $H'_{ij}(\ell m, \ell' m')$ for which either $m' \neq m$ and/or $\ell' = \ell$ disappear, the above secular equation reduces to

$$\begin{vmatrix} -E_n^{(1)} & H'_{12}(00, 10) & 0 & 0 \\ H'_{21}(10, 00) & -E_n^{(1)} & 0 & 0 \\ 0 & 0 & -E_n^{(1)} & 0 \\ 0 & 0 & 0 & -E_n^{(1)} \end{vmatrix} = 0. \tag{10.59}$$

This gives

$$\left(E_n^{(1)}\right)^4 - H'_{12}(00, 10)\, H'_{21}(10, 00)\left(E_n^{(1)}\right)^2 = 0.$$

Therefore,

either
$$\left(E_n^{(1)}\right)^2 = 0, \tag{10.60a}$$

or
$$\left(E_n^{(1)}\right)^2 = H'_{12}(00, 10)\, H'_{21}(10, 00) \tag{10.60b}$$

The matrix elements are

$$H'_{12}(00, 10) = \int \psi_{n_1}^{(0)*}(00)\, H' \psi_{n_2}^{(0)}(10)\, \mathrm{d}\tau \tag{10.61}$$

and

$$H'_{21}(10, 00) = \int \psi_{n_2}^{(0)*}(10)\, H' \psi_{n_1}^{(0)}(00)\, \mathrm{d}\tau. \tag{10.62}$$

Problem

10.6. Show that

$$H'_{12}(00,\ 10) = H'^{*}_{21}(10,\ 00). \tag{10.63}$$

If we take one of the matrix elements as real, then by virtue of the above equation, the other should also be real and of course equal to it. Thus, we have to evaluate only one of these elements, say $H'_{12}(00,\ 10)$. This is given by

$$H'_{12}(00,\ 10) = \int \psi_{n_1}^{(0)*}(00) H' \psi_{n_2}^{(0)}(10)\, \mathrm{d}\tau$$

$$= \frac{1}{32\pi a_0^3} \int_0^\infty \int_0^\pi \int_0^{2\pi} \mathrm{e}^{-r/a_0}(2 - r/a_0) e\mathcal{E} r \cos\theta \frac{r}{a_0} \cos\theta\, r^2\, \mathrm{d}r \sin\theta\, \mathrm{d}\theta\, \mathrm{d}\phi$$

$$= \frac{e\mathcal{E}}{6a_0{}^3} \int_\infty^0 \mathrm{e}^{-r/a_0}(2 - r/a_0) \frac{r^4}{a_0} \mathrm{d}r$$

$$= -3e\mathcal{E}a_0, \tag{10.64}$$

where the length a_0 is the Bohr radius of the hydrogen atom. Substituting this value of $H'_{12}(00,\ 10)$ in (10.60*b*) and simplifying, we obtain

$$E_n^{(1)} = \pm 3e\mathcal{E}a_0.$$

Hence, to first order, for the hydrogen atom in the first excited state, the degeneracy of two energy levels is removed by applying an external electric field. The energy levels are raised or depressed by the same value $3e\mathcal{E}a_0$. The degeneracy of the remaining two energy levels is not removed. The results are shown in figure 10.2. The spin and the relativistic effects also produce a further splitting of these energy levels. This is called the *fine structure*, but we will not consider it here.

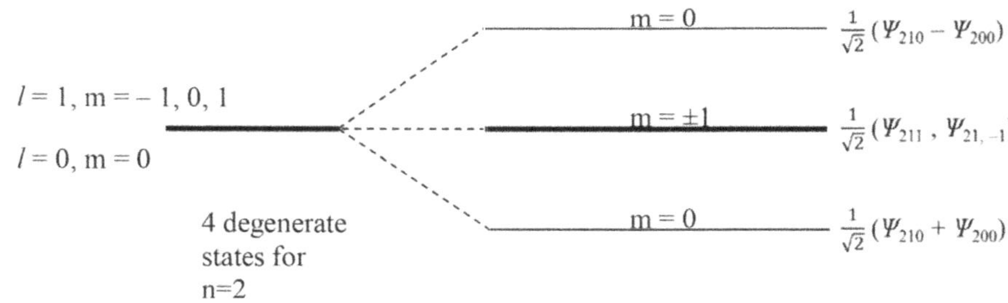

Figure 10.2. Splitting of the degenerate energy level of the hydrogen atom for $n = 2$ due to the linear Stark effect.

Problem

10.7. Show that the states corresponding to the four energy corrections 0, 0, $3e\mathcal{E}a_0$ and $-3e\mathcal{E}a_0$ are ψ_{211}, $\psi_{21,-1}$, $1/\sqrt{2}\,(\psi_{210} - \psi_{200})$, $1/\sqrt{2}\,(\psi_{210} + \psi_{200})$.

10.11 Time-dependent perturbation theory

Let us now consider a physical system whose Hamiltonian $H(\mathbf{r}, t)$ is explicitly dependent upon space and time coordinates. As we know, in this case there are no stationary state solutions of the Schrödinger wave equation. The perturbation theory developed for stationary states has to be modified. We *assume* that the Hamiltonian $H(\mathbf{r}, t)$ is such that the time-dependent Schrödinger equation

$$H(\mathbf{r}, t)\Psi(\mathbf{r}, t) = \mathrm{i}\hbar\frac{\partial\Psi(\mathbf{r}, t)}{\partial t} \tag{10.65}$$

for this system *cannot* be solved exactly. However, this Hamiltonian can be written as

$$H(\mathbf{r}, t) = H_0(\mathbf{r}) + H'(\mathbf{r}, t) \tag{10.66}$$

where the exact solution of the Schrödinger wave equation for the Hamiltonian H_0 can be obtained:

$$\mathrm{i}\hbar\frac{\partial\Psi^{(0)}}{\partial t} = H_0(\mathbf{r})\Psi^{(0)}. \tag{10.67}$$

Moreover, the part $H'(\mathbf{r}, t)$ of the Hamiltonian for the physical system is very small compared to $H_0(\mathbf{r})$ at all times. Our problem is to obtain an approximate solution of differential equation (10.65).

The eigenfunctions of $H_0(\mathbf{r})$ are obtained from equation

$$H_0(\mathbf{r})\psi_n(\mathbf{r}) = E_n\psi_n(\mathbf{r}). \tag{10.68}$$

The functions $\psi_n(\mathbf{r})$, being the eigenfunctions of a Hermitian operator $H_0(\mathbf{r})$, form a complete set so that $\Psi(\mathbf{r}, t)$ can be expanded as a linear combination of them, the time dependence being contained in the coefficients of $\psi(\mathbf{r})$:

$$\Psi(\mathbf{r}, t) = \sum_n c_n(t)\psi_n(\mathbf{r})\mathrm{e}^{-\mathrm{i}E_n t/\hbar}. \tag{10.69}$$

In order to simplify the ensuing arguments, the exponential factor $\mathrm{e}^{-\mathrm{i}E_n t/\hbar}$ has been separated from $c_n(t)$. If we know all $c_n(t)$, the solution $\Psi(\mathbf{r}, t)$ can be obtained. For this purpose, substituting the expressions for H and $\Psi(\mathbf{r}, t)$ from (10.66) and (10.69) in (10.65) and changing sides, we obtain

$$\mathrm{i}\hbar\frac{\partial}{\partial t}\sum_n c_n(t)\psi_n(\mathbf{r})\mathrm{e}^{-\mathrm{i}E_n t/\hbar} = [H_0(\mathbf{r}) + H'(\mathbf{r}, t)]\sum_n c_n(t)\psi_n(\mathbf{r})\mathrm{e}^{-\mathrm{i}E_n t/\hbar},$$

or

$$\left[i\hbar\sum_n \frac{dc_n(t)}{dt}\, e^{-iE_nt/\hbar} - c_n(t)\frac{iE_n}{\hbar}e^{-iE_nt/\hbar}\right]\psi_n(\mathbf{r})$$
$$= \sum_n c_n(t)E_n\psi_n(\mathbf{r})e^{-iE_nt/\hbar} + \sum_n H'(\mathbf{r},\, t)c_n(t)\psi_n(\mathbf{r})e^{-iE_nt/\hbar}$$

where, in writing the first term on the right-hand side of the above equation, we have made use of relation (10.68). Cancelling the equal terms on the two sides of the above equation, writing $\dot{c}_n(t)$ for dc_n/dt, and *assuming* that the Hamiltonian operator of the physical system does not involve any derivative with respect to time, we obtain

$$\sum_n [i\hbar\dot{c}_n(t) - c_n(t)H'(\mathbf{r},\, t)]\psi_n(\mathbf{r})e^{-iE_nt/\hbar} = 0.$$

Multiplying on the left by $\psi_m^*(\mathbf{r})$ $\psi_m^*(\mathbf{r})$ and integrating over all space, we obtain

$$\sum_n [i\hbar\dot{c}_n(t)e^{-iE_nt/\hbar}\int\psi_m^*(\mathbf{r})\psi_n(\mathbf{r})d\tau - c_n(t)e^{-iE_nt/\hbar}\int\psi_m^*(\mathbf{r})H'(\mathbf{r},\, t)\psi_n(\mathbf{r})d\tau] = 0,$$

or

$$\sum_n\left[i\hbar\dot{c}_n(t)e^{-iE_nt/\hbar}\delta_{mn} - c_n(t)e^{-iE_nt/\hbar}H'_{mn}\right] = 0,$$

where the definition of the matrix element H'_{mn} has been extended so as to include time-dependent perturbations. Summing the first term over n, the above equation takes the form

$$i\hbar\dot{c}_m(t)e^{-iE_mt/\hbar} - \sum_n c_n(t)e^{-iE_nt/\hbar}H'_{mn} = 0,$$

or

$$\dot{c}_m(t) = -\frac{i}{\hbar}\sum_n c_n(t)H'_{mn}e^{i(E_m - E_n)t/\hbar},$$

or

$$\dot{c}_m(t) = -\frac{i}{\hbar}\sum_n c_n(t)H'_{mn}e^{i\omega_{mn}t}, \tag{10.70}$$

where

$$\omega_{mn} = \left(E_m - E_n\right)/\hbar. \tag{10.71}$$

Equation (10.70) are a set of first-order, linear, homogeneous, ordinary differential equations. *Everything up to these equations has been exact* but, in general, these differential equations cannot be solved exactly. We therefore have to resort to the perturbation technique just as in the case of the time-independent perturbation theory. The difference is that *this time we expand c(t) instead of ψ_m*. Thus, we write

$$H = H_0(\mathbf{r}) + \lambda H'(\mathbf{r},\, t) \tag{10.72}$$

and

$$c_m(t) = c_m^{(0)}(t) + \lambda c_m^{(1)}(t) + \cdots. \tag{10.73}$$

Differentiating the last equation with respect to time, we obtain

$$\dot{c}_m = \dot{c}_m^{(0)} + \lambda \dot{c}_m^{(1)} + \cdots,$$

where, for convenience, we have dropped the argument t in the c, as we would be using c and $\dot{c}$ with or without the argument t interchangeably. Substituting the expressions for c_n and $c._m$ in (10.70) and replacing, by virtue of (10.72), H'_{mn} by $\lambda H'_{mn}$, we obtain

$$\dot{c}_m^{(0)} + \lambda \dot{c}_m^{(1)} + \cdots = -\frac{i\lambda}{\hbar}\sum\nolimits_n \left(c_n^{(0)} + \lambda c_n^{(1)} + \cdots\right) H'_m \, e^{i\omega_{mn}t}. \tag{10.74}$$

Comparing the coefficients of various powers of λ on the two sides of this equation, we obtain

$$\dot{c}_m^{(0)} = 0, \tag{10.75}$$

$$\dot{c}_m^{(1)} = -\frac{i}{\hbar}\sum\nolimits_n c_n^{(0)} H'_{mn} e^{i\omega mnt}, \tag{10.76}$$

$$\cdots$$

These differential equations can be solved in principle to obtain $c_m^{(0)}$, $c_m^{(1)}$,... and consequently approximate solutions of the Schrödinger wave equation to any order in perturbation. It should be noted that if the perturbation is switched off by putting $\lambda = 0$, then (10.74) reduces to

$$\dot{c}_m^{(0)} = 0, \tag{10.75}$$

i.e. to (10.75). Integrating (10.75) with respect to t, we obtain

$$c_m^{(0)} = \text{const.} \tag{10.77}$$

That is, the coefficients for the unperturbed Hamiltonian $H_0(\mathbf{r})$ are independent of time. The values of these coefficients are the initial conditions of the problem and specify the unperturbed state of the system. The first-order contributions to $c_m(t)$, i.e. $c_m^{(1)}$ (t), are obtained from (10.76) by integrating it with respect to t. We have

$$c_m^{(1)}(t) = -\frac{i}{\hbar}\sum\nolimits_n c_n^{(0)} \int_0^t H'_{mn} e^{i\omega_{mn}t} dt. \tag{10.78}$$

Similarly, we can write the equations for the coefficients of higher order.

Let us consider the *special case* when, at time $t = 0$, *the system is in an unperturbed eigenstate* $\psi_j(\mathbf{r})$ so that $c_j^{(0)} = 1$ and $c_n^{(0)} = 0$ for all $n \neq j$. Then (10.78) gives

$$c_m^{(1)}(t) = -\frac{i}{\hbar}\int_0^t H'_{mj} e^{i\omega_{mj}t} dt. \tag{10.79}$$

Therefore, we have

$$c_m(t) = c_m^{(0)}(t) + c_m^{(1)}(t) = c_m^{(1)}(t), \quad m \neq j \tag{10.80}$$

because for $m \neq j$, $c_m^{(0)}(t) = 0$. Consequently, the probability of finding the system, initially in a state $\psi_j(\mathbf{r})$, in a state $\psi_m(\mathbf{r})$, where $m \neq j$, is $|c_m|^2 = |c_m^{(1)}(t)|^2$, where $c_m^{(1)}(t)$ is given by (10.79). Hence, by using (10.79) and (10.80), in principle all c can be calculated up to the first order and consequently the solution $\psi(\mathbf{r}, t)$ of the Schrödinger wave equation is determined provided $H'(\mathbf{r}, t)$ is known.

10.12 Harmonic perturbation

So far, we have analysed the situation when the perturbation Hamiltonian $H'(\mathbf{r}, t)$, at all times, is small compared to the unperturbed Hamiltonian $H_0(\mathbf{r})$ *without specifying what it is.* Further calculations can be made only if the Hamiltonian of the physical system is known. Let us now consider a system with perturbation part $H'(\mathbf{r}, t)$ of the Hamiltonian given by

$$H'(\mathbf{r}, t) = H'(\mathbf{r}) \cos \omega t, \tag{10.81}$$

i.e. the perturbation varies harmonically in time *with angular frequency* ω, except that it is turned on at time $t = 0$ and turned off at a later time t. Then by (10.79), we have

$$c_m^{(1)}(t) = -\frac{\mathrm{i}}{\hbar}\int_0^t H_{mj} \cos \omega t \, \mathrm{e}^{\mathrm{i}\omega_{mj}t} \, \mathrm{d}t.$$

Since H'_{mj} *now does not involve time*, it can be taken outside the integral so that we have

$$c_m^{(1)}(t) = -\frac{\mathrm{i}}{\hbar}H'_{mj}\int_0^t \cos \omega t \, \mathrm{e}^{\mathrm{i}\omega_{mj}t}\mathrm{d}t. \tag{10.82}$$

But $\cos \omega t$ can be written as a sum of two exponentials:

$$\cos \omega t = \frac{1}{2}\left(\mathrm{e}^{-\mathrm{i}\omega t} + \mathrm{e}^{\mathrm{i}\omega t}\right).$$

Substituting this expression in (10.82), we obtain

$$\begin{aligned}
c_m^{(1)}(t) &= -\frac{\mathrm{i}}{2\hbar}H'_{mj}\int_0^t (\mathrm{e}^{-\mathrm{i}\omega t} + \mathrm{e}^{\mathrm{i}\omega t})\mathrm{e}^{\mathrm{i}\omega_{mj}t} \, \mathrm{d}t \\
&= -\frac{\mathrm{i}}{2\hbar}H'_{mj}\int_0^t \left(\mathrm{e}^{-\mathrm{i}(\omega-\omega_{mj})t} + \mathrm{e}^{\mathrm{i}(\omega+\omega_{mj})t}\right) \mathrm{d}t \\
&= -\frac{\mathrm{i}}{2\hbar}H'_{mj}\int_0^t \left(\mathrm{e}^{\mathrm{i}(\omega_{mj}-\omega)t} + \mathrm{e}^{\mathrm{i}(\omega_{mj}+\omega)t}\right) \mathrm{d}t \\
&= -\frac{1}{2\hbar}H'_{mj}\left[\frac{\mathrm{e}^{\mathrm{i}(\omega_{mj}-\omega)t} - 1}{\omega_{mj} - \omega} + \frac{\mathrm{e}^{\mathrm{i}(\omega_{mj}+\omega)t} - 1}{\omega_{mj} + \omega}\right].
\end{aligned} \tag{10.83}$$

If $\omega \approx \omega_{mj}$ or $\approx -\omega_{mj}$, the denominator in the first or the second term will become very small, making that term very large and the other one negligible. For instance, if $\omega \approx \omega_{mj}$, the first term becomes very large and the second term can be ignored as compared to it. Then, we have

$$c_m^{(1)}(t) = -\frac{1}{2\hbar}H'_{mj}\left[\frac{e^{i(\omega_{mj}-\omega)t} - 1}{\omega_{mj} - \omega}\right].$$

Writing x for $(\omega_{mj} - \omega)$, the above relation takes the form

$$c_m^{(1)}(t) = -\frac{1}{2\hbar}H'_{mj}\left[\frac{e^{ixt} - 1}{x}\right].$$

Therefore, the probability of finding the system in the state $\psi_m(\mathbf{r})$ at time t is given by

$$|c_m|^2 = \left|c_m^{(1)}\right|^2 = \frac{1}{4\hbar^2}\left|H'_{mj}\right|^2\left[\frac{\left(e^{-ixt} - 1\right)\left(e^{ixt} - 1\right)}{x^2}\right]. \tag{10.84}$$

The numerator in the last factor can be written as

$$\begin{aligned}
&\left(e^{-ixt} - 1\right)\left(e^{ixt} - 1\right)\\
&= 1 - e^{-ixt} - e^{ixt} + 1\\
&= 2 - [\cos xt + i\sin xt + \cos xt - i\sin xt]\\
&= 2 - 2\cos xt\\
&= 2(1 - \cos xt)\\
&= 2\left[1 - \left(1 - 2\sin^2\frac{xt}{2}\right)\right]\\
&= 4\sin^2\frac{xt}{2}.
\end{aligned}$$

Substituting this expression in relation (10.84), we obtain the transition probability $P(t)_{(n\to m)}$ as given by

$$\underset{(n\to m)}{P(t)} = \left|c_m^{(1)}\right|^2 = \frac{1}{4\hbar^2}\left|H'_{mj}\right|^2\frac{4\sin^2\frac{xt}{2}}{x^2} = \frac{1}{\hbar^2}\left|H'_{mj}\right|^2\frac{\sin^2\frac{xt}{2}}{x^2}. \tag{10.85}$$

It may be noted that the transition probability oscillates sinusoidally with time. The factor $[\sin^2 xt/2]/x^2$ is plotted as a function of x in figure 10.3. The height of the *central peak* increases in proportion to t^2 while its width decreases as proportional to t^{-1}. Therefore, the area under the curve is proportional to t. The figure shows that, for large t, the factor $[\sin^2 xt/2]/x^2$ and consequently the probability of finding the

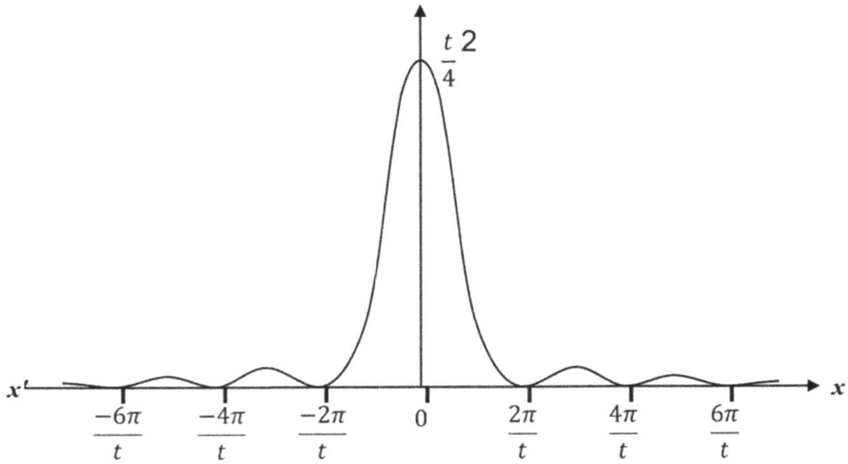

Figure 10.3. The function $[\sin^2(xt/2)]/x^2$ plotted against x.

system in the m-state is strongly peaked around $x = 0$, i.e. around $\omega_{mj} = \omega$, and away from $x = 0$, it oscillates very rapidly. This is the kind of behaviour we associate with a delta function. Thus, after a long time compared to the period of perturbation, the transition probability will be extremely small unless the condition $\omega \approx \omega_{mj}$, i.e. near $x \approx 0$, is fulfilled.

10.13 Fermi's golden rule

It often happens that energy levels in the final state are so close together that they form or approximate to a continuum. Then, instead of considering the transition to a single level, we have to consider the transitions to a *group of final states* whose energies lie within a short range ΔE, say from $E - \Delta E/2$ to $E + \Delta E/2$ centered about E. This means that the energy states within the range ΔE are available for transition. As we are interested in finding the *total* probability for a transition of the system to any one of the available final states in the range ΔE, we must sum the individual probabilities over all available final states in this range. Thus, instead of computing merely $|c_m|^2$, we are interested in calculating $\sum_m |c_m|^2$, where the summation is over all the states in the energy range ΔE. As the states are so closely packed that these form or approximate a continuum, we may make the calculations by considering the density of final states. This density of final states may be defined as the number of final states per unit frequency interval ω_{mj}. Let us denote it by $\rho(\omega_{mj})$. We may then make the replacement

$$\sum_m |c_m|^2 \rightarrow \int_{\Delta E} |c_m|^2 \rho(\omega_{mj})\, \mathrm{d}\omega_{mj}. \tag{10.86}$$

We have seen that $|c_m|^2$ is sharply peaked around the value $\omega_{mj} = \omega$. If we assume that $\rho(\omega_{mj})$, the density of final states, and H'_{mj}, the matrix element, are slowly varying functions of ω_{mj} in the range ΔE, then it is a good approximation to replace these functions by their *values* about $\omega_{mj} = \omega$ and to move them outside the integral so that, by virtue of (10.85), the above transition takes the form

$$\sum_m |c_m|^2 \to \frac{1}{4\hbar^2} \left|H'_{mj}(\omega)\right|^2 \rho(\omega) \int_{\Delta E} \frac{\sin^2 \frac{xt}{2}}{x^2} \, \mathrm{d}\omega_{mj}. \tag{10.87}$$

Since the significant contribution to this integral comes only from the region for which $-2\pi/t < x < 2\pi/t$, we can *now* take the limits as $-\infty$ to $+\infty$, without significantly changing the value of the same integral when the limits were from $E - \Delta E/2$ to $E + \Delta E/2$. Hence, we may write the total transition probability as

$$\frac{1}{4\hbar^2} \left|H'_{mj}(\omega)\right|^2 \rho(\omega) \int_{\infty}^{-\infty} \frac{\sin^2 \frac{xt}{2}}{\left(\frac{x}{2}\right)^2} \, \mathrm{d}\omega_{mj}.$$

Denoting the integral in the above expression by I, we have

$$I = \int_{-\infty}^{\infty} \frac{\sin^2 \frac{xt}{2}}{\left(\frac{x}{2}\right)^2} \, \mathrm{d}\omega_{mj}$$

$$= t^2 \int_{-\infty}^{\infty} \frac{\sin^2 \frac{xt}{2}}{\left(\frac{x}{2}\right)^2} \, \mathrm{d}x$$

as $\mathrm{d}\omega = \mathrm{d}(\omega_{mj} - \omega) = \mathrm{d}x$. Writing

$$\frac{x}{2} t = y,$$

we obtain

$$I = t^2 \int_{\infty}^{-\infty} \frac{\sin^2 y}{y^2} \frac{2}{t} \, \mathrm{d}y = 2t\pi.$$

The total transition probability may therefore be written as

$$\frac{\pi}{2\hbar^2} \left|H'_{mj}(\omega)\right|^2 \rho(\omega) t. \tag{10.88}$$

Next, we name the transition probability per unit time as the *transition rate* and denote it by w. From (10.88), this is

$$w = \frac{\pi}{2\hbar^2} \left|H'_{mj}(\omega)\right|^2 \rho(\omega). \tag{10.89}$$

This equation expresses what is known as Fermi's golden rule. The transition rate is proportional to the product of the square of the magnitude of $H'_{mj}(\omega)$ and $\rho(\omega)$.

Problem

10.8. A charged linear harmonic oscillator is placed in a uniform electric field. Find the first- and second-order corrections to the energy.

Additional problems

10.9. Find the fourth-order correction to the ground state energy of the hydrogen atom.

10.10. A perturbation given by $axF(t)$, where

$$
\begin{aligned}
&= 0 \quad t < 0, \\
F(t) &= 1, \quad 0 < t < T, \\
&= 0, \quad t > T,
\end{aligned}
$$

is applied to a one-dimensional harmonic oscillator in its first excited state. The wave function for the linear harmonic oscillator in this state is

$$\psi_1(x) = \left(\frac{\alpha}{\pi}\right)^{1/4} \frac{1}{\sqrt{2}} \exp\left(-\frac{\alpha x^2}{2}\right) H_1\left(\sqrt{\alpha}\, x\right).$$

What is the probability in the first order that the oscillator will be found in the ground state for $t > T$?

10.11. Discuss the time-independent perturbation theory for a non-degenerate system and find the first- and second-order perturbation energies.

10.12. Determine to the first order the probability of an atom making a transition from a state k to a state m under a time-dependent perturbation $H'(t)$.

A step function perturbation

$$
\begin{aligned}
H' &= 0, \quad t < 0, \\
&= H'(x), \quad t > 0,
\end{aligned}
$$

is applied to a system. Calculate the probability that the system goes from the state k to the state m.

IOP Publishing

Quantum Mechanics

Mohammad Saleem

Chapter 11

Theory of elastic scattering

11.1 Introduction

The deflection of a beam of particles from its normal path when passing close to another particle or nucleus, called the *target* or *centre of force*, is known as *scattering* or *collision*. It does not mean that the colliding particles must come into contact in the sense the word 'collision' is understood in everyday life, because a particle in the incident beam under the influence of the force exerted by the other particle may change its path before having a chance to come in contact with it. If the nature of the colliding particles and their kinetic energies remain unchanged, the collision is said to be *elastic*, otherwise it is called *inelastic*. Thus, if two particles A and B collide such that $A + B \rightarrow A + B$, and the kinetic energies of A and B do not change, then the collision is elastic. The importance of the theory of scattering stems from the fact that it plays a significant role in understanding atomic and nuclear structure and in studying the interactions of elementary particles. For instance, Rutherford's experiments on the scattering of α-particles provided the experimental basis for the nuclear atom model. We will confine ourselves to the theory of *elastic scattering*.

Let us consider the idealised problem of the scattering of an incident beam of *monoenergetic, non-interacting particles* by a target which is initially at rest in the laboratory. The colliding particles would be assumed to be of *different species*. To simplify matters, we will neglect the spins of the particles and will not take their internal structure into account. We will now define certain terms which will be used throughout the discussion on scattering. The number of particles in the incident beam crossing the unit area normal to the direction of the beam in unit time is called the *flux of the beam* or the *incident flux* and is denoted by I. The dimensions of the flux of the beam are L^2T^{-1}. The number of incident particles scattered per unit time into a unit solid angle in the direction (θ, ϕ) of the incident beam is denoted by $S(\theta)$. Its dimension is T^{-1}. The ratio $S(\theta)/I$ is called the

doi:10.1088/978-0-7503-1206-6ch11

differential cross section and is denoted by $d\sigma/d\Omega$. It evidently has the dimension of area L^2. We may write

$$\frac{d\sigma}{d\Omega} = \frac{S(\theta)}{I}. \tag{11.1}$$

Since the particles in the incident beam are assumed to be non-interacting and monoenergetic, it is sufficient to discuss the scattering and behaviour of a single particle. Thus, we have a two-body problem. This can be reduced to a one-body problem as was done in the case of a hydrogen atom.

The integral of differential cross section $\frac{d\sigma}{d\Omega}$ over the entire solid angle is called the *total cross section*. It is denoted by σ_{Total} or σ_{T}, and is given by

$$\begin{aligned}\sigma_{\text{Total}} &= \int \frac{d\sigma}{d\Omega} d\Omega = \int_0^{\pi} \int_0^{2\pi} \frac{d\sigma}{d\Omega} \frac{r \sin\theta \, d\phi r \, d\theta}{r^2} \\ &= \int_0^{\pi} \int_0^{2\pi} \frac{d\sigma}{d\Omega} \sin\theta \, d\theta \, d\phi. \end{aligned} \tag{11.2}$$

For a *spherically symmetric potential*, the differential cross section is independent of ϕ and the expression for total cross section becomes

$$\begin{aligned}\sigma_{\text{Total}} \equiv \sigma_{\text{T}} &= \int_0^{\pi} \frac{d\sigma}{d\Omega} \sin\theta \, d\theta \int_0^{2\pi} d\phi \\ &= 2\pi \int_0^{\pi} \frac{d\sigma}{d\Omega} \sin\theta \, d\theta. \end{aligned} \tag{11.3}$$

It may be pointed out that there is an important difference between the reduced mass problem for the hydrogen atom and the scattering of a single particle by a target. The hydrogen atom was a bound state system and therefore the energy spectrum was discrete. Moreover, while considering its behaviour we were interested only in the energy differences of its various energy levels. The problem under consideration is that of the scattering of particles. Classically the system is unbound. The energy spectrum is continuous. An experimenter makes measurements of the scattering cross sections and a theoretical physicist has to compute these in order to test the theory. But we should not forget that while the experimental measurements are made in the laboratory (the frame of reference) in which the observer is taken to be at rest at the origin, calculations, because of the reduced mass problem, are performed in the centre-of-mass frame (in which the centre-of-mass of the system is taken to be at rest at the origin). Hence in order that a comparison between theory and experiment may be possible, we must know how the cross sections, etc, in the two frames of reference are related to each other. We shall establish these relations before taking up the theory of scattering.

11.2 Centre-of-mass and laboratory frames of reference

Let m_1, m_2 be the masses of the incident particle and the target, respectively, $\mathbf{r}_1$, $\mathbf{r}_2$ their position vectors with respect to a fixed origin and $\mathbf{v}_1$, $\mathbf{v}_2$ their velocities in the laboratory frame of reference before collision. Let **R** be the position vector and **V** the velocity of their centre-of-mass of these two particles in the laboratory frame before collision. Then, by definition, the position vector **R** of the centre-of-mass in the laboratory frame is given by

$$\mathbf{R} = \frac{m_1\mathbf{r}_1 + m_2\mathbf{r}_2}{m_1 + m_2}.$$

Differentiating with respect to time t, we obtain

$$\frac{\mathrm{d}\mathbf{R}}{\mathrm{d}t} = \frac{m_1\dfrac{\mathrm{d}\mathbf{r}_1}{\mathrm{d}t} + m_2\dfrac{\mathrm{d}\mathbf{r}_2}{\mathrm{d}t}}{m_1 + m_2},$$

or

$$\mathbf{V} = \frac{m_1\mathbf{v}_1 + m_2\mathbf{v}_2}{m_1 + m_2}. \tag{11.4}$$

This gives the velocity of the centre-of-mass in the laboratory frame. Usually the target is initially at rest so that $\mathbf{v}_2 = 0$. Equation (11.4) then reduces to

$$\mathbf{V} = \frac{m_1\mathbf{v}_1}{m_1 + m_2}. \tag{11.5}$$

That is, in the laboratory frame, before collision, **V** and $\mathbf{v}_1$ are in the same direction.

If we denote the before-collision velocity of m_1 in the centre-of-mass frame by $\mathbf{u}_1$, then

$$\begin{aligned} \mathbf{u}_1 &= \text{before-collision velocity of } m_1 \text{ with respect to the centre-of-mass} \\ &= \mathbf{v}_1 - \mathbf{V} = \mathbf{v}_1 - \frac{m_1\mathbf{v}_1}{m_1 + m_2} = \frac{m_2}{m_1 + m_2}\mathbf{v}_1, \end{aligned} \tag{11.6}$$

where, while making calculations, the result in (11.5) has been used. Equations (11.5) and (11.6) show that velocities $\mathbf{u}_1$, $\mathbf{v}_1$ and **V** are all in the same direction.

If $\mathbf{u}_2$ is the velocity of the target of mass m_2 in the centre-of-mass frame, then (11.5) shows that

$$\mathbf{u}_2 = \mathbf{v}_2 - \mathbf{V} = -\mathbf{v} = -\frac{m_1}{m_1 + m_2}\mathbf{v}_1, \tag{11.7}$$

where we have made use of the fact that the target is stationary in the laboratory ($\mathbf{v}_2 = 0$) and have also used (11.5). Equations (11.6) and (11.7) show that, before the collision, the two particles are moving in opposite directions in the centre-of-mass frame. Moreover, before collision, the momentum of the system in the centre-of-mass frame is

$$m_1\mathbf{u}_1 + m_2\mathbf{u}_2 = m_1\left[\frac{m_2\mathbf{v}_1}{m_1 + m_2}\right] + m_2\left[-\frac{m_1\mathbf{v}_1}{m_1 + m_2}\right] = 0.$$

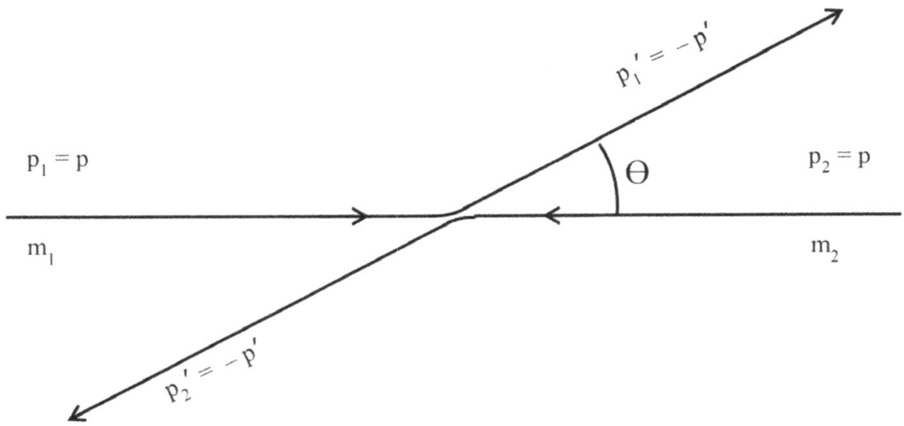

Figure 11.1. Collision of two particles in the centre-of-mass frame.

The law of conservation of linear momentum requires that on collision the momentum of the system should remain zero, i.e. the momenta of m_1 and m_2 should be equal and opposite: the particles recede from each other after the collision. Furthermore, the speeds of the two particles are not affected by collision. This is because if the speed of one particle increases or decreases by collision, the speed of the other particle should also increase or decrease so that the linear momentum of the system may remain zero. But this would increase or decrease the kinetic energy of the system which is against the very definition of elastic collisions according to which the kinetic energy must be conserved in an elastic collision. Therefore, the speeds of the two particles remain unchanged. Hence, in the centre-of-mass frame, the result of the collision is simply to rotate the line joining the two particles m_1 and m_2 through *the centre-of-mass scattering angle* Θ. This is shown in figure 11.1.

11.3 The effect of collision on the velocity of the centre-of-mass in the laboratory frame

Let us next consider the situation when the collision has occurred. Let $\mathbf{v}_1'$ and $\mathbf{v}_2'$ be the velocities of particles with masses m_1 and m_2, and $\mathbf{V}'$ that of the centre-of-mass in the laboratory frame of reference, after the collision. Then

$$\mathbf{V}' = \frac{m_1\mathbf{v}_1' + m_2\mathbf{v}_2'}{m_1 + m_2} = \frac{m_1\mathbf{v}_1 + m_2\mathbf{v}_2}{m_1 + m_2} = \mathbf{V} \tag{11.8}$$

as the momentum of the system is always conserved. Equation (11.8) shows that on elastic collision the velocity of the centre-of-mass in the laboratory frame remains unchanged.

Now, if $\mathbf{u}_1'$ denotes the after-collision velocity of m_1 in the centre-of-mass frame, then

$$\begin{aligned} \mathbf{u}_1' &= \text{after-collision velocity of } m_1 \text{ with respect to the centre-of-mass} \\ &= \mathbf{v}_1' - \mathbf{V}. \end{aligned} \tag{11.9a}$$

Similarly, the velocity $\mathbf{u}_2'$ of m_2 in the centre-of-mass frame is given by:

$$\mathbf{u}_2' = \mathbf{v}_2' - \mathbf{V}. \tag{11.9b}$$

11.4 Relation between scattering angles in the laboratory and centre-of-mass frames

In order to determine the relation between the scattering angles in the laboratory and centre-of-mass frames, let us draw a diagram with the help of the above equations. Let us represent the before-collision velocity $\mathbf{u}_1$ of m_1 in the centre-of-mass frame by the vector **OL**. Since $\mathbf{v}_1$ and $\mathbf{V}$ are in the same direction, these would also be in the direction of OL. Let **OB** represent $\mathbf{u}_1'$ and let Θ be the angle between the initial and final velocities of m_1 in the centre-of-mass frame. Then the angle $LOB = \Theta$. Let **OA** represent $\mathbf{v}_1'$ and let θ be the angle between initial and final directions of motion of m_1 in the laboratory frame. Then $AOL = \theta$. Join BA. Then, from figure 11.2, we have

$$\mathbf{BA} = \mathbf{OA} - \mathbf{OB} = \mathbf{v}_1' - \mathbf{u}_1' = \mathbf{V},$$

where the last result has been obtained using (11.9*a*). Since the direction of **V** is along **OL**, the line BA must be parallel to **OL**. From A and B, draw AP and BQ normal to OL. Then, from figure 11.2, we have

$$\tan\theta = \frac{AP}{OP} = \frac{BQ}{OQ + QP} = \frac{OB\sin\Theta}{OB\cos\Theta + BA}$$

$$= \frac{u_1'\sin\Theta}{u_1'\cos\Theta + V}. \tag{11.10}$$

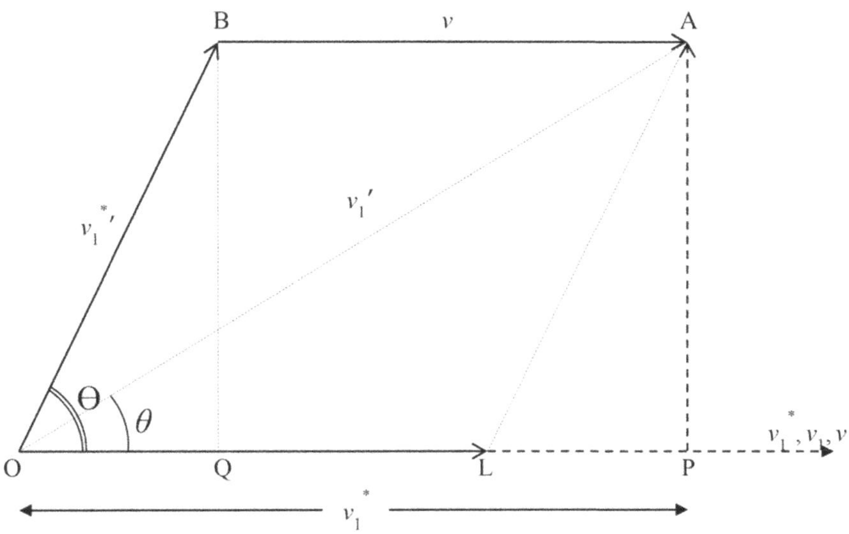

Figure 11.2. Relation between scattering angles in the laboratory and centre-of-mass frames.

Now, by virtue of (11.5) and (11.6), and the fact that, in elastic collisions, the speeds do not change on collision, we have

$$V = \frac{m_1}{m_1 + m_2}\mathbf{v}_1 \tag{11.5'}$$

and

$$u_1' = u_1 \frac{m_2}{m_1 + m_2}\mathbf{v}_1. \tag{11.6'}$$

Substituting these expressions in (11.10), we obtain

$$\tan\theta = \frac{\dfrac{m_2}{m_1 + m_2} v_1 \sin\Theta}{\dfrac{m_2}{m_1 + m_2} v_1 \cos\Theta + \dfrac{m_1}{m_1 + m_2} v_1},$$

or

$$\tan\theta = \frac{\sin\Theta}{\cos\Theta + \dfrac{m_1}{m_2}}. \tag{11.11}$$

This equation shows how the scattering angles θ and Θ in the laboratory and centre-of-mass frames are related to each other.

Suppose that $m_2 \gg m_1$. Then $m_1/m_2 \to 0$ and (11.11) yields

$$\tan\theta \approx \frac{\sin\Theta}{\cos\Theta} = \tan\Theta,$$

or

$$\theta \approx \Theta. \tag{11.12}$$

This shows that if the target mass m_2 is much greater than the mass m_1 of the incident particle, then the scattering angles in the two frames of reference are almost equal.

If $m_1 = m_2$, then by virtue of (11.11), we have

$$\begin{aligned}\tan\theta &= \frac{\sin\Theta}{\cos\Theta + 1} \\ &= \frac{2\sin\dfrac{\Theta}{2}\cos\dfrac{\Theta}{2}}{2\cos^2\dfrac{\Theta}{2}} \\ &= \tan\frac{\Theta}{2},\end{aligned}$$

or

$$\theta = \frac{\Theta}{2}. \tag{11.13}$$

This equation shows that if the incident particle and the target have equal masses, the scattering angle in the laboratory frame is half of the scattering angle in the centre-of-mass frame.

11.5 Relation between differential cross sections in the laboratory and centre-of-mass frames

Let us next find the relation between the differential cross sections in the two frames of reference. Since the number of particles passing within a given solid angle must be the same in both frames, we must have

$$I\sigma_{\rm L}(\theta)\,{\rm d}\Omega(\theta, \phi) = I\sigma_{\rm CM}(\Theta){\rm d}\Omega(\Theta, \phi),$$

where, for convenience, $\sigma_{\rm L}(\theta)$ and $\sigma_{\rm CM}(\theta)$ have been used for differential cross sections in the laboratory and centre-of-mass frames, respectively,

$$2\pi I\sigma_{\rm L}(\theta)\sin\theta\,{\rm d}\theta = 2\pi I\sigma_{\rm CM}(\Theta)\sin\Theta\,{\rm d}\Theta,$$

or

$$\sigma_{\rm L}(\theta) = \sigma_{\rm CM}\frac{\sin\Theta}{\sin\theta}\frac{{\rm d}\Theta}{{\rm d}\theta}. \tag{11.14}$$

To obtain the value of $\frac{{\rm d}\Theta}{{\rm d}\theta}$, we differentiate (11.11) with respect to θ:

$$\begin{aligned}
\sec^2\theta &= \left[\frac{\cos\Theta}{\left(\cos\Theta + \frac{m_1}{m_2}\right)} - \sin\Theta\frac{(-\sin\Theta)}{\left(\cos\Theta + \frac{m_1}{m_2}\right)^2}\right]\frac{{\rm d}\Theta}{{\rm d}\theta} \\
&= \frac{\cos\Theta\left(\cos\Theta + \frac{m_1}{m_2}\right) + \sin^2\Theta}{\left(\cos\Theta + \frac{m_1}{m_2}\right)^2}\frac{{\rm d}\Theta}{{\rm d}\theta} \\
&= \frac{1 + \frac{m_1}{m_2}\cos\Theta}{\left(\cos\Theta + \frac{m_1}{m_2}\right)^2}\frac{{\rm d}\Theta}{{\rm d}\theta}.
\end{aligned} \tag{11.15}$$

Substituting the expression for $d\Theta/d\theta$ from (11.15) in (11.14), we obtain

$$\sigma_L(\theta) = \sigma_{CM}(\theta)\frac{\sin\Theta}{\sin\theta}\frac{1}{\cos^2\theta}\frac{\left(\cos\Theta + \frac{m_1}{m_2}\right)^2}{1 + \frac{m_1}{m_2}\cos\Theta}. \tag{11.16}$$

Moreover, squaring (11.11), we obtain

$$\tan^2\theta = \frac{\sin^2\Theta}{\left(\cos\Theta + \frac{m_1}{m_2}\right)^2}, \tag{11.17}$$

or

$$\frac{\sin^2\theta}{\cos^2\theta}\left\{\cos^2\Theta + \left(\frac{m_1}{m_2}\right)^2 + 2\left(\frac{m_1}{m_2}\right)\cos\Theta\right\} = \sin^2\Theta,$$

or

$$\begin{aligned}\sin^2\theta\left\{\cos^2\Theta + \left(\frac{m_1}{m_2}\right)^2 + 2\left(\frac{m_1}{m_2}\right)\cos\Theta\right\} &= \sin^2\Theta\cos^2\theta \\ &= \sin^2\Theta\left(1 - \sin^2\theta\right) \\ &= \sin^2\Theta - \sin^2\Theta\sin^2\theta,\end{aligned}$$

or

$$\sin^2\theta\left\{1 + \left(\frac{m_1}{m_2}\right)^2 + 2\left(\frac{m_1}{m_2}\right)\cos\Theta\right\} = \sin^2\Theta,$$

or

$$\sin\Theta = \frac{\sin\theta}{\left\{1 + \left(\frac{m_1}{m_2}\right)^2 + 2\left(\frac{m_1}{m_2}\right)\cos\Theta\right\}^{1/2}}. \tag{11.18}$$

Now, from (11.17), we have

$$\frac{1 - \cos^2\theta}{\cos^2\theta} = \frac{\sin^2\Theta}{\left(\cos\Theta + \frac{m_1}{m_2}\right)^2},$$

because

$$\tan^2\theta = \frac{\sin^2\theta}{\cos^2\theta} = \frac{1-\cos^2\theta}{\cos^2\theta}.$$

The above equation yields

$$\left(\cos\Theta + \frac{m_1}{m_2}\right)^2\left(1-\cos^2\theta\right) = \sin^2\Theta\cos^2\theta,$$

or

$$\left(\cos\Theta + \frac{m_1}{m_2}\right)^2 - \left(\cos\Theta + \frac{m_1}{m_2}\right)^2\cos^2\theta = \sin^2\Theta\cos^2\theta$$

or

$$\cos^2\theta = \frac{\left(\cos\Theta + \dfrac{m_1}{m_2}\right)^2}{1+\left(\dfrac{m_1}{m_2}\right)^2} + 2\left(\frac{m_1}{m_2}\right)\cos\Theta. \tag{11.19}$$

Substituting this expression for $\cos^2\theta$ in (11.16), we obtain

$$\sigma_L(\theta) = \sigma_{CM}(\theta)\frac{\sin\theta}{\sin\theta}\frac{\left\{1+\left(\dfrac{m_1}{m_2}\right)^2 + 2\left(\dfrac{m_1}{m_2}\right)\cos\Theta\right\}}{\left(1+\dfrac{m_1}{m_2}\cos\theta\right)}. \tag{11.20}$$

Substituting the expression for $\sin\Theta/\sin\theta$ from (11.18) in the above equation, we obtain

$$\sigma_L(\theta) = \sigma_{CM}(\theta)\frac{\left\{1+\left(\dfrac{m_1}{m_2}\right)^2 + 2\left(\dfrac{m_1}{m_2}\right)\cos\Theta\right\}^{3/2}}{\left(1+\dfrac{m_1}{m_2}\cos\theta\right)}. \tag{11.21}$$

This gives the relation between differential cross sections in the laboratory and centre-of-mass frames in terms of the masses of colliding particles and the scattering angle Θ in the centre-of-mass frame.

The kinetic energy of the system in the laboratory frame is given by

$$T_{LAB} = \frac{1}{2}m_1 v_1^2$$

while in the centre-of-mass frame it is

$$T_{\mathrm{CM}} = \frac{1}{2}m_1 u_1^2 + \frac{1}{2}m_1 u_2^2.$$

Problem

11.1. Show that

$$T_{\mathrm{CM}} = \frac{m_2}{m_1 + m_2} T_{\mathrm{LAB}}.$$

11.6 Scattering by a stationary target

Let us now consider the scattering of an incident particle by a target stationary in the laboratory. Initially the particle is far away from the centre of force, i.e. the target. Therefore the force acting on it may be taken as zero so that its potential, related to the force by the equation $\mathbf{F} = -\nabla V$, is constant. The value of this constant is determined by our choice of the zero of energy scale, just as the height of a roof is determined from the zero of the measurement scale; if we measure the height from the roof itself, its height will be zero. Here, we choose the zero of the energy scale so that the potential energy V of the particle is zero. Then the Schrödinger amplitude equation for the incoming particle is

$$\nabla^2 \psi_{\mathrm{in}} + \frac{2mE}{\hbar^2}\psi_{\mathrm{in}} = 0 \tag{11.22}$$

where ψ_{in} represents the incoming particle and E is the total energy of the particle. Writing k^2 for $2mE/\hbar^2$, we obtain

$$\nabla^2 \psi_{\mathrm{in}} + k^2 \psi_{\mathrm{in}} = 0. \tag{11.23}$$

If the direction of motion of the particle is taken as the z-axis, then the above equation takes the form

$$\frac{\mathrm{d}^2 \psi_{\mathrm{in}}}{\mathrm{d}z^2} + k^2 \psi_{\mathrm{in}} = 0. \tag{11.24}$$

This is a linear, homogeneous, second-order, ordinary differential equation. Its general solution is

$$\psi_{\mathrm{in}} = Ae^{ikz} + Be^{-ikz}. \tag{11.25}$$

Multiplying the above equation by $\exp(-\mathrm{i}Et/\hbar)$, we obtain

$$\psi_{\mathrm{in}} \mathrm{e}^{-\mathrm{i}Et/\hbar} = A\mathrm{e}^{\mathrm{i}kz}\mathrm{e}^{-\mathrm{i}Et/\hbar} + B\mathrm{e}^{-\mathrm{i}kz}\mathrm{e}^{-\mathrm{i}Et/\hbar}, \tag{11.26}$$

or

$$\Psi_{\text{in}} = A e^{ikz} e^{-iEt/\hbar} + B e^{-ikz} e^{-iEt/\hbar}. \tag{11.27}$$

The first term on the right-hand side of this equation represents a particle moving towards the right while the second term describes the particle moving towards the left. Since no particles are moving to the left, the term Be^{-ikz} is not physically acceptable. Thus the incoming particle is represented by a plane wave:

$$\Psi_{\text{in}} = A e^{ikz} e^{-iEt/\hbar}. \tag{11.28}$$

As is customary in the literature, the time-factor will normally be dropped in the following discussion. Thus, for instance, we will write the plane wave as

$$\psi_{\text{in}} = A e^{ikz} \tag{11.29}$$

whereas, actually, it is only the spatial part of the wave function.

A very interesting point is worth careful consideration at this stage. Can a plane wave actually represent the particle under consideration? We note from equation $\psi = Ae^{ikz}$ that $\psi^*\psi = A^*A$. Thus, the probability of finding the particle anywhere in the plane wave extending from $-\infty$ to $+\infty$ is not only independent of x and y which are not present in (11.29) but also of z. Hence, the position of the particle is completely indeterminate; its momentum must be sharply defined. This is impossible because the scattering experiment is actually performed in the laboratory in a well-defined region of space. For that reason its position and momentum should both be uncertain by certain amounts. However, the size of the apparatus used in the scattering experiment is extremely large as compared to the dimensions of the particles in the incident beam. Therefore, the uncertainty in the momentum is negligible. Hence, to a good approximation a plane wave can represent the incoming particle in this problem.

When the particle approaches the centre of force, it is deflected from its normal path under the influence of the force which is exerted by the target. Let us then find the form of the scattered wave at a large distance from the centre of force. We shall assume that the potential $V(r)$ is such that

$$\lim_{r\to\infty}(rV(r)) = 0, \tag{11.30}$$

i.e. $V(r)$ decreases more rapidly than $1/r$. This will simplify the further analysis. Except for this restriction, the potential $V(r)$ may have any form. The Schrödinger wave equation for the scattered particle is

$$\nabla^2\psi_{\text{sc}} + \frac{2m}{\hbar^2}[(E - V(r))]\psi_{\text{sc}} = 0. \tag{11.31}$$

In spherical polar coordinates (r, θ, ϕ), this differential equation takes the form

$$\frac{1}{r}\frac{\partial^2}{\partial r^2}(r\psi_{\text{sc}}) + \frac{1}{r^2}\Omega\psi_{\text{sc}} + \left\{k^2 - \frac{2m}{\hbar^2}V(r)\right\}\psi_{\text{sc}} = 0, \tag{11.32}$$

where Ω is the angular part of the differential equation and is given by

$$\Omega = \frac{1}{\sin\theta}\frac{\partial}{\partial\theta} - \left(\sin\theta\frac{\partial}{\partial\theta}\right) + \frac{1}{\sin^2\theta}\frac{\partial^2}{\partial\phi^2}. \tag{11.33}$$

Since $1/r^2$ and $V(r)$ decrease more rapidly than $1/r$, as r tends to infinity, i.e. at large distances from the centre of force, the terms involving these may be neglected as compared to the rest. Therefore the asymptotic form of (11.32) would be

$$\frac{1}{r}\frac{\partial^2}{\partial r^2}(r\psi_{sc}) + k^2\psi_{sc} = 0. \tag{11.34}$$

The two solutions of this differential equation are

$$f(\theta, \phi)\frac{e^{\pm ikr}}{r}.$$

The function $f(\theta, \phi)e^{-ikr}/r$ represents radially incoming waves. Since no such waves exist, this solution is not physically acceptable. The only acceptable solution is $f(\theta,\phi)$ e^{ikr}/r which represents the *scattered wave outgoing radially from the centre of force.* We denote it by ψ_{sc}. Then, we may write

$$\psi_{sc} = f(\theta, \phi)\frac{e^{ikr}}{r} \tag{11.35}$$

where $f(\theta,\phi)$ is called the *scattering amplitude.* Therefore, the total *asymptotic solution* of the Schrödinger wave equation is

$$\psi_{asymptotic} = \psi_{in} + \psi_{sc} = e^{ikz} + f(\theta, \phi)\frac{e^{ikr}}{r}. \tag{11.36}$$

We have omitted the normalisation factor A because in the scattering process our interest lies in the *relative probability* rather than the absolute one. Equation (11.36) may be considered as a boundary condition for the solution of the Schrödinger wave equation for the scattering problem, i.e.

$$\nabla^2\psi + \frac{2m}{\hbar^2}[E - V(r)]\psi = 0. \tag{11.37}$$

It may be emphasised that this is the Schrödinger wave equation for the complete interacting system which is assumed to be in a stationary state. This scattering process is illustrated in figure 11.3.

11.7 Relation between the scattering amplitude and differential cross section

We have shown that the incident beam is represented by $\psi_{in} = Ae^{ikz}$. Let $\mathbf{v}$ be the velocity of the beam. Then the number of particles in the incident beam crossing unit area normal to its direction of motion in unit time is

$$v\psi_{in}{}^*\psi_{in} = ve^{-ikz}e^{ikz} = v. \tag{11.38}$$

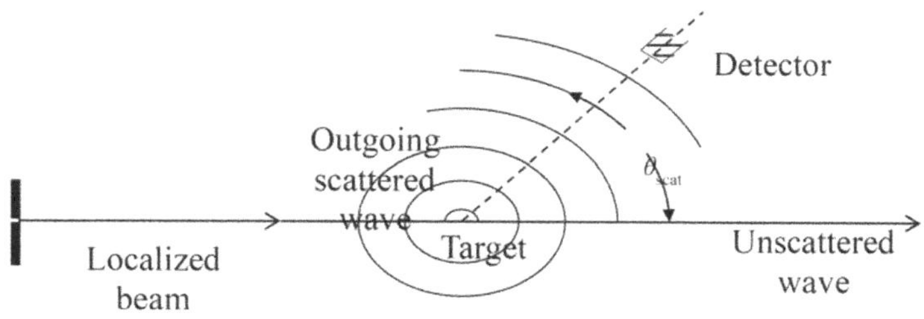

Figure 11.3. Scattering by a stationary target.

This is the *incident flux*. The number of scattered particles crossing the unit area normal to the direction of motion of the scattered beam in unit time is given by

$$v\psi_{sc}{}^{*}\psi_{sc} = vf^{*}(\theta,\,\phi)\frac{e^{-ikr}}{r}f(\theta,\,\phi)\frac{e^{ikr}}{r} = \frac{vf^{*}(\theta,\,\phi)f(\theta,\,\phi)}{r^2}. \tag{11.39}$$

The number of scattered particles crossing area dS normal to its direction of motion in unit time is

$$\frac{vf^{*}(\theta,\,\phi)f(\theta,\,\phi)}{r^2}\mathrm{d}S = vf^{*}(\theta,\,\phi)f(\theta,\,\phi)\mathrm{d}\Omega, \tag{11.40}$$

where dΩ is the solid angle subtended by the area dS at the centre of the potential. This expression therefore gives the number of particles scattered per unit time into the solid angle dΩ (subtended by the area normal to the direction of scattered motion at the centre of force) in the direction $(\theta,\,\phi)$ of the incident beam in unit time. The probability that a particle will be scattered per unit time into a unit solid angle in the direction $(\theta,\,\phi)$ of the incident beam is called the *differential cross section* and is given by

$$\frac{\mathrm{d}\sigma}{\mathrm{d}\Omega} = \frac{vf^{*}(\theta,\,\phi)f(\theta,\,\phi)}{v} = f^{*}(\theta,\,\phi)f(\theta,\,\phi) = \left|f(\theta,\,\phi)\right|^2. \tag{11.41}$$

Thus, the computation of the differential cross section depends upon a knowledge of the scattering amplitude $f(\theta,\,\phi)$. Since the dimension of the differential cross section is L^2, the dimension of the scattering amplitude is that of length. It may be noted that in calculating the differential cross section, the terms $\psi_{in}(= e^{ikz})$ and $\psi_{sc}(= f(\theta,\,\phi)e^{ikr}/r)$, representing, respectively, the incident and scattered beams, were considered separately, thus ignoring the interference terms between them which would arise if the wave function ψ $(= \psi_{in} + \psi_{sc})$ itself was used in computing the fluxes. This is justified for the reason that the incident and the scattered beams are experimentally detected in the regions which are far away from the target so that the two beams do not overlap. The interference terms are effective only in the neighbourhood of the target where both the incident and the scattered beams are present. Since no measurements are made in the vicinity of the target, one need not bother about the interference terms.

It should be pointed out that in the scattering problem, the energy spectrum is continuous because the scattered particle is asymptotically free.

11.8 Computation of the scattering amplitude

We have shown that the differential cross section for a scattering problem can be calculated if the scattering amplitude $f(\theta, \phi)$ is known. To calculate $f(\theta, \phi)$, we proceed as follows.

The time-independent Schrödinger wave equation for the scattering problem is

$$\nabla^2\psi(r) + \frac{2m}{\hbar^2}\{E - V(r)\}\psi(r) = 0. \tag{11.42}$$

Writing

$$\frac{2mE}{\hbar^2} = k^2, \quad \frac{2m}{\hbar^2}V(r) = U(r) \quad \text{and} \quad U(r)\psi(\mathbf{r}) = F(\mathbf{r}), \tag{11.43}$$

(11.42) becomes

$$\left(\nabla^2 + k^2\right)\psi(\mathbf{r}) = F(\mathbf{r}). \tag{11.44}$$

At large distances from the centre of force where interference between ψ_{in} and ψ_{sc} can be ignored, this gives

$$\left(\nabla^2 + k^2\right)\left[\psi_{in}(\mathbf{r}) + \psi_{sc}(\mathbf{r})\right] = F(\mathbf{r}), \tag{11.45}$$

where $\psi_{\text{in}}(\mathbf{r})$ stands for the incident beam while $\psi_{\text{sc}}(\mathbf{r})$ stands for the scattered beam. Since the incident particle, being far away from the target, is moving freely, the potential $V(r) = 0$ and therefore

$$\left(\nabla^2 + k^2\right)\psi_{in}(\mathbf{r}) = 0, \tag{11.46}$$

and differential equation (11.45) reduces to

$$\left(\nabla^2 + k^2\right)\psi_{sc}(\mathbf{r}) = F(\mathbf{r}). \tag{11.47}$$

This differential equation is of type

$$Ly(\mathbf{r}) = f(\mathbf{r}) \tag{11.48}$$

where L is an operator. Its solution in terms of the Green's function $G(\mathbf{r}, \mathbf{r}')$ is given by

$$y(\mathbf{r}) = \int G(\mathbf{r}, \mathbf{r}')f(\mathbf{r}')\,d\mathbf{r}', \tag{11.49}$$

where $G(\mathbf{r}, \mathbf{r}')$, the Green's function for the operator L, is a solution of the equation

$$LG(\mathbf{r}, \mathbf{r}') = \delta(\mathbf{r} - \mathbf{r}'). \tag{11.50}$$

Problem

11.2. Find the Green's function $G(\mathbf{r}, \mathbf{r}')$ for the operator $(\nabla^2 + k^2)$ which satisfies the equation

$$\left(\nabla^2 + k^2\right)G(\mathbf{r}, \mathbf{r}') = \delta(\mathbf{r} - \mathbf{r}')$$

such that the asymptotic solution is of the form

$$f(\theta, \phi)\frac{e^{ikr}}{r}.$$

The Green's function in this case is given by

$$G(\mathbf{r}, \mathbf{r}') = -\frac{1}{4\pi}\frac{e^{ik|\mathbf{r}-\mathbf{r}'|}}{|\mathbf{r} - \mathbf{r}'|}. \tag{11.51}$$

Of course, the boundary condition is incorporated.

Hence the solution of the differential equation

$$\left(\nabla^2 + k^2\right)\psi_{sc} = F(\mathbf{r})$$

is

$$\begin{aligned}\psi_{sc} &= \int G(\mathbf{r}, \mathbf{r}')F(\mathbf{r}')d\mathbf{r}' \\ &= -\frac{1}{4\pi}\int \frac{e^{ik|\mathbf{r}-\mathbf{r}'|}}{|\mathbf{r} - \mathbf{r}'|}F(\mathbf{r}')d\mathbf{r}'.\end{aligned} \tag{11.52}$$

For large values of r, this reduces to

$$\psi_{sc} = -\frac{1}{4\pi}\frac{e^{ikr}}{r}\int e^{-i\mathbf{k}'\cdot\mathbf{r}'}F(\mathbf{r}')d\mathbf{r}' \tag{11.53}$$

where $\mathbf{k}' = k\mathbf{n}$ is a vector in the direction of the outgoing wave and is called the *wave vector* of the scattered beam.

Problem

11.3. Show that for large values of r, (11.52) reduces to (11.53).

The total asymptotic wave function is

$$\psi(\mathbf{r}) = e^{ikz} + \frac{e^{ikr}}{r}\int\left(-\frac{1}{4\pi}\right)e^{-i\mathbf{k}'\cdot\mathbf{r}'}F(\mathbf{r}')\,d\mathbf{r}'. \tag{11.54}$$

Comparing (11.54) and (11.36), we note that

$$f(\theta, \phi) = -\frac{1}{4\pi}\int e^{-i\mathbf{k}'\cdot\mathbf{r}'} F(\mathbf{r}')d\mathbf{r}'$$
$$= -\frac{1}{4\pi}\int e^{-i\mathbf{k}'\cdot\mathbf{r}'} U(\mathbf{r}')\psi(\mathbf{r}')d\mathbf{r}' \tag{11.55}$$

where $\psi(\mathbf{r}')$ is the total wave function at $\mathbf{r}'$. Equation (11.55) is exact. It may be thought that (11.54) gives the solution $\psi(\mathbf{r})$ which may be used in (11.55) to evaluate $f(\theta, \phi)$. But the difficulty arises because $F(\mathbf{r}') = U(\mathbf{r}')\psi(\mathbf{r}')$ occurs under the integral sign. Hence the scattering amplitude $f(\theta, \phi)$ cannot be calculated from (11.55) as it stands. Some additional constraint has to be imposed. However, this equation becomes the basis to obtain an approximate solution of the problem.

11.9 The Born approximation

We can get around the difficulty and solve the equation to a *first approximation* if the scattering potential is *localised*, i.e. it is effective to a small region *and* is *sufficiently weak*, i.e. small compared to the kinetic energy of the incident beam. In this case it is reasonably good to assume that the incident plane wave representing the incoming beam of particles is not very much distorted by the scattering potential. Therefore, to a first approximation, we may replace $\psi(\mathbf{r}')$ in (11.55) by $\psi_{\text{in}}(\mathbf{r}')$, the incident wave. Equation (11.55), then becomes

$$f(\theta, \phi) = -\frac{1}{4\pi}\int e^{-i\mathbf{k}'\mathbf{r}'} U(\mathbf{r}')\psi_{\text{in}}(\mathbf{r}')d\mathbf{r}'. \tag{11.56}$$

But $\psi_{in}(\mathbf{r}) = e^{ikz}$. Changing to spherical polar coordinates r, θ, ϕ, taking the origin at the centre of the potential and the polar axis ($\theta = 0$) as the z-axis, we have

$$\psi_{\text{in}}(\mathbf{r}) = e^{ikz} = e^{ikr\cos\theta} = e^{i\mathbf{k}\cdot\mathbf{r}}, \tag{11.57}$$

where $\mathbf{k}$ is a vector of magnitude $k(=|\mathbf{k}|)$ directed along the beam of incident particles which is in the direction of the z-axis and $\mathbf{r}$ is a vector drawn from the origin to the point (r, θ, ϕ). Replacing $\mathbf{r}$ by $\mathbf{r}'$ throughout (11.57), we obtain

$$\psi_{\text{in}}(\mathbf{r}') = e^{i\mathbf{k}\cdot\mathbf{r}'}. \tag{11.58}$$

Substituting this expression in (11.56), we obtain

$$f(\theta, \phi) = -\frac{1}{4\pi}\int e^{i(\mathbf{k}-\mathbf{k}')\cdot\mathbf{r}'} U(\mathbf{r}')d\mathbf{r}'. \tag{11.59}$$

For convenience, we may now drop the prime from $\mathbf{r}'$ and write the above equation as

$$f(\theta, \phi) = -\frac{1}{4\pi}\int e^{i(\mathbf{k}-\mathbf{k}')\cdot\mathbf{r}} U(r)d\mathbf{r}. \tag{11.60}$$

This is called the *Born approximation formula*. The differential cross section is given by

$$\frac{d\sigma}{d\Omega} = |f(\theta, \phi)|^2$$

$$= \left| -\frac{1}{4\pi} \int e^{i(\mathbf{k}-\mathbf{k}')\cdot\mathbf{r}} U(r) d\mathbf{r} \right|^2 \tag{11.61}$$

Equation (11.60), the Born approximation formula, is further simplified if the *scattering potential is spherically symmetric*, i.e. a function of radial coordinates only: $V = V(r)$. Let

$$\mathbf{k} - \mathbf{k}' = \mathbf{X} \tag{11.62}$$

and define a new set of spherical polar coordinates r, θ_N, ϕ_N (these should not be confused with the spherical polar coordinates r, θ, ϕ whose polar axis lies along the direction of $\mathbf{k}$) with an origin at the centre of the potential and polar axis $\theta_N = 0$ along the direction of $\mathbf{X}$. Then

$$(\mathbf{k} - \mathbf{k}') \cdot \mathbf{r} = \mathbf{X} \cdot \mathbf{r} = Xr \cos\theta_N$$

and $d\mathbf{r} = r^2 \sin\theta_N dr d\theta_N d\phi_N$. The solution should then show cylindrical symmetry about the direction of incidence ($\theta = 0$), i.e. about the polar axis. Then $f(\theta, \phi) \to f(\theta)$. Substituting these expressions in (11.60), we obtain the scattering amplitude as given by

$$f(\theta) = -\frac{1}{4\pi} \int e^{i(\mathbf{k}-\mathbf{k}')\cdot\mathbf{r}} U(r) d\mathbf{r}$$

$$= -\frac{1}{4\pi} \int_0^\infty \int_0^\pi \int_0^{2\pi} U(r) e^{iXr\cos\theta_N} r^2 \sin\theta_N dr d\theta_N d\phi_N$$

$$= -\frac{1}{4\pi} \int_0^\infty \int_0^\pi U(r) e^{iXr\cos\theta_N} 2\pi r^2 dr \sin\theta_N d\theta_N. \tag{11.63}$$

Let $\cos\theta_N = u$. Then $-\sin\theta_N d\theta_N = du$ and (11.63) yields

$$f(\theta) = -\frac{1}{2} \int_0^\infty \int_1^{-1} U(r) e^{iXru} (-du) r^2 dr$$

$$= \frac{1}{2} \int_0^\infty \int_1^{-1} U(r) r^2 dr \left[\frac{e^{iXru}}{iXr} \right]$$

$$= \frac{1}{2} \int_0^\infty U(r) r^2 \frac{\left(e^{-iXr} - e^{iXr}\right)}{iXr} dr$$

$$= \frac{1}{2} \int_0^\infty U(r) r^2 \frac{(-2i \sin Xr)}{iXr} dr$$

$$= -\int_0^\infty U(r) \frac{r \sin(Xr)}{X} dr$$

$$= -\frac{2m}{\hbar^2 X} \int_0^\infty V(r) r \sin(Xr) dr \tag{11.64}$$

where in deriving the last equation, we have made use of the relation

$$\frac{2m}{\hbar^2}V(r) = U(r). \tag{11.43'}$$

The expression for X in terms of k and θ can be obtained from the equation

$$\mathbf{X} = \mathbf{k} - \mathbf{k}'. \tag{11.62'}$$

Squaring, we obtain

$$X^2 = k^2 + k'^2 - 2kk'\cos\theta. \tag{11.65}$$

Since $k' = k$ for elastic scattering, (11.65) can be written as

$$\begin{aligned} X^2 &= 2k^2 - 2k^2\cos\theta \\ &= 2k^2(1 - \cos\theta) \\ &= 4k^2\sin^2\frac{\theta}{2}. \end{aligned}$$

Hence,

$$X = 2k\sin\frac{\theta}{2}. \tag{11.66}$$

It may be noted that

$$\hbar X = \hbar\mathbf{k} - \hbar\mathbf{k}' = \mathbf{p} - \mathbf{p}' \tag{11.67}$$

and therefore represents the difference in momenta of incident and scattered particles; $\hbar X$ is the momentum transferred from the incident particle to the scattering field during the collision.

Problem

11.4. Show that if $V(-r) = V(r)$, the Born scattering amplitude is real.

11.10 Scattering of high energy electrons by a screened Coulomb potential

An important example of the application of the Born approximation formula is the scattering of high energy electrons each of charge $-e$ by a screened Coulomb potential. An electron of charge $-e$ incident on an atom of nuclear charge Ze sees the nuclear charge as if it were screened by the surrounding atomic electrons. Just outside the atom, which is behaving as a neutral entity, the potential is zero. And when it approaches very close to the nucleus, the potential becomes $-Ze^2/r$, where r is the distance of the electron from the centre of force. In between these states, there is a variation in potential. Such a potential is usually represented by

$$V(r) = -\frac{Ze^2}{r}\mathrm{e}^{-\alpha r} \tag{11.68}$$

where α has the dimensions of $1/r$, and is called a *screened Coulomb potential.* $1/\alpha$ is called the *screening radius.*

Substituting this expression for $V(r)$ in (11.64), we obtain

$$\begin{aligned} f(\theta) &= \frac{2m}{\hbar^2 X}\int_0^\infty \frac{Ze^2}{r}\mathrm{e}^{-\alpha r} r\sin(Xr)\,\mathrm{d}r \\ &= \frac{2mZe^2}{\hbar^2 X}\int_0^\infty e^{-\alpha r}\sin(Xr)\,\mathrm{d}r \\ &= b\int_0^\infty \mathrm{e}^{-\alpha r}\sin(Xr)\,\mathrm{d}r, \end{aligned} \tag{11.69}$$

where

$$b = \frac{2mZe^2}{\hbar^2 X}. \tag{11.70}$$

The integral in (11.69) does not possess a definite value because $\cos(Xr)$ is periodic and the upper limit is ∞. Now (11.69) can be written as

$$\begin{aligned} f(\theta) &= b\,\mathrm{Im}\int_0^\infty \mathrm{e}^{-\alpha r}\mathrm{e}^{\mathrm{i}Xr}\mathrm{d}r \\ &= b\,\mathrm{Im}\int_0^\infty \mathrm{e}^{(-\alpha+\mathrm{i}X)r}\mathrm{d}r \\ &= b\int_\infty^0 \mathrm{Im}\left[\frac{\mathrm{e}^{(-\alpha+\mathrm{i}X)r}}{(-\alpha+\mathrm{i}X)}\right]. \end{aligned} \tag{11.71}$$

When we substitute the upper limit $r \to \infty$, we find that $\mathrm{e}^{-\alpha r} \to 0$ while $\mathrm{e}^{\mathrm{i}Xr}$, being periodic, although changing with r, remains oscillating between finite values. Therefore, for $r \to \infty$, the function

$$\mathrm{e}^{(-\alpha+\mathrm{i}X)r} = \mathrm{e}^{-\alpha r}\mathrm{e}^{\mathrm{i}Xr} \to 0.$$

Consequently, (11.71) yields

$$f(\theta) = b\,\mathrm{Im}\left[-\frac{1}{(-\alpha + \mathrm{i}X)}\right]$$

$$= b\,\mathrm{Im}\left[-\frac{(-\alpha - \mathrm{i}X)}{\left(\alpha^2 + X^2\right)}\right]$$

$$= b\,\mathrm{Im}\left[\frac{\alpha + \mathrm{i}X}{\alpha^2 + X^2}\right]$$

$$= b\left[\frac{X}{\left(\alpha^2 + X^2\right)}\right]. \tag{11.72}$$

Therefore, the differential cross section is given by

$$\frac{\mathrm{d}\sigma}{\mathrm{d}\Omega} = \left|f(\theta)\right|^2 = \frac{b^2X^2}{\left(\alpha^2 + X^2\right)^2}. \tag{11.73}$$

For the scattering of high energy α-particles by heavy nuclei, the potential energy is $2Ze^2/r$. The scattering amplitude is obtained from (11.72) by replacing b by $-2b$, and taking the limit as $\alpha \to 0$. Then, we have

$$f(\theta) = -\frac{2b}{X}. \tag{11.74}$$

Substituting the expressions for b and X from (11.70) and (11.66) in (11.74), we obtain

$$f(\theta) = -\frac{Ze^2}{2E}\mathrm{cosec}^2\frac{\theta}{2}. \tag{11.75}$$

Therefore, we have

$$|f(\theta)|^2 = \frac{Z^2e^4}{4E^2}\mathrm{cosec}^4\frac{\theta}{2},$$

or

$$\frac{\mathrm{d}\sigma}{\mathrm{d}\Omega} = |f(\theta)|^2 = \frac{Z^2e^4}{4E^2}\mathrm{cosec}^4\frac{\theta}{2}. \tag{11.76}$$

This is the well-known *Rutherford scattering formula* that has been verified experimentally. The result obtained using the Born approximation formula is the same as the classical Rutherford scattering formula. Why? It should be pointed out that the exact quantum mechanical calculations lead to the same result as obtained using the Born approximation. Why?

11.11 Partial wave analysis

The Born approximation can be used to solve only those scattering problems for which the kinetic energy of the incident particles is much greater than the scattering potential. We will now describe a method which *in principle* is valid for incident particles of all energies but *in practice* is useful only when the incident beam is of low energy. This is the method of *partial wave analysis* because it employs a decomposition of the wave function representing the particle into partial waves each corresponding to a particular value of ℓ. We will confine our analysis to the case when the scattering potential is spherically symmetric, $V = V(r)$. Suppose that the incident particle is coming freely from a long distance towards the target, its direction of motion being taken as the z-axis. If there is no potential, i.e. $V(r) = 0$, then the Schrödinger amplitude equation is

$$\frac{d^2\psi_{in}}{dz^2} + \frac{2m}{\hbar^2}E\psi_{in} = 0, \qquad (11.77a)$$

$$\frac{d^2\psi_{in}}{dz^2} + k^2\psi_{in} = 0. \qquad (11.77b)$$

This differential equation has

$$\psi_{in} = e^{ikz} \qquad (11.78)$$

as a physically acceptable solution. We have shown that at large distances, the physically acceptable complete solution ψ of Schrödinger's differential equation for the scattering process is

$$\psi = \psi_{in} + \psi_{sc} = e^{ikz} + f(\theta)\frac{e^{ikr}}{r}. \qquad (11.79)$$

The first term is in Cartesian coordinates while the second term is in spherical polar coordinates. It would be more appropriate if both the terms were expressed in the same system of coordinates. Since the potential is spherically symmetric, it will be convenient to use spherical polar coordinates. We will therefore express the first term in spherical polar coordinates.

11.12 The incident particle wave in terms of partial waves

The incident particle is represented by a plane wave moving in the positive direction of the z-axis:

$$\psi_{in} = e^{ikz}. \qquad (11.78')$$

In terms of spherical polar coordinates, this can be written as

$$\psi_{in} = e^{ikr\cos\theta} \qquad (11.80)$$

where ψ_{in} is a solution of

$$\nabla^2 \psi_{\text{in}} + k^2\psi_{\text{in}} = 0, \tag{11.23'}$$

the origin being taken at the centre of the scattering potential and the polar axis chosen as the z-axis. Note that as the target is away from the incident particle, we have taken the potential energy as zero. Let us express ψ_{in} as a product of $R(r)$, $P(\theta)$ and $F(\phi)$. Then we write

$$\mathrm{e}^{ikr\cos\theta} = R(r)P(\theta)F(\phi), \tag{11.81}$$

where, as seen in the case of a centrally symmetric field ($V \equiv V(r)$), the functions R, P and F satisfy the differential equations

$$\frac{1}{r^2}\frac{\mathrm{d}}{\mathrm{d}r}\left(r^2\frac{\mathrm{d}R}{\mathrm{d}r}\right) + \left[k^2 - \frac{\ell(\ell+1)}{r^2}\right]R = 0, \quad \text{where } k^2 = \frac{2m}{\hbar^2}E, \tag{11.82}$$

$$\frac{1}{\sin^2\theta}\frac{\mathrm{d}}{\mathrm{d}\theta}\left(\sin\theta\frac{\mathrm{d}P(\theta)}{\mathrm{d}\theta}\right) + \ell(\ell+1)P(\theta) - \frac{m^2}{\sin^2\theta}P(\theta) = 0, \tag{11.83}$$

$$\frac{\mathrm{d}^2F(\phi)}{\mathrm{d}\phi^2} + m^2F(\phi) = 0. \tag{11.84}$$

Now the left-hand side of (11.81) does not depend upon ϕ. Therefore, the function $F(\phi)$ on the right-hand side of this equation must be a constant:

$$F(\phi) = \text{const.}$$

Note that this constant cannot be zero because in that case the wave function $\psi_{\text{in}} = R(r)P(\theta)F(\phi)$ will be zero for all values of r and θ, and the particle will disappear! Since F cannot be zero, (11.84) gives $m = 0$. Substituting this value of m in (11.83), we obtain

$$\frac{1}{\sin^2\theta}\frac{\mathrm{d}}{\mathrm{d}\theta}\left(\sin\theta\frac{\mathrm{d}P}{\mathrm{d}\theta}\right) + \ell(\ell+1)P = 0. \tag{11.85}$$

This is a second-order ordinary differential equation and therefore has, in general, two linearly independent solutions valid for even and odd values of ℓ. The two solutions are:

(1) a polynomial with even values of ℓ and
(2) a polynomial with odd values of ℓ.

When put together, these constitute a Legendre polynomial of order ℓ, $P_\ell(\cos\theta)$.

By introducing the variable $\rho = kr$, the radial equation (11.82) with $V(r) = 0$ takes the form

$$\frac{\mathrm{d}^2R}{\mathrm{d}\rho^2} + \frac{2}{\rho}\frac{\mathrm{d}R}{\mathrm{d}\rho} + \left[1 - \frac{\ell(\ell+1)}{\rho^2}\right]R = 0. \tag{11.86}$$

This is called the *spherical Bessel differential equation.* It has two linearly independent solutions $j_\ell(\rho)$ and $n_\ell(\rho)$ which can be expressed in terms of Bessel functions of half-odd integral order:

$$j_\ell(\rho) = \left(\frac{\pi}{2\rho}\right)^{1/2} j_{\ell+1/2}(\rho) = (-1)^\ell \left(\frac{d}{d\rho}\right)^\ell \frac{\sin\rho}{\rho} \tag{11.87}$$

and

$$n_\ell(\rho) = (-1)^{\ell+1} \left(\frac{\pi}{2\rho}\right)^{1/2} n_{-\ell-1/2}(\rho) = (-1)^{\ell+1} \left(\frac{d}{d\rho}\right)^\ell \frac{\cos\rho}{\rho}. \tag{11.88}$$

The functions $j_\ell(\rho)$ and $n_\ell(\rho)$ are, respectively, called the *spherical Bessel function* and the *spherical Neumann function.*

The first few spherical Bessel functions are

$$j_0(kr) = \frac{\sin kr}{kr}, \tag{11.89a}$$

$$j_1(kr) = \frac{\sin kr}{(kr)^2} - \frac{\cos kr}{kr}, \tag{11.89b}$$

$$j_2(kr) = \left[\frac{3}{(kr)^3} - \frac{1}{kr}\right] \sin kr - \frac{3}{(kr)^2} \cos kr. \tag{11.89c}$$

The other spherical Bessel functions can be obtained from the formula

$$j_{\ell+1}(kr) = \frac{2\ell + 1}{kr} j_\ell(kr) - j_{\ell-1}(kr). \tag{11.90}$$

The asymptotic, i.e. the large kr values of the spherical Bessel functions are given by

$$j_\ell(kr) = \frac{1}{kr} \sin\left(kr - \frac{\ell\pi}{2}\right). \tag{11.91}$$

This may also be written as

$$j_\ell(kr) \xrightarrow[kr\to\infty]{} \frac{1}{kr} \sin\left(kr - \frac{\ell\pi}{2}\right). \tag{11.92}$$

For small kr, by expanding $(i/kr)\sin kr$ and $(1/kr)\cos kr$ in power series in kr, it can be shown that

$$j_\ell(kr) \xrightarrow[kr\to 0]{} \frac{(kr)^\ell}{1, 3, 5, \ldots, (2\ell + 1)}. \tag{11.93}$$

The explicit expressions for the first few spherical Neumann functions are

$$n_0(kr) = -\frac{\cos kr}{kr}, \tag{11.94a}$$

$$n_1(kr) = -\frac{\cos kr}{(kr)^2} - \frac{\sin kr}{kr}, \tag{11.94b}$$

$$n_2(kr) = -\left[\frac{3}{(kr)^3} - \frac{1}{kr}\right]\cos kr - \frac{3}{(kr)^2}\sin kr. \tag{11.94c}$$

The other spherical Neumann functions can be obtained from the formula

$$n_{\ell+1}(kr) = \frac{2\ell}{kr}n_\ell(kr) - n_{\ell-1}(kr). \tag{11.95}$$

The asymptotic, i.e. the large kr values of spherical Neumann functions are given by

$$n_\ell(kr) = -\frac{1}{kr}\sin\left(kr - \frac{\ell\pi}{2}\right). \tag{11.96}$$

This may also be written as

$$n_\ell(kr) \xrightarrow[kr\to\infty]{} -\frac{1}{kr}\sin\left(kr - \frac{\ell\pi}{2}\right). \tag{11.97}$$

For small kr, by expanding $(i/kr)\sin kr$ and $(1/kr)\cos kr$ in a power series in kr, it can be shown that

$$n_\ell(kr) \xrightarrow[kr\to 0]{} -\frac{1, 3, 5, \ldots, (2\ell - 1)}{(kr)^{\ell+1}}. \tag{11.98}$$

Since according to (11.98), the function $n_\ell(\rho = kr)$ is *singular at the origin,* i.e. at $r = 0$, it is not physically acceptable, and is therefore rejected. The other solution $j_\ell(\rho)$, according to (11.93), is finite at the origin. In fact, it is finite throughout the region, and is therefore physically acceptable. Hence, a physically acceptable solution of differential equation (11.82)

$$\nabla^2\psi_{\text{in}} + k^2\psi_{\text{in}} = 0 \tag{11.82'}$$

for the incident particle, is

$$a_\ell P_\ell(\cos\theta)j_\ell(kr), \tag{11.99}$$

where a_ℓ is an arbitrary constant. Since the differential equation (11.82) is independent of ℓ, the general solution is

$$\psi_{in} = \sum\nolimits_\ell a_\ell P_\ell(\cos\theta)j_\ell(kr), \tag{11.100a}$$

or

$$e^{ikr\cos\theta} = \sum_{\ell} a_\ell P_\ell(\cos\theta) j_\ell(kr). \tag{11.100b}$$

This expresses the incident plane wave as a sum of partial waves with $\ell = 0, 1, 2, \ldots$.

11.13 Phase shift and scattering

For the scattering process, the radial equation in the presence of the potential $V(r)$ is given by

$$\frac{d^2}{dr^2} + \frac{2}{r}\frac{d}{dr} - \frac{\ell(\ell+1)}{r^2} - \frac{2\mu}{\hbar^2}V(r) + k^2 R_\ell(k, r) = 0 \tag{11.86'}$$

We do not know the potential but we assume that it is such that

$$\lim_{r\to\infty}\left\{r^2 V(r)\right\} = 0,$$

i.e. at the origin, it is less singular than $1/r^2$. As we have to make the experimental measurements at sufficiently large distances from the target, say for $r > a$, therefore the potential $V(r)$ may be neglected. Then for each ℓ, the two linearly independent solutions are $j_\ell(\rho)$ and $n_\ell(\rho)$ but this time both the solutions are physically acceptable because the matching condition at the origin makes it imperative that the solution $R_\ell(k, r)$ remains regular. The general solution is of the form

$$R_\ell(\rho) = B_\ell j_\ell(\rho) + C_\ell n_\ell(\rho). \tag{11.101a}$$

Now as the detector is placed at a large distance from the target, we have to consider the asymptotic expressions for the functions $j_\ell(\rho)$ and $n_\ell(\rho)$. These are

$$j_\ell(\rho) \xrightarrow[\rho\to\infty]{} \frac{1}{\rho}\sin\left(\rho - \frac{\ell\pi}{2}\right),$$

$$n_\ell(\rho) \xrightarrow[\rho\to\infty]{} \frac{1}{\rho}\cos\left(\rho - \frac{\ell\pi}{2}\right).$$

Then for $\rho \to \infty$, relation (11.101a) may be expressed as

$$R_\ell(\rho) \xrightarrow[\rho\to\infty]{} B_\ell \frac{1}{\rho}\sin\left(\rho - \frac{\ell\pi}{2}\right) - C_\ell \frac{1}{\rho}\cos\left(\rho - \frac{\ell\pi}{2}\right). \tag{11.101b}$$

Let us write

$$\mathrm{B}_\ell(k) = A_\ell(k)\cos\delta_\ell$$

and

$$C_\ell(k) = -A_\ell(k)\sin\delta_\ell.$$

Substituting these expressions in (11.101b), we obtain

$$R_\ell(\rho) \underset{\rho\to\infty}{\longrightarrow} A_\ell(k)\cos\delta_\ell \frac{1}{\rho}\sin\left(\rho - \frac{\ell\pi}{2}\right) + A_\ell \sin\delta_\ell \frac{1}{\rho}\cos\left(\rho - \frac{\ell\pi}{2}\right),$$

or

$$R_\ell(\rho) \underset{\rho\to\infty}{\longrightarrow} A_\ell(k)\,\frac{1}{\rho}\sin\left(\rho - \frac{\ell\pi}{2} + \delta_\ell\right).$$

We may choose $A_r(k) = 1$ in normalisation. Then

$$R_\ell(\rho) \underset{\rho\to\infty}{\longrightarrow} \frac{1}{\rho}\sin\left(\rho - \frac{\ell\pi}{2} + \delta_\ell\right).$$

The above analysis shows that when the particle is free, the physically acceptable asymptotic radial solution is

$$j_\ell(\rho) = \frac{1}{\rho}\sin\left(\rho - \frac{\ell\pi}{2}\right).$$

When the potential has been applied, the *asymptotic* radial solution transforms to

$$R_\ell(\rho) = \frac{1}{\rho}\sin\left(\rho - \frac{\ell\pi}{2} + \delta_\ell\right).$$

Hence, the potential introduces a *phase shift δ_ℓ in the asymptotic solution.* The phase shift δ_ℓ characterises the strength of the scattering in the ℓth partial wave by the potential $V(r)$ and energy $E = \hbar^2k^2/2m$. It can be computed either analytically or numerically, subject to appropriate matching conditions at the origin, i.e. at $r = 0$. Knowledge of δ_ℓ may give us some idea about the potential energy that causes the scattering of incident particles.

Example 11.1.

Problem

Determine the vaues of the constants a_ℓ in (11.100b).

Solution

To determine the value of a_ℓ, we proceed as follows. Writing μ (it should not be confused with reduced mass) for $\cos\theta$, (11.100b) takes the form

$$e^{ikr\mu} = \sum_\ell a_\ell P_\ell(\mu) j_\ell(kr). \tag{11.102}$$

Multiplying by $P_\ell(\mu)$ and integrating with respect to μ over the range (−1, 1) because θ varies from 0 to π, and interchanging the lower and upper limits on both sides of this equation, we obtain

$$\int_{-1}^{1} P_{\ell'}(\mu) e^{ikr\mu} d\mu = \int_{-1}^{1} P_{\ell'}(\mu) \sum_\ell a_\ell P_\ell(\mu) j_\ell(kr) d\mu. \tag{11.103}$$

Evaluating the integral on the left by parts, we obtain

$$\int_{-1}^{1}\left[P_{\ell'}(\mu)\frac{e^{ikr\mu}}{ikr}\right] - \int_{-1}^{1} P'_{\ell'}(\mu)\frac{e^{ikr\mu}}{ikr}d\mu$$
$$= \sum_{\ell} a_\ell j_\ell(kr)\int_{-1}^{1} P_{\ell'}(\mu)P_\ell(\mu)d\mu. \quad (11.104)$$

Problem

11.5. Integrating by parts, show that

$$\int_{-1}^{1} P_{\ell'}(\mu)e^{ikr\mu}d\mu = O\left(\frac{1}{r^2}\right). \quad (11.105)$$

By virtue of the above result, (11.104) can be written as

$$\left[P_{\ell'}(1)\frac{e^{ikr}}{ikr} - P_{\ell'}(-1)\frac{e^{-ikr}}{ikr}\right] - O\left(\frac{1}{r^2}\right)$$
$$= \sum_{\ell} a_\ell j_\ell(kr)\frac{2}{2\ell+1}\delta_{\ell\ell'}$$

where in writing the value of the last integral, we have made use of the orthonormalisation properties of the Legendre polynomials $P_\ell(\mu)$. Summing over ℓ on the right-hand side, the above equation reduces to

$$\left[P_{\ell'}(1)\frac{e^{ikr}}{ikr} - P_{\ell'}(-1)\frac{e^{-ikr}}{ikr}\right] - O\left(\frac{1}{r^2}\right) = a_{\ell'} j_{\ell'}(kr)\frac{2}{2\ell'+1}.$$

Dropping the 'prime' throughout this equation, we can write it as:

$$a_\ell\, j_\ell(kr)\frac{2}{2\ell+1} = \left[P_\ell(1)\frac{e^{ikr}}{ikr} - P_\ell(-1)\frac{e^{-ikr}}{ikr}\right] - O\left(\frac{1}{r^2}\right). \quad (11.106)$$

Since a_ℓ are constants, they do not depend upon r; they have the same value for all r. For convenience, we will calculate them when $r \to \infty$. Then $O(1/r^2)$ can be neglected compared to the first term on the right-hand side of the above equation which reduces to

$$a_\ell j_\ell(kr)\frac{2}{2\ell+1} = \left[P_\ell(1)\frac{e^{ikr}}{ikr} - P_\ell(-1)\frac{e^{-ikr}}{ikr}\right]$$
$$= \frac{1}{ikr}\left[e^{ikr} - (-1)^\ell e^{-ikr}\right] \quad (11.107)$$

because $P_\ell(1) = 1$ and $P_\ell(-1) = (-1)^\ell$. Since $(-1)^\ell = (\cos\pi + \mathrm{i}\sin\pi)^\ell = e^{i\pi\ell}$, equation (11.107) takes the form

$$a_\ell j_\ell(kr)\frac{2}{2\ell+1} = \frac{1}{\mathrm{i}kr}\left[\mathrm{e}^{\mathrm{i}kr} - \mathrm{i}^{2\ell}\mathrm{e}^{-\mathrm{i}kr}\right]$$
$$= \frac{1}{\mathrm{i}kr}\mathrm{i}^\ell\left[\mathrm{i}^{-\ell}\mathrm{e}^{\mathrm{i}kr} - \mathrm{i}^\ell\mathrm{e}^{-\mathrm{i}kr}\right],$$

but

$$\mathrm{e}^{\mathrm{i}\ell\pi/2} = \left(\mathrm{e}^{\frac{\mathrm{i}\pi}{2}}\right)^\ell = \left(\cos\frac{\pi}{2} + \mathrm{i}\sin\frac{\pi}{2}\right)^\ell = \mathrm{i}^\ell.$$

Similarly

$$\mathrm{e}^{-\mathrm{i}\ell\pi/2} = (\mathrm{i})^{-\ell}.$$

Therefore,

$$a_\ell j_\ell(kr)\frac{2}{2\ell+1} = \frac{1}{\mathrm{i}kr}\mathrm{i}^\ell\left[\mathrm{e}^{\mathrm{i}\left(kr-\frac{\ell\pi}{2}\right)} - \mathrm{e}^{-\mathrm{i}\left(kr-\frac{\ell\pi}{2}\right)}\right]$$
$$= \frac{2\mathrm{i}^\ell}{kr}\sin(kr - \ell\pi/2). \tag{11.108}$$

For $r \to \infty$, the spherical Bessel function $j_\ell(kr) \to \frac{1}{kr}\sin(kr - \ell\pi/2)$. Equation (11.108), therefore, gives

$$a_\ell = \mathrm{i}^\ell(2\ell + 1). \tag{11.109}$$

This gives the values of a_ℓ.

Substituting this expression for a_ℓ in (11.100*b*), we obtain

$$\mathrm{e}^{\mathrm{i}kr\cos\theta} = \sum_\ell \mathrm{i}^\ell(2\ell+1)j_\ell(kr)P_\ell(\cos\theta). \tag{11.110}$$

In this way it has been possible to expand the incoming plane wave in terms of partial waves.

Our next step will be to expand the *total asymptotic wave function* in terms of partial waves. This will enable us to compare two sides of (11.79) and consequently determine an expression for the scattering amplitude $f(\theta)$ for a spherically symmetric potential.

11.14 A general solution in terms of partial waves

Let $\psi(r, \theta)$ be the physically acceptable complete solution of the Schrödinger wave equation for a scattering process for a spherically symmetric field in terms of spherical polar coordinates. Then, we may write

$$\psi(r, \theta) = \sum_\ell b_\ell R_\ell(kr)P_\ell(\mu) \tag{11.111}$$

where b_ℓ are constants.

We note that R_ℓ and j_ℓ differ essentially only for small values of r, i.e. in the vicinity of the target where the spherically symmetric potential $V(r) \neq 0$. For large r, $V(r) \to 0$, and both R_ℓ and j_ℓ satisfy the same differential equation. This does not mean that for large r the two solutions R_ℓ and j_ℓ are identical. Since for small r, the differential equations differ, spherically symmetric potential causes a phase shift for large r.

Example 11.2.

Problem

Determine the values of the constants b_ℓ.

Solution

To determine the values of b_ℓ, we proceed as follows. Since b_ℓ are constants, they do not depend upon the spherical polar coordinates r, θ, ϕ. Therefore, the value of any b_ℓ will not change with a change in r, θ or ϕ. For our convenience, we will calculate b_ℓ for $r \to \infty$. Then equation $\psi = \psi_{\text{in}} + \psi_{\text{sc}}$ gives

$$\sum_\ell b_\ell \frac{\sin\left(kr - \frac{\ell\pi}{2} + \delta_\ell\right)}{kr} P_\ell(\mu)$$

$$= \sum_\ell \mathrm{i}^\ell(2\ell + 1)\frac{\sin\left(kr - \frac{\ell\pi}{2}\right)}{kr} P_\ell(\mu) + f(\theta)\frac{\mathrm{e}^{ikr}}{r}.$$

Expressing the sine functions in terms of complex exponentials, simplifying and separately grouping together the coefficients of $\mathrm{e}^{ikr}$and e^{-ikr}, we obtain

$$\left[\sum_\ell\left\{(2\ell + 1)\frac{\mathrm{i}^\ell}{2\mathrm{i}}\mathrm{e}^{-\mathrm{i}\ell\pi/2} - \frac{b_\ell}{2\mathrm{i}}\mathrm{e}^{-\mathrm{i}\ell\pi/2}\mathrm{e}^{\mathrm{i}\delta_\ell}\right\}\right]P_\ell(\mu) + f(\theta)k\mathrm{e}^{ikr}$$

$$+\left[\sum_\ell\left\{-(2\ell + 1)\frac{\mathrm{i}^\ell}{2\mathrm{i}}\mathrm{e}^{\,\mathrm{i}\ell\mu/2} + \frac{b_\ell}{2\mathrm{i}}\mathrm{e}^{\mathrm{i}\ell\mu/2}\mathrm{e}^{-\mathrm{i}\delta_\ell}\right\}\right]P_\ell(\mu)\mathrm{e}^{-ikr} = 0.$$

Since e^{ikr} and e^{-ikr} are linearly independent, the coefficient of each one of them must be separately zero, i.e. each of the expressions in square brackets must vanish. Thus,

$$\sum_\ell\left\{(2\ell + 1)\frac{\mathrm{i}^\ell}{2i}\mathrm{e}^{-\mathrm{i}\ell\pi/2} - \frac{b_\ell}{2i}\mathrm{e}^{-\mathrm{i}\ell\pi/2}\mathrm{e}^{\mathrm{i}\delta_\ell} P_\ell(\mu) + kf(\theta) = 0 \qquad (11.112)$$

and

$$\sum_\ell\left\{-(2\ell + 1)\frac{\mathrm{i}^\ell}{2\mathrm{i}}\mathrm{e}^{\mathrm{i}\ell\pi/2} + \frac{b_\ell}{2\mathrm{i}}\,\mathrm{e}^{\mathrm{i}\ell\pi/2}\mathrm{e}^{-\mathrm{i}\delta_\ell}\right\}P_\ell(\mu) = 0. \qquad (11.113)$$

Equation (11.113) must hold for all values of θ. This is possible only if the coefficient of each $P_\ell(\mu = \cos\theta)$ is separately zero:

$$-(2\ell+1)\frac{i^\ell}{2i}e^{i\ell\pi/2} + \frac{b_\ell}{2i}e^{i\ell\pi/2}e^{-i\delta\ell} = 0,$$

or

$$-(2\ell+1)i^\ell + b_\ell e^{-i\delta\ell} = 0,$$

or

$$b_\ell = i^\ell(2\ell+1)e^{i\delta\ell}. \tag{11.114}$$

This gives the values of b_ℓ.

We have not yet made use of (11.112). It gives the value of $f(\theta)$ and consequently that of the differential cross section. Substituting the above expression for b_ℓ in (11.112), we obtain

$$\sum_\ell \left\{(2\ell+1)\frac{i^\ell}{2i}e^{-i\ell\pi/2} - \frac{1}{2i}(2\ell+1)e^{-i\ell\pi/2}e^{2i\delta\ell}\right\}P_\ell(\mu) + kf(\theta) = 0,$$

or

$$f(\theta) = \frac{1}{k}\sum_\ell \frac{1}{2i}(2\ell+1)i^\ell e^{-i\ell\pi/2}\left(e^{2i\delta\ell} - 1\right)P_\ell(\mu),$$

or

$$f(\theta) = \frac{1}{k}\sum_\ell (2\ell+1)e^{i\delta\ell}\sin\delta_\ell P_\ell(\cos\theta) \tag{11.115}$$

where we have replaced μ by $\cos\theta$. This is the basic equation for studying the effects of scattering by partial wave analysis, the potential being spherically symmetric. The scattering effect may be regarded as a summation of the effects of all possible momenta of the incident particle. The differential cross section $d\sigma/d\Omega$ is given by

$$\frac{d\sigma}{d\Omega} = |f(\theta)|^2 = \frac{1}{k^2}\left|\sum_\ell (2\ell+1)e^{i\delta\ell}\sin\delta\ell P\ell(\cos\theta)\right|^2$$

$$= \frac{1}{k^2}\sum_\ell\sum_{\ell'}(2\ell+1)(2\ell'+1)e^{i(\delta\ell-\delta\ell')}\sin\delta_\ell\sin\delta_{\ell'}P_\ell(\cos\theta)P_{\ell'}(\cos\theta). \tag{11.116}$$

The total cross section σ_{Total} is given by

$$\sigma_T = \int |f(\theta)|^2 \, d\Omega = 2\pi \int_0^{\pi} |f(\theta)|^2 \sin\theta \, d\theta$$
$$= \frac{4\pi}{k^2} \sum\nolimits_\ell (2\ell + 1)\sin^2\delta_\ell \qquad (11.117)$$

11.15 Optical theorem

For forward scattering, i.e. for $\theta = 0$, (11.115) gives

$$f(0) = \frac{1}{k} \sum\nolimits_\ell (2\ell + 1)e^{i\delta_\ell} \sin \delta_\ell, \qquad (11.118)$$

because $P_\ell(1) = 1$. Therefore, the imaginary part of the forward scattering amplitude is given by

$$\text{Im} f(0) = \frac{1}{k} \sum\nolimits_\ell (2\ell + 1)\sin^2 \delta_\ell. \qquad (11.119)$$

Substituting the expression for $\sum_\ell (2\ell + 1)\sin^2 \delta_\ell$ from (11.119) in (11.117), we obtain

$$\sigma_T = \frac{4\pi}{k} \text{Im } f(0). \qquad (11.120)$$

This is known as the *optical theorem*. This shows that the total cross section is $4\pi/k$ times the imaginary part of the forward scattering amplitude.

The following remark is in order. We have proved that

$$f(\theta) = \frac{1}{k} \sum\nolimits_\ell (2\ell + 1)e^{i\delta_\ell} \sin \delta_\ell P_\ell(\cos\theta). \qquad (11.115')$$

This formula gives the scattering amplitude $f(\theta)$ in terms of phase shifts δ_ℓ. Note that each term in the series on the right-hand side of the above equation corresponds to a definite value of angular momentum ℓ. If we know V and *can* solve the corresponding Schrödinger wave equation, we may calculate all δ_ℓ and hence determine the scattering amplitude $f(\theta)$ and consequently the differential and total cross sections. In nuclear and particle physics, the situation is quite different. V is *not* known! Thus, if we know δ_ℓ and σ_T, we may try various expressions for the potential energy V. This may give us some idea about the nature of that force.

Example 11.3.

Problem

Show that e^{2iA} can be expressed as

$$e^{2iA} = \left[1 - \tan^2 A + 2i \tan A\right] / \left[1 + \tan^2 A\right].$$

Solution

We can write

$$\begin{aligned} e^{2iA} &= \cos 2A + i \sin 2A \\ &= 1 - 2\sin^2 A + 2i \sin A \cos A \\ &= 1 - 2\frac{\tan^2 A}{1 + \tan^2 A} + 2i\frac{\tan A}{1 + \tan^2 A} \\ &= \frac{1 - \tan^2 A + 2i \tan A}{1 + \tan^2 A}. \end{aligned}$$

11.16 Scattering by a hard sphere

Consider a particle incident on a sphere of radius a. Suppose that the scattering potential has a finite range a such that

$$\begin{aligned} V(r) &= \infty, \quad \text{for } r < a, \\ &= 0, \quad \text{for } r > a, \end{aligned}$$

the distance being measured from the centre of the sphere. This is shown in figure 11.4. Since for $r < a$, the potential $V(r) = \infty$, the incident particle cannot penetrate into the sphere. The sphere is therefore called a *hard* or *rigid sphere*. Thus, a is the distance of closest approach to the sphere. For the region $r > a$ for which $V(r) = 0$, the radial part of the Schrödinger wave equation is

$$\frac{d^2R}{dr^2} + \frac{2}{r}\frac{dR}{dr} + \left[k^2 - \frac{\ell(\ell+1)}{r^2}\right]R = 0, \qquad (11.82a')$$

where

$$k^2 = \frac{2mE}{\hbar^2}. \qquad (11.82b')$$

It has two linearly independent solutions, $j_\ell(kr)$ and $n_\ell(kr)$. The Neumann function $n_\ell(kr)$ is singular at the origin. But the incident particle is prevented by the infinite repulsive potential to reach the origin. Therefore, in the region *accessible* to the

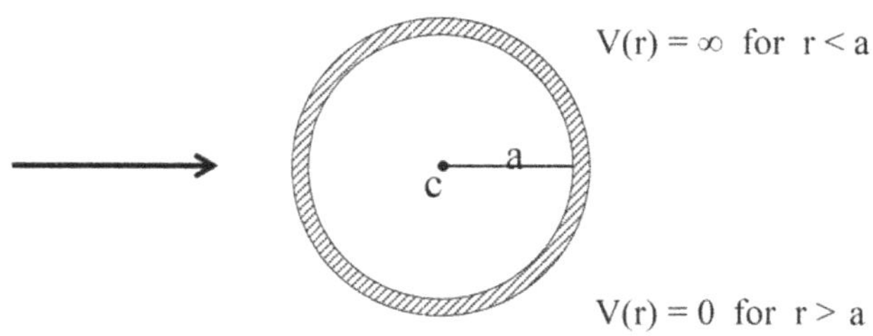

Figure 11.4. Scattering by a hard sphere.

incident particle, the function $n_\ell(kr)$ is not singular. Consequently, both the solutions are physically acceptable for the region $r > a$. Thus, in this region, the general solution of the radial differential equation may be written as

$$R_\ell(k, r) = B_\ell(k) j_\ell(kr) + C_\ell(k) n_\ell(kr). \tag{11.124}$$

Writing

$$B_\ell(k) = A_\ell(k) \cos\delta_\ell(k), \tag{11.125a}$$

$$C_\ell(k) = -A_\ell(k) \sin\delta_\ell(k), \tag{11.125b}$$

(11.124) yields

$$R_\ell(kr) = A_\ell(k) \cos\delta_\ell(k) j_\ell(kr) - A_\ell(k) \sin\delta_\ell(k) n_\ell(kr). \tag{11.126}$$

Dropping the arguments in the parameters, we can write this equation as

$$R_\ell(k, a) = A_\ell \cos\delta_\ell j_\ell(ka) - A_\ell \sin\delta_\ell(k) n_\ell(ka). \tag{11.127}$$

The condition for the continuity of the wave function at the boundary requires that at $r = a$, where the wave function *cannot* penetrate the rigid sphere (the region $r < a$) further due to its infinite repulsive potential, the function $R_\ell(k, r)$ must vanish: $R_\ell(k, a) = 0$. Equation (11.127) therefore gives

$$A_\ell\left[\cos\delta_\ell j_\ell(kr) - \sin\delta_\ell(k) n_\ell(kr)\right] = 0. \tag{11.128}$$

But A_ℓ cannot be zero because in that case, the radial function $R_\ell(k, r)$ and consequently the wave function will vanish for all values of r; the particle would disappear! Therefore, we must have

$$\left[\cos\delta_\ell j_\ell(ka) - \sin\delta_\ell(k) n_\ell(ka)\right] = 0,$$

or

$$\frac{\sin\delta_\ell}{\cos\delta_\ell} = \frac{j_\ell(ka)}{n_\ell(ka)},$$

or

$$\tan\delta_\ell = \frac{j_\ell(ka)}{n_\ell(ka)}. \tag{11.129}$$

Squaring both sides of this equation, we obtain

$$\tan^2\delta_\ell = \frac{j_\ell^2(ka)}{n_\ell^2(ka)},$$

or

$$\sin^2\delta_\ell = \frac{1}{1 + \cot^2\delta_\ell} = \frac{j_\ell^2(ka)}{j_\ell^2(ka) + n_\ell^2(ka)}. \qquad (11.130)$$

The total cross section σ_T is given by

$$\sigma_T = \frac{4\pi}{k^2}\sum\nolimits_\ell (2\ell + 1)\sin^2\delta_\ell.$$

Substituting the expression for $\sin^2\delta_\ell$ from (11.130) in (11.117′), we obtain

$$\sigma_T = \frac{4\pi}{k^2}\sum\nolimits_\ell (2\ell + 1)\frac{j_\ell^2(ka)}{j_\ell^2(ka) + n_\ell^2(ka)}. \qquad (11.131)$$

In the low energy limit, i.e. for $ka \approx 1$, dividing (11.93) by (11.98), we obtain

$$\frac{j_\ell(ka)}{n_\ell(ka)} = -\frac{(ka)^{2\ell+1}}{(2\ell + 1)!!(2\ell - 1)!!'} \qquad (11.132)$$

where

$$(2\ell + 1)!! \equiv 1, 3, 5, \ldots, (2\ell + 1) \qquad (11.133a)$$

and

$$(2\ell - 1)!! \equiv 1, 3, 5, \ldots, (2\ell - 1). \qquad (11.133b)$$

Substituting the expression for $j_\ell(k, a)/n_\ell(k, a)$ from (11.129) in (11.132), we obtain

$$\tan\delta_\ell = -\frac{(ka)^{2\ell+1}}{(2\ell + 1)!!(2\ell - 1)!!}. \qquad (11.134)$$

This equation shows that $|\tan\delta_\ell|$ and consequently δ_ℓ *decreases very rapidly* as ℓ increases. Therefore, the series in (11.117) converges so rapidly that only the first few terms in the series are significant. Specifically, *at very low energy*, the scattering process is dominated by an $\ell = 0$ partial wave, i.e. by the s-wave. Now, for $\ell = 0$, we obtain

$$\tan\delta_0 = \frac{j_0(ka)}{n_0(ka)}, \qquad (11.135)$$

but

$$j_0(ka) = \frac{\sin ka}{ka} \qquad (11.136)$$

and

$$n_0(ka) = -\frac{\cos ka}{ka}. \qquad (11.137)$$

Dividing (11.136) by (11.137), we obtain

$$\frac{j_0(ka)}{n_0(ka)} = -\tan ka. \tag{11.138}$$

Comparing (11.135) and (11.138), we obtain

$$\delta_0 = -ka. \tag{11.139}$$

We know that the total cross section σ_T is given by

$$\sigma_T = \frac{4\pi}{k^2}\sum_\ell (2\ell + 1)\sin^2\delta_\ell.$$

Since at low energies, i.e. for $k \approx 1$, σ_T converges very rapidly, the $\ell = 0$ contribution dominates and therefore we may write

$$\sigma_T = \frac{4\pi}{k^2}\sin^2\delta_0 = \frac{4\pi}{k^2}\delta_0^2 = \frac{4\pi}{k^2}k^2a^2 \approx 4\pi a^2. \tag{11.140}$$

i.e.

$$\sigma_T \to 4\pi a^2.$$
$$kr \to 0 \tag{11.141}$$

Thus, σ_T is four times the geometrical cross section of the hard sphere. This is four times the classical value πa^2. This also reflects that in such a situation the scattering process is isotropic.

Problems

11.6. Show that at very low energy, the differential cross section is a^2.
11.7. Show that the total cross section at high energies is given by

$$\sigma_T = 2\pi a^2.$$

This is twice the classical value. How can this apparently anomalous result be explained?

11.17 Scattering from a potential square well

Let us now consider scattering by an attractive square-well potential of length a and depth V_0 such that

$$V(r) = -V_0, \quad \text{for } r < a, (V_0 > 0),$$
$$= 0, \quad \text{for } r > a. \tag{11.142}$$

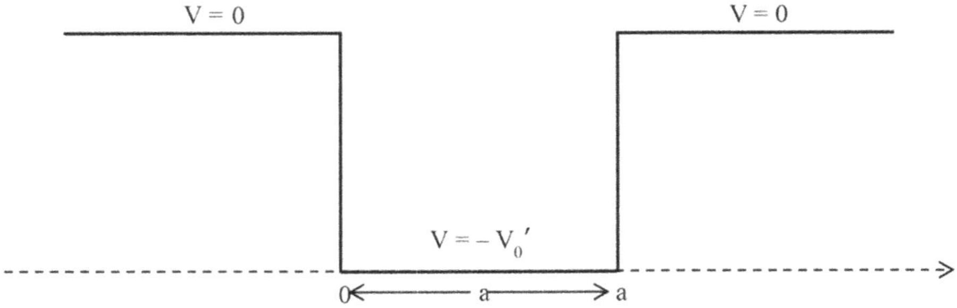

Figure 11.5. Scattering from a potential square well.

This is shown in figure 11.5. Suppose that every particle in the incident beam has energy E. For $r < a$, the potential $V(r) = -V_0$ and the radial equation takes the form

$$\frac{d^2R}{dr^2} + \frac{2}{r}\frac{dR}{dr} + \left[K^2 - \frac{\ell(\ell+1)}{r^2}\right]R = 0, \quad r < a, \tag{11.143a}$$

where

$$K^2 = \frac{2m}{\hbar^2}(E + V). \tag{11.143b}$$

For $r > a$, for which $V(r) = 0$, the radial equation becomes

$$\frac{d^2R}{dr^2} + \frac{2}{r}\frac{dR}{dr}\left[k^2 - \frac{\ell(\ell+1)}{r^2}\right]R = 0, \quad r > a, \tag{11.144}$$

where

$$k^2 = \frac{2m}{\hbar^2}E. \tag{11.145}$$

The differential equation (11.144) for the region $r > a$ has two linearly independent solutions, $j_\ell(kr)$ and $n_\ell(kr)$. The solution $n_\ell(kr)$ is singular at the origin. But the origin is not included in the external region for which $r > a$, and therefore in this region both solutions are physically acceptable. Thus, we may write the general solution in the *external region* as

$$R_\ell(kr) = B_\ell j_\ell(kr) + C_\ell n_\ell(kr), \quad r > a. \tag{11.146}$$

For the region inside the well, i.e. for $r < a$, the second-order ordinary differential equation (11.143) has two linearly independent solutions, $j_\ell(kr)$ and $n_\ell(kr)$. The solution $n_\ell(kr)$ is singular at the origin which is included in this region, and is therefore rejected. Consequently, we write the solution for the *internal region* as

$$R_\ell(Kr) = A_\ell j_\ell(Kr), \quad r > a. \tag{11.147}$$

The continuity condition requires that at the boundary of the square well at $r = a$, the two solutions and their derivatives should be equal. Therefore, we must have

$$A_\ell j_\ell(Ka) = B_\ell j_\ell(ka) + C_\ell n_\ell(ka), \tag{11.148}$$

$$kA_\ell j_\ell'(Ka) = kB_\ell j_\ell'(ka) + kC_\ell n_\ell'(ka), \tag{11.149}$$

where

$$j_\ell'(x) = \frac{\mathrm{d}j_\ell(x)}{\mathrm{d}x},$$
$$n_\ell'(x) = \frac{\mathrm{d}n_\ell(x)}{\mathrm{d}x}.$$

Dividing (11.49) by (11.48), we obtain

$$k\frac{j_\ell'(Ka)}{j_\ell(Ka)} = k\frac{B_\ell j_\ell'(ka) + C_\ell n_\ell'(ka)}{B_\ell j_\ell(ka) + C_\ell n_\ell(ka)},$$

or

$$k\frac{j_\ell'(Ka)}{j_\ell(Ka)} = k\frac{j_\ell'(ka) + C_{\ell/B_\ell} n_\ell'(ka)}{j_\ell(ka) + C_\ell/B_\ell n_\ell(ka)}. \tag{11.150}$$

The asymptotic form of (11.146) is given by

$$R_\ell(k, r) \xrightarrow[kr\to\infty]{} B_\ell\frac{1}{kr}\sin\left(kr - \frac{\ell\pi}{2}\right) + C_\ell\frac{1}{kr}\cos\left(kr - \frac{\ell\pi}{2}\right).$$

Writing

$$B_\ell = A_\ell \cos\delta_\ell, \tag{11.125a'}$$

$$C_\ell = -A_\ell \sin\delta_\ell, \tag{11.125b'}$$

we obtain

$$R_\ell(k, r) \xrightarrow[\mathrm{kr}\to\infty]{} A_\ell \cos\delta_\ell\frac{1}{kr}\sin\left(kr - \frac{\ell\pi}{2}\right) + A_\ell \sin\delta_\ell\frac{1}{kr}\cos\left(kr - \frac{\ell\pi}{2}\right),$$

or

$$R_\ell(k, r) \xrightarrow[kr\to\infty]{} A_\ell\frac{1}{kr}\sin\left(kr - \frac{\ell\pi}{2} + \delta_\ell\right). \tag{11.151}$$

This is the asymptotic form of $R_\ell(k, r)$. Now, from (11.125a') and (11.125b'), we have

$$-\frac{C_\ell}{B_\ell} = \tan\delta_\ell(k). \tag{11.152}$$

Substituting this expression for C_ℓ/B_ℓ in (11.150) and simplifying, we obtain

$$\tan\delta_\ell(k) = \frac{kj'_\ell(ka)j_\ell(Ka) - Kj'_\ell(Ka)j_\ell(ka)}{kn'_\ell(ka)j_\ell(Ka) - Kn_\ell(ka)j'_\ell(Ka)}. \tag{11.153}$$

If we know the potential energy in the well, the phase shifts can be calculated in principle. On the other hand, if the phase shifts can be determined, we can have an idea about the nature of the scattering potential in the well. This is actually a complicated expression for tan δ_ℓ. We will therefore confine our attention to low energies in which the dominant contributions come from the s-wave. Consequently, we may concentrate on $\ell = 0$ state. We could proceed further using (11.139) but it is simpler to determine the shifts directly. We proceed as follows.

11.18 s-wave scattering for a square-well potential

For s-waves, $\ell = 0$ and therefore the radial differential equation yields

$$\frac{\mathrm{d}^2R}{\mathrm{d}r^2} + \frac{2}{r}\frac{\mathrm{d}R}{\mathrm{d}r} + \left[k^2 - \frac{2m}{\hbar^2}V(r)\right]R = 0, \tag{11.154a}$$

where

$$k^2 = \frac{2mE}{\hbar^2}. \tag{11.82b'}$$

Writing

$$R = \frac{X}{r}$$

and differentiating it once and twice, we obtain

$$\frac{\mathrm{d}R}{\mathrm{d}r} = -\frac{X}{r^2} + \frac{1}{r}\frac{\mathrm{d}X}{\mathrm{d}r}$$

and

$$\frac{\mathrm{d}^2R}{\mathrm{d}r^2} = \frac{2X}{r^3} - \frac{1}{r^2}\frac{\mathrm{d}X}{\mathrm{d}r} - \frac{1}{r^2}\frac{\mathrm{d}X}{\mathrm{d}r} + \frac{1}{r}\frac{\mathrm{d}^2X}{\mathrm{d}r^2}.$$

Substituting these expressions for R and its derivatives in (11.154a), we obtain

$$\frac{2X}{r^3} - \frac{2}{r^2}\frac{\mathrm{d}X}{\mathrm{d}r} + \frac{1}{r}\frac{\mathrm{d}^2X}{\mathrm{d}r^2} - \frac{2X}{r^3} + \frac{2}{r^2}\frac{\mathrm{d}X}{\mathrm{d}r} + \left\{k^2 - \frac{2m}{\hbar^2}V(r)\right\}\frac{X}{r} = 0,$$

or

$$\frac{1}{r}\frac{\mathrm{d}^2X}{\mathrm{d}r^2} + \left\{k^2 - \frac{2m}{\hbar^2}V(r)\right\}\frac{X}{r} = 0,$$

or

$$\frac{d^2X}{dr^2} + \left\{k^2 - \frac{2m}{\hbar^2}V(r)\right\}X = 0,$$

or

$$\frac{d^2X}{dr^2} + \frac{2m}{\hbar^2}\{E - V(r)\}X = 0. \tag{11.155}$$

Inside the well, the potential energy is $-V_0$. Therefore, the above equation reduces to

$$\frac{d^2X}{dr^2} + \frac{2m}{\hbar^2}\left(E + V_0\right)X = 0. \tag{11.156}$$

Writing

$$\frac{2m}{\hbar^2}\left(E + V_0\right) = K^2,$$

(11.142) takes the form

$$\frac{d^2X}{dr^2} + K^2X = 0.$$

The general solution of this differential equation is

$$X = A\sin Kr + C\cos Kr. \tag{11.157}$$

In order that the function X/r may be *finite* at the origin i.e. at $r = 0$, the function X must vanish at $r = a$. Therefore, according to (11.143), at $r = 0$, the function $X = 0$ so that

$$0 = C$$

and (11.43) reduces to

$$X = A\sin Kr. \tag{11.158}$$

Outside the well, for which $V = 0$, the differential equation (11.41) takes the form

$$\frac{d^2X}{dr^2} + \frac{2m}{\hbar^2}EX = 0,$$

or

$$\frac{d^2X}{dr^2} + k^2X = 0. \tag{11.159}$$

The general solution of this differential equation is

$$\begin{aligned} X &= F \sin kr + G \cos kr \\ &= B \cos\delta_0 \sin kr + B \sin\delta_0 \cos kr \\ &= B \sin(kr + \delta_0), \end{aligned} \tag{11.160}$$

where

$$F = B \cos\delta_0, \tag{11.161a}$$

$$G = B \sin\delta_0 \tag{11.161b}$$

The two solutions should be continuous at $r = a$. Therefore, we must have

$$A \sin Ka = B \sin(kr + \delta_0). \tag{11.162}$$

The derivatives of the two solutions are:

$$\frac{\mathrm{d}X}{\mathrm{d}r} = AK \cos Kr$$

and

$$\frac{\mathrm{d}X}{\mathrm{d}r} = Bk \cos(kr + \delta_0).$$

Since the derivatives should also be continuous at $r = a$, we obtain

$$AK \cos Ka = Bk \cos(kr + \delta_0). \tag{11.163}$$

Equations (11.162) and (11.163) can be used to determine δ_0. Dividing (11.163) by (11.162), we obtain

$$K \cot Ka = k \cot(ka + \delta_0), \tag{11.164}$$

or

$$k \cot(ka + \delta_0) = c, \tag{11.165}$$

where

$$c = K \cot\ Ka,$$

or

$$\tan(ka + \delta_0) = \frac{k}{c},$$

or

$$\frac{\tan ka + \tan\delta_0}{1 - \tan ka\ \tan\delta_0} = \frac{k}{c},$$

or

$$c \tan ka + c \tan \delta_0 = k - k \tan ka \tan \delta_0,$$

or

$$(c + k \tan ka) \tan \delta_0 = k - c \tan ka,$$

or

$$\tan \delta_0 = \frac{k - c \tan ka}{c + k \tan ka},$$

or

$$\tan \delta_0 = \frac{k/c - \tan ka}{1 + \frac{k}{c} \tan ka}.$$

Substituting the expression for c, i.e. $\frac{K}{\tan Ka}$ in the above equation, we obtain

$$\tan \delta_0 = \frac{\frac{k}{K} \tan Ka - \tan ka}{1 + \frac{k}{K} \tan Ka \tan ka}. \tag{11.166}$$

From (11.164), we have

$$K \cot\ Ka = k \cot\left(ka + \delta_0\right),$$

or

$$K \tan\left(ka + \delta_0\right) = k \tan Ka,$$

or

$$\tan\left(ka + \delta_0\right) = \frac{k}{K} \tan Ka. \tag{11.167}$$

But for low energy:

$$\tan\left(ka + \delta_0\right) \approx ka + \delta_0.$$

Therefore, (11.153) takes the form

$$ka + \delta_0 = \frac{k}{K} \tan Ka,$$

or

$$\delta_0 = \frac{k}{K} \tan Ka - ka,$$

or

$$\delta_0 = \frac{ka}{Ka}\tan Ka - ka,$$

or

$$\delta_0 = ka\left(\frac{\tan Ka}{Ka} - 1\right). \tag{11.168}$$

The s-wave cross section σ_0 is given by

$$\sigma_0 = \frac{4\pi}{k^2}\sin^2\delta_0 \approx \frac{4\pi}{k^2}\delta_0^2.$$

Substituting the expression for δ_0 from (11.168) in the above relation, we obtain

$$\sigma_0 = \frac{4\pi}{k^2}k^2a^2\left(\frac{\tan Ka}{Ka} - 1\right)^2,$$

or

$$\sigma_0 = 4\pi a^2\left(\frac{\tan Ka}{Ka} - 1\right)^2. \tag{11.169}$$

That is, the partial cross section for $\ell = 0$ is approximately isotropic.

Equations (11.141) and (11.169) for the rigid sphere and the potential square well show that at low energies the scattering is substantially isotropic and independent of energy E. This is almost always true for a short range potential. However, large variation in the cross section over a small interval of energy can also arise if, say, the denominator in the expression for tan δ_ℓ given in (11.153) is very small. If a graph is drawn between energy and the scattering cross section, then a peak in the graph is called a *resonance*. It is such a partial wave which dominates the scattering behaviour. For σ_ℓ, if the phase shift suddenly enhances through an odd multiple of $\pi/2$, the total cross section becomes maximum. The scattering phenomenon near resonance is called *resonance scattering*.

11.19 Resonance scattering

Let us examine the scattering near a resonance in detail. We have seen that the total cross section σ_T is given by

$$\sigma_T = \frac{4\pi}{k^2}\sum_\ell(2\ell + 1)\sin^2\delta_\ell, \tag{11.117'}$$

but
$$\sin^2\delta_\ell = \frac{\sin^2\delta_\ell}{\sin^2\delta_\ell + \cos^2\delta_\ell} = \frac{1}{1 + \cot^2\delta_\ell}.$$

Therefore, the expression for the ℓth partial cross section can be written as

$$\sigma_\ell = \frac{4\pi}{k^2}(2\ell + 1)\frac{1}{1 + \cot^2\delta_\ell(E)}. \tag{11.170}$$

Let us examine the behaviour of the cross section near a resonance. At resonance energy E_r, i.e. at the energy at which δ_ℓ rises rapidly through an odd multiple of $\frac{\pi}{2}$, we have

$$\cot\delta_\ell(E_r) = 0.$$

Therefore, at E_r, we must have

$$\sin\delta_\ell(E_r) = \sin\frac{n\pi}{2} = 1, \quad \text{where } n \text{ is an odd integer,}$$

$$\cos\delta_\ell(E_r) = \cos\frac{n\pi}{2} = 0, \quad \text{where } n \text{ is an odd integer.}$$

Now at the resonance, i.e. for $E = E_r$, we can expand $\cos\delta_\ell(E)$ and $\sin\delta_\ell(E)$ in terms of $\cos\delta_\ell(E_r)$ and $\sin\delta_\ell(E_r)$. By using Taylor's theorem, we have

$$\sin\delta_\ell(E) \approx \sin\delta_\ell(E_r) + (E - E_r)\left[\cos\delta_\ell(E)\frac{d\delta_\ell(E)}{dE}_{E=E_r}\right] = 1 \tag{11.171}$$

and

$$\cos\delta_\ell(E) \approx \cos\delta_\ell(E_r) + (E - E_r)\left[-\sin\delta_\ell(E)\frac{d\delta_\ell(E)}{dE}_{E=E_r}\right]$$

$$= -(E - E_r)\left[\frac{d\delta_\ell(E)}{dE}_{E=E_r}\right] \tag{11.172}$$

Denoting the expression in square brackets at $E = E_r$ by $\frac{-2}{\Gamma}$, we obtain

$$\cos\delta_\ell(E) = -\frac{2(E - E_r)}{\Gamma(E_r)} \tag{11.173}$$

We may then write

$$\cot\ \delta_\ell(E) = \frac{\cos\delta_\ell(E)}{\sin\delta_\ell(E)}$$

$$= -\frac{2(E - E_r)}{\Gamma(E_r)},$$

where we have made use of (11.171) and (11.172). Substituting this expression in (11.170), we obtain an expression for the ℓth partial-wave cross section σ_ℓ :

$$\sigma_\ell = \frac{4\pi}{k^2}(2\ell + 1)\frac{1}{1 + \left(\frac{4}{\Gamma^2(E_r)}\right)(E - E_r)^2}$$

$$= \frac{4\pi}{k^2}(2\ell + 1)\frac{\Gamma^2(E_r)}{\Gamma^2(E_r) + 4(E - E_r)^2}$$

$$= \frac{4\pi}{k^2}(2\ell + 1)\frac{\Gamma^2(E_r)/4}{\Gamma^2(E_r)/4 + (E - E_r)^2} \tag{11.174}$$

This is known as the *Breit–Wigner resonance formula.* It is illustrated in figure 11.6. The partial cross section attains its maximum value $\sigma_\ell^{\max}$ at the resonance energy $E = E_r$; it is

$$\alpha_\ell^{\max} = \frac{4\pi}{k^2}(2\ell + 1). \tag{11.175}$$

Substituting this expression for $\sigma_\ell^{\max}$ in (11.174), we obtain

$$\frac{\sigma_\ell}{\sigma_\ell^{\max}} = \frac{\Gamma^2(E_r)}{\Gamma^2(E_r) + 4(E - E_r)^2}. \tag{11.176}$$

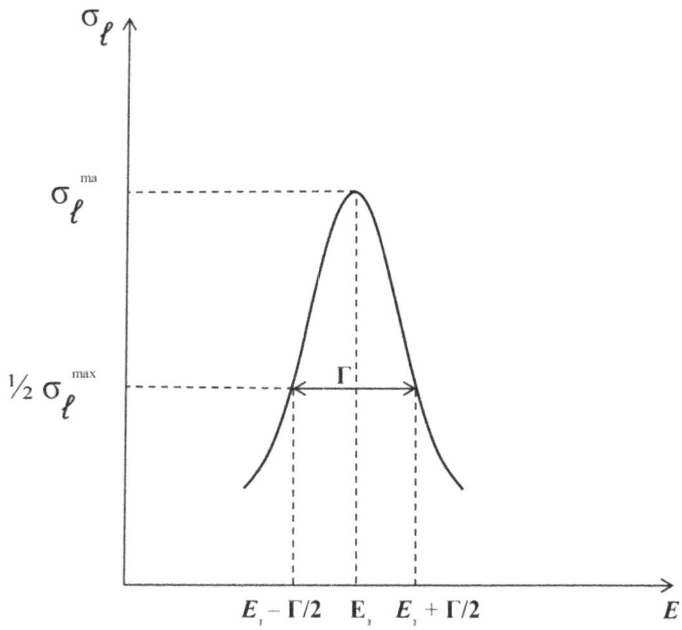

Figure 11.6. The resonance cross section and the resonance width.

Let us see what the value of Γ at $\sigma_\ell = \frac{1}{2}\sigma_\ell^{\max}$ will be. Substituting this expression for σ_ℓ in (11.170), and dropping for convenience the argument of Γ, we obtain

$$\frac{1}{2} = \frac{\Gamma^2}{\Gamma^2 + 4(E - E_r)^2},$$

or

$$2\Gamma^2 = \Gamma^2 + 4(E - E_r)^2,$$

or

$$\Gamma^2 = 4(E - E_r)^2,$$

or

$$\Gamma = 2(E - E_r).$$

The width Γ of the resonance curve at $\sigma_\ell = \frac{1}{2}\,\sigma_\ell^{\max}$ is called the *resonance width* and is equal to $2(E - E_r)$.

The above analysis shows that at resonance energy, the scattering amplitude within a potential well is very large. Therefore, the probability of finding the scattered particle inside the well is very high.

11.20 Zero-energy scattering and the scattering length

The technique of partial wave analysis is used only if the series

$$f(\theta) = \frac{1}{k}\sum_\ell (2\ell + 1)e^{i\delta_\ell}\sin\delta_\ell P_\ell(\cos\theta) \qquad (11.115')$$

for the scattering amplitude and the series

$$\sigma_T = \frac{4\pi}{k^2}\sum_\ell (2\ell + 1)\sin^2\delta_\ell \qquad (11.117')$$

for the total cross section converge so rapidly that only a few terms of each series make a significant contribution. That is, *in practice*, the partial wave analysis is mostly suitable only for low energy scattering. This is found to be the case for short range potentials.

If the energy is so low that only the $\ell = 0$ partial wave makes a significant contribution, then the analysis becomes very much simplified as only δ_0, the s-wave phase-shift, will be different from zero. Then, from (11.107), we have

$$f_k(\theta) = f_k^{(0)}(\theta) = \frac{1}{k}e^{i\delta_0}\sin\delta_0 \qquad (11.177)$$

as $P_0(\cos\theta) = 1$. The scattering amplitude is therefore independent of the scattering angle, i.e. the scattering is isotropic. The parameter k is given by

$$k^2 = \frac{2m}{\hbar^2}E.$$

The *limiting value of energy* for which (11.177) holds is called *zero energy*. The scattering amplitude at the zero-energy is constant and its negative is called the *scattering length*. Denoting it by a, we may write

$$a = \lim_{k\to 0}\left[-f_k^{(0)}(\theta)\right] = -\frac{1}{k}e^{i\delta 0}\sin\delta_0. \tag{11.178}$$

If the interaction is weak and the phase shift is small, then (11.178) shows that

$$a = -\lim_{k\to 0}\frac{\delta_0}{k}.$$

That is, at very small energy, the phase shift is given by $\delta_0 = -ka$.

Problem

11.8. Discuss the various aspects of the low energy proton–neutron scattering within the deuteron.

11.21 Identical particles

11.21.1 Introduction

Let us next examine the behaviour of two identical particles, i.e. of particles which belong to the same species: for instance, two electrons. These particles have the same intrinsic properties such as mass, charge and spin. Classically, if we consider the motion of these two electrons, *in principle* we can keep a track of their trajectories and can always distinguish one electron from the other. Quantum mechanically, the situation is quite different. The position and momentum of an electron cannot be measured precisely at the same time and in the classical sense the state of the electron cannot be determined. The trajectory of an electron has no meaning at the quantum mechanical level and, therefore, even in principle we cannot keep track of the electrons. Thus, the two electrons cannot be distinguished from each other: quantum mechanically these are *indistinguishable*. This indistinguishability of identical particles gives results which, although strange from the classical point of view, are consistent with experiments.

Let us compare, for instance, the classical and quantum mechanical results of the *elastic collision* of two identical particles with the collision of two different particles, the particles having the same interaction between them. Will there be any difference between the results of these two experiments? Classically, the answer is no. This is because in principle we can distinguish between the particles even when they are identical. *In practice,* this distinction is made when the colliding particles are *not* identical. The cross section for the system is the sum of independent cross sections for the two particles. And the same result would be valid for identical particles as, in principle, these can be distinguished from each other. Quantum mechanically, the situation is entirely different. As the identical particles cannot be distinguished, the result computed classically will not be valid.

11.21.2 Symmetric and antisymmetric wave functions

Before making calculations for the cross section of identical particles, let us determine the nature of the wave function representing a system of two identical particles. Let us denote this wave function by $\phi(1, 2)$, where 1 denotes the spin and the space coordinates of one of the two particles and 2 denotes the spin and space coordinates of the other particle. We will show that the wave function $\phi(1, 2)$ should be either symmetric or antisymmetric. We proceed as follows. Since the particles are indistinguishable, the interchange of two particles should not change the wave function, except for a physically unimportant phase factor. Therefore, we may write

$$\phi(1, 2) = e^{i\alpha}\phi(2, 1), \qquad (11.179)$$

where α is real. By repeating the interchange, we obtain

$$\phi(2, 1) = e^{i\alpha}\phi(1, 2). \qquad (11.180)$$

Substituting the expression for $\phi(2, 1)$ from (11.180) in (11.179), we obtain

$$\phi(1, 2) = e^{2i\alpha}\phi(1, 2).$$

But such a repetition should reproduce the original function. Therefore, we must have

$$e^{2i\alpha} = 1.$$

This yields

$$e^{i\alpha} = \pm 1.$$

Hence,

$$\phi(1, 2) = \pm\phi(2, 1).$$

This relation shows that there are two possibilities: either $\phi(1, 2) = \phi(2, 1)$ or $\phi(1, 2) = -\phi(2, 1)$. In the first case, the wave function is said to be *symmetric*; interchanging the identical particles does not change the wave function at all. In the second case, when $\phi(1, 2) = -\phi(2, 1)$, the wave function is said to be *antisymmetric*; interchanging the identical particles changes only the sign of the wave function. To sum up: two identical particles can be represented either by a symmetric or by an antisymmetric wave function.

11.21.3 Exchange operator

The same result can be derived by following a *formal approach*. Let us introduce an *exchange operator* P_{12} which acting on the wave function $\phi(1, 2)$ interchanges the identical particles 1 and 2. Then, we may write

$$P_{12}\phi(1, 2) = \phi(2, 1).$$

Operating again by P_{12}, we obtain

$$P_{12}^2\phi(1, 2) = P_{12}\phi(2, 1) = \phi(1, 2).$$

This equation shows that 1 is the eigenvalue of P_{12}^2 so that

$$P_{12} = \pm 1.$$

This in turn means that either $P_{12}\phi(1, 2) = \phi(2, 1)$ or $P_{12}\phi(1, 2) = -\phi(2, 1)$. That is, the eigenfunction $\phi(1, 2)$ of a system of two identical particles should be either symmetric or antisymmetric with respect to interchange of 1 and 2.

11.21.4 Construction of symmetric and antisymmetric wave functions

Every solution of the Schrödinger wave equation for a system of two identical particles does not represent the system. The acceptable wave functions must be either symmetric or antisymmetric. We will now show how to construct such wave functions. Let $H(1,2)$ be the Hamiltonian of a system of two identical particles. Since interchanging 1 and 2 does not affect the system, we have

$$H(1, 2) = H(2, 1).$$

The time-dependent Schrödinger wave equation for the system is

$$i\hbar\frac{\partial\Psi(1, 2)}{\partial t} = H(1, 2)\Psi(1, 2)$$

where $\Psi(1,2)$ is a mathematical solution of this differential equation. Interchanging 1 and 2, we obtain

$$i\hbar\frac{\partial\Psi(2, 1)}{\partial t} = H(2, 1)\Psi(2, 1).$$

As $H(2,1) = H(1,2)$, this differential equation reduces to

$$i\hbar\frac{\partial\Psi(2, 1)}{\partial t} = H(1, 2)\Psi(2, 1).$$

This shows that if $\Psi(1,2)$ is a solution corresponding to $H(1,2)$, then $\Psi(2,1)$ is also a solution corresponding to the same Hamiltonian. According to the principle of superposition, a linear combination of these solutions is also a solution of the same differential equation. In particular, we note that

$$\frac{1}{\sqrt{2}}[\Psi(1, 2) + \Psi(2, 1)],$$

$$\frac{1}{\sqrt{2}}[\Psi(1, 2) - \Psi(2, 1)].$$

are also solutions of this differential equation. These two solutions are, respectively, symmetric and antisymmetric. The factor $1/\sqrt{2}$ has been introduced for the sake of normalisation. Denoting these solutions by ϕ_S and ϕ_A, respectively, we may write

$$\phi_S = \frac{1}{\sqrt{2}}[\Psi(1, 2) + \Psi(2, 1)],$$

$$\phi_A = \frac{1}{\sqrt{2}}[\Psi(1, 2) - \Psi(2, 1)].$$

We conclude that every solution of the Schrödinger wave equation with $H(1,2)$ as the Hamiltonian of a system of two identical particles does not represent the system of identical particles. For such a system, only symmetric or antisymmetric solutions are acceptable.

Remark

If the wave function $\phi(1, 2)$ for two identical particles is different from zero only when 1 is in some region A, 2 is in some region B, and A and B do not have any common domain, then the identical particles can be considered as distinguishable.

11.21.5 Spin and statistics

Which of these two wave functions are we going to choose? The choice of the symmetric and antisymmetric wave functions for a system of identical particles is determined by the type of a particle. Particles of half odd-integral spin obey Fermi–Dirac statistics, are called *fermions* and are represented by an antisymmetric wave function while the particles with zero or integral spin obey Bose–Einstein statistics, are called *bosons* and are represented by symmetric wave functions. Thus, there is an intimate connection between the spins of the particles and their statistics. Electrons, protons and neutrons are fermions while pi-mesons are bosons. Aggregates of particles which are sufficiently tightly bound so as to behave as 'particles' can be treated as fermions or bosons,

11.21.6 Symmetry character of a wave function

We have seen that the wave function $\phi(1, 2)$ representing a system of identical particles can be either symmetric or antisymmetric. We will now prove that a wave function maintains its symmetry character with time; if it is symmetric, it remains symmetric with time and if it is antisymmetric, it remains antisymmetric with time.

The time-dependent Schrödinger wave equation for a system of two identical particles is

$$i\hbar\frac{\partial\Psi(1, 2)}{\partial t} = H(1, 2)\Psi(1, 2).$$

Suppose that $\phi_S(1, 2)$ represents a symmetric solution of the above differential equation. Then

$$i\hbar\frac{\partial\phi_S(1, 2)}{\partial t} = H(1, 2)\phi_S(1, 2).$$

The Hamiltonian $H(1,2)$, being that of two identical particles, is symmetric with respect to interchange of 1 and 2. Therefore, $H(1, 2)\phi_S(1, 2)$ is also symmetric. This being the right-hand side of the above equation, the expression $\partial\phi_S(1, 2)/\partial t$ on the left-hand side of this equation must also be symmetric. After an infinitesimal time dt, we have

$$\phi_S(1, 2) \rightarrow \phi_S(1, 2) + \frac{\partial\phi_S(1, 2)}{\partial t}\mathrm{d}t.$$

This is again symmetric. If this process is continued, we find that the symmetric function retains its character even after finite intervals of time. The same is true for an antisymmetric wave function.

11.21.7 Exchange operator: a constant of the motion

We will show that if the Hamiltonian H of a system of two identical particles does not contain the time variable t explicitly, then the exchange operator is a constant of the motion. In this case, the energy eigenvalue equation is given by

$$H(1, 2)u(1, 2) = Eu(1, 2),$$

where $u(1,2)$ is the eigenfunction of $H(1,2)$ corresponding to the eigenvalue E. Interchanging 1 and 2, we obtain

$$H(2, 1)u(2, 1) = Eu(2, 1).$$

But as $H(2,1) = H(1,2)$, the above equation reduces to

$$H(1, 2)u(2, 1) = Eu(2, 1).$$

Operating both sides from the left by P_{12}, we obtain

$$\begin{aligned} P_{12}H(1, 2)u(2, 1) &= P_{12}Eu(2, 1) \\ &= EP_{12}u(2, 1) \\ &= Eu(1, 2) \\ &= H(1, 2)u(1, 2) \\ &= H(1, 2)P_{12}u(2, 1). \end{aligned}$$

Since this equation is valid for all $u(1, 2)$, we have

$$P_{12}H(1, 2) = H(1, 2)P_{12},$$

or

$$[P_{12}, H(1, 2)] = 0.$$

This shows that the exchange operator P_{12} is a constant of the motion and hence its eigenvalues do not change in the course of temporal development. This makes it possible to restrict the allowed states of any system of two identical particles to be either symmetric or antisymmetric with respect to the exchange of particles.

11.21.8 Non-interacting identical particles and Pauli's exclusion principle

Let us next consider a system of two *non-interacting* identical particles. If $H(1)$ and $H(2)$ denote the Hamiltonians of the two particles and $H(1, 2)$ denotes the Hamiltonian of the system, then we write

$$H(1, 2) = H(1) + H(2).$$

Let $\Psi(1, 2)$ be the solution corresponding to $H(1, 2)$. Suppose that it can be expressed as a product of wave functions $\Psi(1)$ and $\Psi(2)$ representing the two identical particles constituting the system. Then, we have

$$\Psi_{\alpha\beta}(1, 2) = \Psi_\alpha(1)\Psi_\beta(2)$$

where $\Psi_\alpha(1)$ denotes the state α of particle 1, etc. Operating from the left by $H(1, 2)$, we obtain

$$\begin{aligned}
H(1, 2)\Psi_{\alpha\beta}(1, 2) &= H(1, 2)\left[\Psi_\alpha(1)\Psi_\beta(2)\right] \\
&= [H(1) + H(2)]\left[\Psi_\alpha(1)\Psi_\beta(2)\right] \\
&= H(1)\left[\Psi_\alpha(1)\Psi_\beta(2)\right] + H(2)\left[\Psi_\alpha(1)\Psi_\beta(2)\right] \\
&= \Psi_\beta(2)\left[H(1)\Psi_\alpha(1)\right] + \Psi_\alpha(1)\left[H(2)\Psi_\beta(2)\right] \\
&= \Psi_\beta(2)\left[E_1\Psi_\alpha(1)\right] + \Psi_\alpha(1)\left[E_2\Psi_\beta(2)\right] \\
&= (E_1 + E_2)\Psi_\alpha(1)\Psi_\beta(2) \\
&= E\Psi_{\alpha\beta}(1, 2).
\end{aligned}$$

Similarly, we can show that

$$\Psi_{\beta\alpha}(1, 2) = \Psi_\beta(1)\Psi_\alpha(2)$$

is an independent eigenfunction of $H(1, 2)$ and belongs to the same energy eigenvalue. Thus, there is an exchange-degeneracy. A linear combination of these eigenfunctions will also be an eigenfunction of $H(1, 2)$. But $\Psi_{\alpha\beta}(1, 2) = \Psi_\alpha(1)\Psi_\beta(2)$ (or $\Psi_{\alpha\beta}(1, 2)$) cannot be an acceptable solution for a system of two identical particles because it is neither symmetric nor antisymmetric. For fermions (electrons, protons, etc), the acceptable solution must be antisymmetric. Therefore, we have to choose the wave function $\Psi_A(1, 2)$ as

$$\Psi_A(1, 2) = \frac{1}{\sqrt{2}}\left[\Psi_\alpha(1)\Psi_\beta(2) - \Psi_\beta(1)\Psi_\alpha(2)\right]$$
$$= \frac{1}{\sqrt{2}}\begin{vmatrix} \Psi_\alpha(1) & \Psi_\beta(1) \\ \Psi_\alpha(2) & \Psi_\beta(2) \end{vmatrix}.$$

If both the particles are in the same state, i.e. $\Psi_\beta(i) = \Psi_\alpha(i)$, $i = 1, 2$, then the above equation yields

$$\Psi_A(1, 2) = \frac{1}{\sqrt{2}}\begin{vmatrix} \Psi_\alpha(1) & \Psi_\alpha(1) \\ \Psi_\alpha(2) & \Psi_\alpha(2) \end{vmatrix} = 0.$$

Consequently, the probability of finding both fermions in the same state is zero: only one fermion can exist in any state. This is known as *Pauli's exclusion principle.* It was postulated by Pauli for an explanation of the periodic system of elements.

Pauli, a Nobel laureate in physics, was a versatile genius with a mastery of satire. He did not believe that parity would be violated in weak interactions and expressed his opinion by stating that God cannot be left-handed! However, when parity violation in weak interactions was confirmed experimentally, he wrote to Weiskopf: 'It is good that I did not make a bet. It would have resulted in a heavy loss of money (which I think I cannot afford to do)...What shocks me is not the fact that God is just left-handed but the fact that in spite of this He exhibits Himself as left–right symmetric when he expresses Himself strongly.'

The analysis can be easily extended to a system of n identical fermions so that we have

$$\psi_A(1, 2, \ldots, n) = \frac{1}{\sqrt{n!}}\begin{vmatrix} \psi_\alpha(1) & \psi_\beta(1) & \ldots & \psi_\mu(1) \\ \psi_\alpha(2) & \psi_\beta(2) & \ldots & \psi_\mu(2) \\ \ldots & \ldots & \ldots & \ldots \\ \psi_\alpha(n) & \psi_\beta(n) & \ldots & \psi_\mu(n) \end{vmatrix}.$$

This is called the Slater determinant and shows that for $\Psi_\alpha = \Psi_\beta$, the antisymmetric wave function $\Psi_A(1, 2, \ldots, n) = 0$. This again leads to Pauli's exclusion principle: there cannot be more than one fermion in any one of the states $\alpha, \beta, \ldots, \mu$.

Problem

11.9. Is it essential that the wave function representing the deuteron ground state be either symmetric or antisymmetric?

For a system of bosons, say photons, we have to construct symmetric wave functions from $\Psi_\alpha(1)$ and $\Psi_\beta(2)$. Indeed, any number of bosons can be in the same state.

Problem

11.10. Starting with $\Psi_\alpha(1)$ and $\Psi_\beta(2)$, construct possible symmetric wave functions for two bosons. How many states can be obtained classically?

11.21.9 Scattering of identical particles

Consider a particle incident on an identical particle called the target, the collision being elastic. In the centre-of-mass frame, the two identical particles will be approaching each other before collision and will be receding from each other after the collision. The deviation of a particle from its normal path is called scattering. Then, as shown in figure 11.7, if a particle, scattered at an angle θ, enters the detector, then the conservation of momentum requires that the other particle scattered at an angle $(\pi - \theta)$ should also enter the same detector.

Quantum mechanically, it is not possible to distinguish one particle from the other so that the two *amplitudes* $f(\theta)$ and $f(\pi - \theta)$ interfere. Let us first examine the case of two identical spinless particles colliding with each other; for instance, two α-particles. Quantum mechanically, the cross section for the scattering of α-particles is given by

$$\frac{d\sigma_{QM}}{d\Omega} = |f(\theta) + f(\pi - \theta)|^2.$$

For $\theta = \frac{\pi}{2}$, it yields

$$\frac{d\sigma_{QM}}{d\Omega} = \left| f\left(\theta = \frac{\pi}{2}\right) + f\left\{\pi - \left(\theta = \frac{\pi}{2}\right)\right\}\right|^2$$

$$= \left| f\left(\frac{\pi}{2}\right) + f\left(\frac{\pi}{2}\right)\right|^2$$

$$= 4\left| f\left(\frac{\pi}{2}\right)\right|^2$$

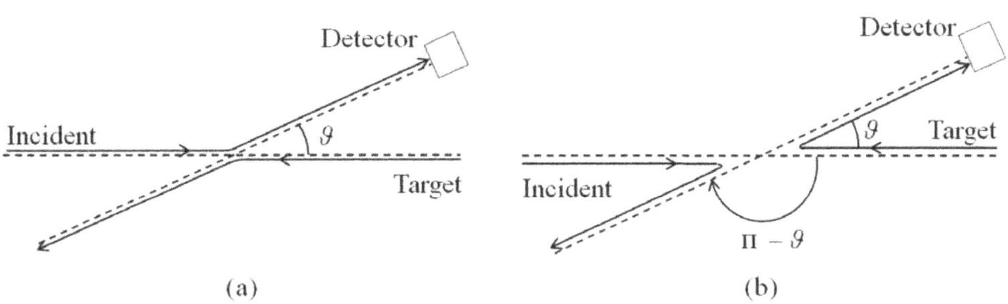

Figure 11.7. Scattering of identical particles in the centre-of-mass frame.

Classically, since the identical particles can be distinguished, we have

$$\frac{d\sigma_{CL}}{d\Omega} = |f(\theta)|^2 + |f(\pi - \theta)|^2.$$

For $\theta = \frac{\pi}{2}$, this yields

$$\frac{d\sigma_{CL}}{d\Omega} = \left|f\left(\frac{\pi}{2}\right)\right|^2 + \left|f\left(\frac{\pi}{2}\right)\right|^2$$

$$= 2\left|f\left(\frac{\pi}{2}\right)\right|^2.$$

The quantum mechanical result, which is twice the classical prediction, is consistent with experiments.

Consider next the collision of two spin $\frac{1}{2}$ identical particles, for instance, proton–proton scattering. The total amplitude of such particles should be antisymmetric with respect to the interchange of two identical particles. Now as each particle has spin $\frac{1}{2}$, the total spin of the system can have values from $\frac{1}{2} + \frac{1}{2}$ to $\frac{1}{2} - \frac{1}{2}$ in unit steps, i.e. two values 1 and 0. If the spins of the identical particles are in the opposite direction, then the spin of the system is zero. Its spin wave function is antisymmetric and it is said to be in the spin singlet state. Since the total wave function for fermions is antisymmetric, the spatial part of the wave function must be symmetric. Hence

$$\frac{d\sigma_s}{d\Omega} = |f(\theta)|^2 + |f(\pi - \theta)|^2.$$

If the spins of two identical particles are parallel, the system has spin 1 so that its third component has the values −1, 0, 1. Then it is said to be in a spin triplet state. The spin wave function is symmetric. The total wave function being antisymmetric, the spatial part of the wave function must be antisymmetric. Hence,

$$\frac{d\sigma_A}{d\Omega} = |f(\theta)|^2 - |f(\pi - \theta)|^2.$$

In the scattering of two unpolarised spin $\frac{1}{2}$ identical particles, all spin states of the system are equally likely. Therefore, the probability of finding the system in the spin triplet state is three times the probability of finding the system in the spin singlet state. Hence, we have

$$\frac{d\sigma}{d\Omega} = \frac{3}{4}\left|f(\theta) - f(\pi - \theta)\right|^2 + \frac{1}{4}\left|f(\theta) + f(\pi - \theta)\right|^2$$

$$= |f(\theta)|^2 + |f(\pi - \theta)|^2 - \frac{1}{2}\left[\left|f^*(\theta)\quad f(\pi - \theta)\right| + \left|f(\theta)\quad f^*(\pi - \theta)\right|\right]$$

This shows that for spin $\frac{1}{2}$ identical particles, there is symmetry for $\theta \to (\pi - \theta)$.

Additional problems

11.11. Define symmetric and anti-symmetric functions. Show that a wave function describing a system of two identical particles must be either symmetric or antisymmetric. Prove that if a system of two identical particles is described by an anti-symmetric wave function, then the probability that the two particles may be found in the same state is zero.

11.12. Define the scattering length and effective range for s-wave scattering. Show that

$$\frac{1}{a(k)} = \frac{1}{a} - \frac{1}{2}k^2 r_0$$

where $1/a(k) = -k \cot \delta$, $a = a(0)$ and r_0 is the effective range in the s-wave scattering phase shift and k is the wave number.

11.13. Outline the partial wave analysis of the two-body scattering problem in quantum mechanics. Using the impact parameter argument, or otherwise, show that this approach is most useful for short-range interactions at low energies. In a typical scattering experiment, a beam of particles is scattered from a thin target containing many scattering centres. Under what conditions are the differential and total scattering cross sections calculated from two-body theory applicable to those measured experimentally?

11.14. Particles of mass m and initial momentum $\hbar k_0$ are scattered from a spherically symmetric potential $V(r)$. Derive the Born approximation formula for the cross section. By using this formula, show that the differential cross section $I(\theta)\mathrm{d}\Omega$ for the non-relativistic elastic scattering of fast electrons by an atom, assuming that the interaction can be represented by the potential

$$V(r) = -\frac{Ze^2}{r}\exp(-\alpha r)$$

is given by

$$I(\theta) = \frac{Z^2 e^4}{16E^2}\left(\sin^2\frac{\theta}{2} + \frac{\alpha^2\hbar^2}{8mE}\right)^{-2}$$

where $E = \hbar^2 k^2/2m$ is the energy of the electron.

IOP Publishing

Quantum Mechanics

Mohammad Saleem

Chapter 12

Dirac's formalism

12.1 Introduction

We will now introduce a formalism devised by Dirac, the famous British physicist, who, with Schrödinger, was awarded the Nobel Prize in physics in 1933 'for the discovery of new productive forms of atomic theory'. This formalism is elegant and concise and is very often used in the literature. Anyone interested in research in high energy physics must be well-versed in it.

We know that, in classical mechanics, the state of a physical system is determined by its position and momentum variables. The temporal development of the system is governed by Newton's second law of motion. If the state of the system at a particular time is known, then its state at any other time can be computed using the dynamical law. However, in quantum mechanics, Heisenberg's uncertainty principle tells us that the position and momentum of even a single particle cannot be measured precisely at the same time. That is, the state of a particle cannot be determined in the classical sense. Consequently, it is *assumed* that all the accessible information about a state of a system is contained in a vector (strictly speaking, a *ray* which is defined as the set of all vectors with different magnitudes but the same direction) in a Hilbert space for the system; the space usually having an infinite number of dimensions. Any vector of the Hilbert space represents a state of the system and vice versa. All possibly available information about the state of the system is contained in the state vector. If the vector is known at any time, its temporal development can be determined using the equation of motion.

But how much information can we obtain about a state of the system? Even in principle, all the observables of a system cannot be measured precisely at the same time. In fact, a simultaneous precise measurement can be made only for those observables of a system whose operators mutually commute. But what is the maximum number of attributes of a system which can be determined precisely at the same time? *Nobody knows*. However, we *assume* that there exists a maximum number of attributes which can be determined precisely at the same time. These

doi:10.1088/978-0-7503-1206-6ch12

quantities, being observables, must be represented by mutually commuting Hermitian operators. Such a complete set of commuting Hermitian operators is called a complete set of commuting Hermitian observables. A complete set of simultaneous eigenvectors of these operators exists and is characterised by the eigenvalues of the complete set of commuting operators. Each of these simultaneous eigenvectors gives the optimum information about a state of the system. These eigenvectors span the entire Hilbert space so that every vector in this space can be expanded in terms of them. These are the basis vectors in the Hilbert space and can always be chosen as orthonormal.

If a physical system has N degrees of freedom, then it may be thought that, quantum mechanically, N is the maximum number of independent dynamical quantities which can be measured precisely at the same time. This is not true. The quantum mechanical system may possess properties, such as spin and parity, which *have no classical analogue*. The system might possess properties which can be measured along with others but have not yet been discovered. Hence, in quantum mechanics, the number of quantities which can be measured precisely at the same time is not known. Nevertheless, we will *assume* that every physical system does possess a *complete set* of compatibles. It may be noted that the complete set of compatibles is *not unique*.

Dirac represented a vector in a Hilbert space by $|\rangle$ or, if we want to label it, say by α, by $|\alpha\rangle$. He called it a *ket vector* or merely *ket*. He defined another type of vector, denoting each vector by the symbol $\langle|$ or $\langle\alpha|$, and called each one a *bra vector* or merely *bra* in another space. The space of bra vectors is said to be a dual space with respect to the space of ket vectors. As the two types of vectors, when placed side by side, with the lines dropped, form an angle bracket $\langle\rangle$, these were labelled with the two parts of *bra*(c)*ket*: bra and ket vectors. The *scalar product* of two such vectors, $|\alpha\rangle$ and $\langle\beta|$, is *defined* by associating the ordered pair with a complex number and writing it as

$$\langle\beta|\alpha\rangle,$$

the bra vector being on the left, and the two vertical lines being contracted to one for the sake of brevity. It is *assumed* that

$$\langle\beta|\alpha\rangle = \langle\alpha|\beta\rangle^*. \tag{12.1}$$

Putting $|\beta\rangle = |\alpha\rangle$, we find that

$$\langle\alpha|\alpha\rangle = \langle\alpha|\alpha\rangle^*$$

and therefore the scalar product of a vector with itself must be real. We also make the *assumption* that

$$\langle\alpha|\alpha\rangle > 0,$$

except when $|\alpha\rangle = 0$. We *assume* that bra vectors have a *one-to-one correspondence* with ket vectors such that

$$|\alpha\rangle^\dagger = \langle\alpha|$$

and

$$c|\alpha\rangle^{\dagger} = c\langle\alpha|,$$

where, by definition, $|\alpha\rangle^{\dagger}$ is called the *Hermitian conjugate* of $\langle\alpha|$. In fact, the algebra of bra and ket vectors was developed by Dirac so that it might be consistent with the mathematical techniques used in quantum mechanics.

If an operator A operates upon a ket vector $|\alpha\rangle$, we write it as

$$A|\alpha\rangle.$$

We *assume* that when an operator A operates upon a ket vector $|\alpha\rangle$, it produces another ket vector $|\gamma\rangle$:

$$A|\alpha\rangle = |\gamma\rangle. \tag{12.2}$$

If an operator A operates upon a bra vector $\langle\beta|$, we write it as $\langle\beta|A$. It is assumed that it produces a bra vector:

$$\langle\beta|A\rangle = \langle\in|.$$

Let us take the scalar product of a bra vector $|\beta\rangle$ with the ket vector $|A\alpha>\rangle(=|\gamma\rangle)$. Then, we have

$$\langle\beta|A|\alpha\rangle = \langle\beta|\gamma\rangle \tag{12.3}$$

In particular, if A is a number equal to c, then

$$\langle\beta|c|\alpha\rangle = c\langle\beta|\alpha\rangle. \tag{12.4}$$

Quantities such as $\langle\beta|A|\alpha\rangle$, being equal to $\langle\beta|\gamma\rangle$, are in general complex numbers and are usually called *matrix elements*.

The *Hermitian conjugate* $A^{\dagger}$ of an operator A is defined by the equation

$$\langle\beta|A|\alpha\rangle^{*} = \langle\alpha|A^{\dagger}|\beta\rangle. \tag{12.5}$$

Also, by definition, we have

$$\langle\beta|\gamma\rangle^{*} = \langle\gamma|\beta\rangle. \tag{12.6}$$

Comparing (12.5) and (12.6), we obtain

$$\langle\gamma\rangle = A|\alpha\rangle \tag{12.7}$$

and

$$\langle\gamma| = \langle\alpha|A\rangle^{\dagger}. \tag{12.8}$$

If $A^{\dagger} = A$, the operator A is said to be *Hermitian*. Thus, from (12.5), for a Hermitian operator A, we have

$$\langle\beta|A|\alpha\rangle^{*} = \langle\alpha|A|\beta\rangle. \tag{12.9}$$

Let us next consider the vectors $|\alpha_j\rangle$ and $\langle\alpha_i|$. Then the *orthonormalisation* condition may be expressed as

$$\langle\alpha_i|\alpha_j\rangle = \delta_{ij}. \tag{12.10}$$

Suppose that an operator A operating upon a ket vector $|\alpha\rangle$ reproduces that ket vector multiplying it with a constant. Then

$$A|\alpha\rangle = \alpha\,|\alpha\rangle. \tag{12.11}$$

The constant α is an *eigenvalue* of the operator A corresponding to the *eigenvector* $|\alpha\rangle$. The equation itself is called an *eigenvalue equation.*

Consider an operator A with eigenvalues α_1, α_2,... corresponding to the eigenvectors $|\alpha_1\rangle$, $|\alpha_2\rangle$,.... Then the eigenvalue equation for $|\alpha_i\rangle$ may be written as

$$A|\alpha_i\rangle = \alpha_i\,|\alpha_i\rangle.$$

If we multiply the above equation by a complex number a, we obtain

$$A(a|\alpha_i\rangle) = \alpha_i(a|\alpha_i\rangle), \quad i = 1, 2, \ldots$$

This shows that the eigenvalues are independent of the length of the corresponding eigenvector. Since only the eigenvalues of a Hermitian operator representing an observable are the physically measurable quantities, the length of a state vector is unimportant. As already stated, the vectors which differ only in their length represent the same state of the system; they are said to constitute a ray. In other words, a state of a physical system is completely represented by a ray in the Hilbert space. The vectors in this space are called *state vectors*. It is usual to normalise the state vectors.

There is another kind of product of ket and bra vectors which deserves careful consideration. It is

$$|\alpha\rangle\langle\beta|. \tag{12.12}$$

Multiplying from the right by an arbitrary ket $|\gamma\rangle$, we obtain

$$|\alpha\rangle\langle\beta|\gamma\rangle.$$

Since $\langle\beta|\gamma\rangle$ is a complex number, say $a_{\beta\gamma}$, the above product may be written as $a_{\beta\gamma}|\alpha\rangle$. Therefore, $|\alpha\rangle\langle\beta|$ operating upon a ket vector $|\gamma\rangle$, produces the ket vector $|\alpha\rangle$ multiplying it with a complex number:

$$|\alpha\rangle\langle\beta|\gamma\rangle = a_{\beta\gamma}\,|\alpha\rangle.$$

This ket vector depends linearly on the ket vector $|\gamma\rangle$. Hence, $|\alpha\rangle\langle\beta|$ *is an operator which operating upon ket vectors produces ket vectors.* A similar statement is valid regarding bra vectors.

Problem

12.1. Show that

$$[|\alpha\rangle\langle\beta|]^* = |\beta\rangle\langle\alpha|. \tag{12.13}$$

Example 12.1.

Problem

Prove that

$$\sum_i |\alpha_i\rangle\langle\alpha_i|,$$

the sum being taken over all the eigenvectors of a Hermitian operator, is a unit operator.

Solution

Multiplying the given expression

$$\sum_i |\alpha_i\rangle\langle\alpha_i|$$

from the right by a basic ket vector $\langle\alpha_i|$, we obtain

$$\sum_i |\alpha_i\rangle|\alpha_i\rangle\langle\alpha_i| = \sum_i |\alpha_i\rangle\delta_{ij} = |\alpha_j\rangle = I|\alpha_j\rangle$$

where I denotes the unit operator. Since $|\alpha_j\rangle$ can be any basic vector, we have

$$\sum_i |\alpha_i\rangle\langle\alpha_i| = I. \tag{12.14}$$

This is said to represent the *completeness relation.*

Problem

12.2. Show that for an operator having continuous eigenvalues

$$\int |\alpha\rangle\langle\alpha| \, \mathrm{d}\alpha = I. \tag{12.15}$$

This is the completeness relation for a system with a continuous spectrum of eigenvalues.

Theorem 12.1.

Prove that the eigenvalues of a Hermitian operator are always real.

Proof

Let A be a Hermitian operator. Let $|\alpha\rangle$ be an eigenvector of A corresponding to the eigenvalue α. Then

$$A\,|\alpha\rangle = \alpha\langle\alpha|. \tag{12.16}$$

Multiplying from the left by a bra vector $\langle\alpha|$, we obtain

$$\langle\alpha|\, A\, |\alpha\rangle = \alpha\langle\alpha|\alpha\rangle.$$

If the vectors are normalised, then $\langle\alpha|\alpha\rangle = 1$, and the above equation reduces to

$$\alpha = \langle\alpha|\, A\, |\alpha\rangle. \tag{12.17}$$

Taking the Hermitian conjugate of both sides of (12.16), we obtain

$$\langle\alpha|\, A^{\dagger} = \alpha^{*}\langle\alpha|.$$

Multiplying from the right by a ket vector $|\alpha\rangle$, we obtain

$$\langle\alpha|A^{\dagger}|\alpha\rangle = \alpha^{*}\langle\alpha|\alpha\rangle = \alpha^{*}, \text{ as the states are normalized.}$$

Since A is a Hermitian operator, we must have $A^{\dagger} = A$, and therefore the above equation yields

$$\langle\alpha|A|\alpha\rangle = \alpha^{*}. \tag{12.18}$$

Comparing (12.17) and (12.18), we obtain

$$\alpha^{*} = \alpha. \tag{12.19}$$

This shows that α, an eigenvalue of a Hermitian operator, is real.

Theorem 12.2.

Prove that the eigenvectors of a Hermitian operator corresponding to distinct eigenvalues are orthogonal.

Proof

Let A be a Hermitian operator. Let $|\alpha_i\rangle$ and $|\alpha_j\rangle$ be two eigenvectors of A corresponding to distinct eigenvalues α_i and α_j. Then, we may write

$$A|\alpha_i\rangle = \alpha_i|\alpha_i\rangle, \tag{12.20}$$

$$A\,|\alpha_j\rangle = \alpha_j\,|\alpha_j\rangle, \quad \alpha_j \neq \alpha_i. \tag{12.21}$$

Multiplying (12.20) from the left by $\langle\alpha_j|$, we obtain

$$\langle\alpha_j|A|\alpha_i\rangle = \alpha_i\langle\alpha_j|\alpha_i\rangle. \tag{12.22}$$

Taking the Hermitian conjugate of (12.20), we have

$$\langle\alpha_j|A^{\dagger} = \alpha_j^{*}\langle\alpha_j|.$$

Since A is Hermitian, $A^\dagger = A$, and the above equation becomes

$$\langle\alpha_j|A = \alpha_j^*\langle\alpha_j|.$$

Moreover, as the eigenvalues of a Hermitian operator are real, we have $\alpha_j^* = \alpha_j$. The above equation therefore becomes

$$\langle\alpha_j|A = \alpha_j\langle\alpha_j|.$$

Multiplying from the right by $|\alpha_i\rangle$, we obtain

$$\langle\alpha_j|A|\alpha_i\rangle = \alpha_j\langle\alpha_j|\alpha_i\rangle. \tag{12.23}$$

Comparing (12.22) and (12.23), we obtain

$$\alpha_i\langle\alpha_j|\alpha_i\rangle = \alpha_j\langle\alpha_j|\alpha_i\rangle$$

or

$$(\alpha_i - \alpha_j)\langle\alpha_j|\alpha_i\rangle = 0. \tag{12.24}$$

As the eigenvalues are distinct, $(\alpha_i - \alpha_j)$ cannot be zero. Therefore, we must have

$$\langle\alpha_j|\alpha_i\rangle = 0, \tag{12.25}$$

i.e. the eigenvectors of a Hermitian operator corresponding to distinct eigenvalues, $\alpha_j \neq \alpha_i$, are orthogonal. If the vectors are normalised, we have

$$\langle\alpha_j|\alpha_i\rangle = 1. \tag{12.26}$$

Combining the last two, (12.25) and (12.26), we can write

$$\langle\alpha_j|\alpha_i\rangle = \delta_{ij}. \tag{12.27}$$

This shows that eigenvectors $|\alpha_i\rangle$ and $|\alpha_j\rangle$ of a Hermitian operator are orthonormal.

If the eigenvalues of A are continuous, the above relation may be written as

$$\langle\alpha|\alpha'\rangle = \delta(\alpha - \alpha'). \tag{12.28}$$

12.2 Unitary operators

An operator U is said to be unitary if

$$U^\dagger U = I = UU^\dagger. \tag{12.29}$$

We shall now show that a unitary operator preserves the norm of a vector.

Consider an arbitrary vector $|\alpha\rangle$. By definition, its *norm* is $\langle\alpha|\alpha\rangle$. If the vector $|\alpha\rangle$ is operated upon by a unitary operator U, we obtain the vector $U|\alpha\rangle$. We have to show that the norm of this vector is also $\langle\alpha|\alpha\rangle$. If we denote this vector by $|\beta\rangle$, we have

$$|\beta\rangle = U|\alpha\rangle.$$

Taking the Hermitian conjugate of both sides of the above equation, we obtain

$$|\beta\rangle^\dagger = (U|\alpha\rangle)^\dagger$$

or

$$\langle\beta| = \langle\alpha|U^\dagger.$$

Hence, the norm of the vector $U|\alpha\rangle$ $(= |\beta\rangle)$ is:

$$\langle\beta|\beta\rangle = \langle\alpha|U^\dagger U|\alpha\rangle = \langle\alpha|I|\alpha\rangle = \langle\alpha|\alpha\rangle.$$

The importance of this property stems from the fact that the preserving norm is equivalent to the conservation of probability.

Problem

12.3 Prove the above statement.

12.3 Unitary transformation

Consider two operators A and $\hat{A}$ which are related together by a unitary operator U such that

$$\hat{A} = U^\dagger A U, \tag{12.30}$$

then the two operators A and $\hat{A}$ are said to be connected by a unitary transformation. Since U is a unitary operator, we have

$$U^\dagger U = I = UU^\dagger,$$

or

$$U^\dagger = U^{-1}.$$

Writing U^{-1} for $U^\dagger$ in (12.30), we obtain

$$\hat{A} = U^{-1}AU. \tag{12.31}$$

Moreover, multiplying (12.31) from the left by U and from the right by U^{-1}, we obtain

$$U\hat{A}U^{-1} = UU^{-1}AUU^{-1} = A,$$

or

$$A = U\hat{A}U^{-1}. \tag{12.32a}$$

Thus, if $\hat{A} = U^{-1}AU$ where U is unitary, then $A = U\hat{A}U^{-1}$.

We know that an operator A acting on a ket vector $|\phi\rangle$ will produce another ket vector $|\psi\rangle$ so that

$$A|\phi\rangle = |\psi\rangle.$$

Operating from the left by U^{-1}, we obtain

$$U^{-1}A|\phi\rangle = U^{-1}|\psi\rangle.$$

Denoting the left-hand side of the above equation by $|\hat{\psi}\rangle$, we obtain

$$|\hat{\psi}\rangle = U^{-1}|\psi\rangle. \tag{12.32b}$$

This shows how the state vectors are connected in such a transformation.

Theorem 12.3.

If an operator A is connected to another operator $\hat{A}$ by a *unitary transformation*, then

(1) $\hat{A}$ has the same eigenvalues as A;
(2) if A is Hermitian, then $\hat{A}$ is also Hermitian;
(3) if the vector $|\psi\rangle$ is normalised, then the vector $|\hat{\psi}\rangle = U^{-1}|\psi\rangle$, where U is a unitary operator, is also normalized;
(4) all algebraic equations remain invariant.

Proof

(1) It is given that two operators A and $\hat{A}$ are connected by a unitary transformation. Then we have

$$\hat{A} = U^{-1}AU, \tag{12.31'}$$

where

$$U^{\dagger}U = I = UU^{\dagger}, \quad \text{i.e. } U^{\dagger} = U^{-1}. \tag{12.29'}$$

Let α be the eigenvalue of A corresponding to the state vector $|\alpha\rangle$. Then

$$A|\alpha\rangle = \alpha|\alpha\rangle. \tag{12.33}$$

Operating from the left by U^{-1}, we obtain

$$U^{-1}A|\alpha\rangle = \alpha U^{-1}|\alpha\rangle.$$

By utilising the relation $UU^{-1} = I$, the above equation can be written as

$$U^{-1}AUU^{-1}|\alpha\rangle = \alpha U^{-1}|\alpha\rangle.$$

By using (12.31′), this can be expressed as

$$\hat{A}U^{-1}|\alpha\rangle = \alpha U^{-1}|\alpha\rangle. \tag{12.34}$$

Writing

$$U^{-1}|\alpha\rangle = |\hat{\alpha}\rangle, \tag{12.35}$$

(12.34) takes the form

$$\hat{A} = \alpha\ |\hat{\alpha}\rangle. \tag{12.36}$$

This shows that α is also an eigenvalue of the operator $\hat{A}$ but corresponds to the eigenvector $|\hat{\alpha}\rangle$.

(2) Since A is Hermitian, we have

$$A^{\dagger} = A. \tag{12.37}$$

Therefore,

$$\begin{aligned}\hat{A}^{\dagger} &= (U^{-1}AU)^{\dagger} = U^{\dagger}A^{\dagger}(U^{-1})^{\dagger} \\ &= U^{-1}A(U^{-1})^{\dagger} = U^{-1}A(U^{\dagger})^{\dagger} \\ &= U^{-1}AU = \hat{A}.\end{aligned} \tag{12.38}$$

This proves that if the operator A is Hermitian, then the operator $\hat{A}$ is also Hermitian.

(3) We know that by using a unitary transformation given by

$$\hat{A} = U^{-1}AU, \tag{12.32a$'$}$$

$$|\hat{\psi}\rangle = U^{-1}|\psi\rangle, \tag{12.32b$'$}$$

we can write

$$\begin{aligned}\langle\hat{\psi}| &= |\hat{\psi}\rangle^{\dagger} = (U^{-1}|\psi\rangle)^{\dagger} = |\psi\rangle^{\dagger}(U^{-1})^{\dagger} \\ &= |\psi|(U^{\dagger})^{\dagger} = \langle\psi|U.\end{aligned} \tag{12.39}$$

Therefore,

$$\langle\hat{\psi}|\hat{\psi}\rangle = \langle\psi|UU^{-1}|\psi\rangle = \langle\psi|\psi\rangle. \tag{12.40}$$

If $\langle\psi|\psi\rangle = 1$, then $\langle\hat{\psi}|\hat{\psi}\rangle = 1$, i.e. if the vector $|\psi\rangle$ is normalised, then the vector $|\hat{\psi}\rangle$ is also normalised.

(4) We will now prove that any algebraic equation remains invariant under a unitary transformation. Consider the algebraic equation

$$A_1A_2|\psi\rangle = |\phi\rangle.$$

Operating throughout by U^{-1} from the left, we obtain

$$U^{-1}A_1A_2|\psi\rangle = U^{-1}|\phi\rangle.$$

Since $UU^{-1} = 1$, the above equation can be written as

$$U^{-1}A_1UU^{-1}A_2UU^{-1}|\psi\rangle = U^{-1}|\phi\rangle,$$

or

$$\hat{A}_1\hat{A}_2U^{-1}|\psi\rangle = U^{-1}|\phi\rangle,$$

or

$$\hat{A}_1\hat{A}_2\psi\rangle = |\phi\rangle. \tag{12.41}$$

Similarly, we can prove that if

$$(A_1 + A_2)|\psi\rangle = |\phi\rangle \tag{12.42}$$

then

$$(\hat{A}_1 + \hat{A}_2)|\hat{\psi}\rangle = |\hat{\phi}\rangle. \tag{12.43}$$

The above analysis shows that an algebraic equation, being built up by repeated application of addition and multiplication, remains invariant under a unitary transformation.

Hence, a unitary transformation does not change the physics. It just connects two *physically equivalent descriptions of a system.*

Conversely, if a physical system is described by two sets of operators, then these two sets are connected by a unitary transformation.

Example 12.2.

Problem

Show that the commutation relation

$$[x, p_x] = i\hbar \tag{12.44}$$

known as the quantum condition, remains invariant under the unitary transformation

$$\hat{x} = U^{-1}xU. \tag{12.45}$$

Solution

We have

$$[x, p_x] = i\hbar,$$

or

$$xp_x - p_xX = i\hbar.$$

Multiplying from the left by U^{-1}, we obtain

$$U^{-1}xp_x - U^{-1}p_x X = i\hbar U^{-1}.$$

Since $UU^{-1} = I$, we can insert it anywhere between two operators. This yields

$$U^{-1}xUU^{-1}p_x UU^{-1} - U^{-1}p_x UU^{-1}xUU^{-1} = i\hbar U^{-1},$$

or

$$\hat{x}\hat{p}_x U^{-1} - \hat{p}_x\hat{x}U^{-1} = i\hbar U^{-1}.$$

Since U^{-1} is arbitrary, we must have

$$\hat{x}\hat{p}_x - \hat{p}_x\hat{x} = i\hbar,$$

or

$$[\hat{x}, \hat{p}_x] = 0.$$

This proves the above assertion. If $\hbar \to 0$, then $[\hat{x}, \hat{p}_x] \to 0$ as it must because, in this limit, quantum mechanics reduces to classical mechanics.

Conversely, if a physical system is described by two sets of operators, then these two sets are connected by a unitary transformation.

12.4 A particular unitary operator

Consider an operator U operating upon a ket vector $|\alpha_i\rangle$ and producing another ket vector $|\beta_i\rangle$. Then, we may write

$$U|\alpha_i\rangle = |\beta_i\rangle. \tag{12.46}$$

Multiplying from the right by a bra vector $\langle\alpha_i|$, we obtain

$$U|\alpha_i\rangle\langle\alpha_i| = |\beta_i\rangle\langle\alpha_i|.$$

Summing over i, we obtain

$$\sum_i U|\alpha_i\rangle\langle\alpha_i| = \sum_i |\beta_i\rangle\langle\alpha_i|,$$

or

$$U\sum_i |\alpha_i\rangle\langle\alpha_i| = \sum_i |\beta_i\rangle\langle\alpha_i|,$$

or

$$UI = \sum_i |\beta_i\rangle\langle\alpha_i|,$$

or

$$U = \sum_i |\beta_i\rangle\langle\alpha_i|.$$

We will now show that the operator U as defined above is a unitary operator. Taking the Hermitian conjugate of (12.46), we obtain

$$\langle\alpha_i|\ U^\dagger = \langle\beta_i|.$$

Multiplying from the left by the ket vector $|\alpha_i\rangle$, we have

$$|\alpha_i\rangle\langle\alpha_i|\ U^\dagger = |\alpha_i\rangle\langle\beta_i|.$$

Summing over i, we obtain

$$\sum_i |\alpha_i\rangle\langle\alpha_i|\ U^\dagger = \sum_i |\alpha_i\rangle\langle\beta_i|.$$

This yields

$$U^\dagger = \sum_i |\alpha_i\rangle\langle\beta_i|$$

$$= \sum_j |\alpha_j\rangle\langle\beta_i|.$$

Therefore, we have

$$U^\dagger U = \left(\sum_i |\alpha_j\rangle\langle\beta_j|\right)\left(\sum_i |\beta_i\rangle\langle\alpha_i|\right)$$

$$= \sum_i \sum_j |\alpha_j\rangle\langle\beta_j||\beta_i\rangle\langle\alpha_i|$$

$$= \sum_i \sum_j |\alpha_j\rangle\delta_{ij}\langle\alpha_i|.$$

Summing over j, we obtain

$$U^\dagger U = \sum_i |\alpha_i\rangle\langle\alpha_i| = I. \tag{12.47}$$

Similarly, we have

$$UU^\dagger = \sum_j |\beta_j\rangle\langle\alpha_j|\sum_i |\alpha_i\rangle\langle\beta_i|$$

$$= \sum_i \sum_j |\beta_j\rangle\langle\alpha_j||\alpha_i\rangle\langle\beta_i|$$

$$= \sum_i \sum_j |\beta_j\rangle\delta_{ij}\langle\beta_i|.$$

Summing over j, we obtain

$$UU^\dagger = \sum_i |\beta_i\rangle\langle\beta_i = I. \tag{12.48}$$

Equations (12.47) and (12.48) show that $U \equiv \sum_i |\beta_i\rangle\langle\alpha_i|$ is a unitary operator as

$$U^\dagger U = I = UU^\dagger. \tag{12.29}$$

12.5 Representations and change of basis

A state vector in a Hilbert space is not a physically measurable quantity, i.e. it is not an observable. It is an abstract quantity that can be put in various mathematical terms. A particular mathematical formalism chosen to describe the state vector is called a *representation*. The temporal development of the physical system can be determined in the following three ways.

1) The state vectors are time-dependent while the operators, *except those which are explicitly time-dependent*, are time-independent.
2) The state vectors are time-independent but the operators do not depend upon time.
3) The state vectors and operators are both time-dependent.

The temporal developments in accordance with the above three modes are said to represent, respectively, the *Schrödinger picture*, the *Heisenberg picture* and the *interaction picture*. The choice of representation depends upon the physical problem at hand. We will develop these three pictures one by one.

Any set of n linearly independent vectors can be used as the set of expansion vectors in an n-dimensional space. Such a set of vectors is called a *basis* and the vectors are called *basis vectors*. A set of base vectors is *not unique*. Any two sets of base vectors, bases, are related to each other. The problem arises: how are two sets of base vectors related to each other? The transformation from one set of basis vectors to another is called a *change of basis* and is accomplished through a unitary transformation. It is *convenient* to have an orthonormal set of basis vectors. The change from one orthonormal basis to another orthonormal basis is accomplished through a unitary transformation.

Now we will consider the Schrödinger picture, the Heisenberg picture and the interaction picture which have different bases of representation and show that these are connected through unitary transformations.

12.5.1 The Schrödinger picture

A representation of the type in which the state vectors are time-dependent and the operators corresponding to observables, *except those which are explicitly time-dependent*, are time-independent, is called the *Schrödinger picture* or the *Schrödinger representation*. We will now find the equation of motion in this representation, i.e. the equation which describes the temporal development of the system. Let $|\psi(t_0)\rangle$ be a state vector representing the state of a physical system at a certain time t_0 and

$A_i(t_0)$ be a *set* of time-independent operators at time t_0. These operators represent observables. Since the operators have been assumed to be time-independent, all the temporal development of the system is ascribed to the time-dependent state vector $|\psi(t)\rangle$. Assuming that the time evolution of states is linear, $|\psi(t_0)\rangle$ and $|\psi(t)\rangle$ should be related by a linear operator, say, $U(t, t_0)$. We therefore have

$$|\psi(t)\rangle = U(t, t_0)|\psi(t_0)\rangle, \tag{12.49}$$

where $U(t, t_0)$ is a time-dependent mathematical operator which describes the evolution of the state vector from time t_0 and t and is such that $U(t_0, t_0) = I$. Our problem therefore reduces to finding the operator $U(t,t_0)$. Taking the Hermitian conjugate of both sides of (12.49), we obtain

$$|\psi(t)\rangle^{\dagger} = (U(t, t_0)|\psi(t_0)\rangle)^{\dagger},$$

or

$$\begin{aligned}\langle\psi(t)| &= |\psi(t_0)\rangle^{\dagger} U^{\dagger}(t, t_0)\\ &= \langle\psi(t_0)| U^{\dagger}(t, t_0).\end{aligned} \tag{12.50}$$

Therefore, by virtue of (12.49) and (12.50), the scalar product of the bra and ket vectors $\langle\psi(t)|$ and $|\psi(t)\rangle$ yields

$$\langle\psi(t)|\psi(t)\rangle = |\psi(t_0)| U^{\dagger}(t, t_0)U(t, t_0)|\psi(t_0)\rangle. \tag{12.51}$$

Since the conservation of total probability requires that the norm of a vector be constant in time, we must have

$$\langle\psi(t)|\psi(t)\rangle = \langle\psi(t_0)|\psi(t_0) = 1, \tag{12.52}$$

as $|\psi\rangle$ is to be normalised at each instant of time. Comparing (12.51) and (12.52), we obtain

$$U^{\dagger}(t, t_0)U(t, t_0) = I. \tag{12.53a}$$

Since we are considering the physical systems whose states are represented by vectors (strictly speaking, rays) in an infinite dimensional Hilbert space, the above equation does not guarantee that U is unitary. For that purpose, we will have to show that the following equation must *also* hold:

$$U(t, t_0)U^{\dagger}(t, t_0) = I. \tag{12.53b}$$

Problem

12.3. Prove the above relation.

Hence, the operator $U(t, t_0)$ is unitary so that

$$U^{\dagger}(t, t_0) = U^{-1}(t, t_0). \tag{12.54}$$

That is, the time evolution of a state is governed by a unitary operator.

Let us consider the case when t and t_0 are infinitesimally separated so that the infinitesimal increment δt in time is $(t - t_0)$. Then to first order in δt, we may write

$$\begin{aligned} U(t_0 + \delta t, t_0) &= U(t_0, t_0) - \mathrm{i}\frac{H}{\hbar}\delta t \\ &= I - \mathrm{i}\frac{H}{\hbar}\delta t, \end{aligned} \tag{12.55}$$

where H is a Hermitian operator of the dimension of energy and the constant $\hbar$ has been introduced for later convenience of interpretation.

Problem

12.4. Show that H has the dimensions of energy and as U is a unitary operator, the operator H must be Hermitian.

Now, by virtue of (12.49), the state vector at time $t_0 + \delta t$ is given by

$$\begin{aligned} |\psi(t_0 + \delta t)\rangle &= U(t_0 + \delta t, t_0)|\psi(t_0)\rangle \\ &= \left(I - \mathrm{i}\frac{H}{\hbar}\delta t\right)|\psi(t_0)\rangle, \text{ by virtue of (12.55)}, \\ &= |\psi(t_0)\rangle - \mathrm{i}\frac{H}{\hbar}\delta t\,|\psi(t_0)\rangle. \end{aligned} \tag{12.56}$$

Taking the term $|\psi(t_0)\rangle$ to the left-hand side of the above equation, we obtain

$$|\psi(t_0 + \delta t)\rangle - |\psi(t_0)\rangle = -\mathrm{i}\frac{H}{\hbar}\delta t\,|\psi(t_0)\rangle,$$

or

$$\frac{|\psi(t_0 + \delta t)\rangle - |\psi(t_0)\rangle}{\delta t} = -\mathrm{i}\frac{H}{\hbar}\,|\psi(t_0)\rangle.$$

Taking the limit as $\delta t \to 0$, the above relation takes the form

$$\frac{\partial|\psi(t)\rangle}{\partial t} = -\mathrm{i}\frac{H}{\hbar}|\psi(t)\rangle,$$

or

$$\mathrm{i}\hbar\frac{\partial|\psi(t)\rangle}{\partial t} = H\,|\psi(t)\rangle. \tag{12.57}$$

This is the equation of motion for the state vector: it is the *Schrödinger equation of motion* provided the Hermitian operator H is taken as the Hamiltonian of the system. For a closed system, the Hamiltonian operator H does not depend upon time explicitly and all time dependence is contained in the vector $|\psi(t)\rangle$. In general, the Hamiltonian may depend upon time explicitly. Equation (12.57) is *assumed* to be valid even in that case. Hence, the formalism which we have been using in this book is in the Schrödinger picture.

The temporal development of the system can *also* be described by an operator equation involving the operator $U(t, t_0)$. We have proved that

$$i\hbar\frac{\partial|\psi(t)\rangle}{\partial t} = H|\psi(t)\rangle, \tag{12.57'}$$

but

$$|\psi(t)\rangle = U(t, t_0)|\psi(t_0)\rangle. \tag{12.49'}$$

Substituting this expression for $|\psi(t)\rangle$ in (12.57′), we obtain

$$i\hbar\frac{\partial}{\partial t}U(t, t_0)|\psi(t_0)\rangle\} = HU(t, t_0)|\psi(t_0)\rangle,$$

or

$$\left[i\hbar\frac{\partial}{\partial t}\{U(t, t_0)\}\right]|\psi(t_0)\rangle = HU(t, t_0)|\psi(t_0)\rangle.$$

Since this equation holds for arbitrary $|\psi(t_0)\rangle$, we can write

$$i\hbar\frac{\partial}{\partial t}\{U(t, t_0)\} = HU(t, t_0). \tag{12.58}$$

This is a differential equation that enables us to obtain $U(t, t_0)$ for any time t provided H is known.

Problem

12.5. Show that for a time-independent Hamiltonian the *formal solution* of (12.58) is

$$U(t, t_0) = \exp[-iH(t - t_0)/\hbar]. \tag{12.59}$$

To summarise, the Schrödinger picture can be represented *either* by the equation of motion

$$i\hbar\frac{\partial|\psi(t)\rangle}{\partial t} = H|\psi(t)\rangle$$

or by the operator equation

$$i\hbar\frac{\partial}{\partial t}\{U(t, t_0)\} = HU(t, t_0).$$

where $U(t_0, t_0) = I$ and $U(t, t_0) = e^{-iH(t-t_0)/\hbar}$, provided H does not depend upon time explicitly.

12.5.2 The Heisenberg picture

In the Schrödinger picture, *except* when an operator is explicitly time-dependent, the temporal development of the system is resident in the state vectors *only*. We will now present an alternative scheme, called the *Heisenberg picture* or the *Heisenberg representation*, in which the temporal development of the system lies in the *operators only*; of course, in this picture, the state vectors do not change with time.

The Heisenberg picture is obtained from the Schrödinger picture using a unitary transformation. We have shown that a Schrödinger state vector is given by the equation

$$|\psi(t)\rangle_S = U(t, t_0)|\psi(t_0)\rangle_S, \tag{12.49'}$$

the subscript indicating the picture we are considering. Now we *define* a vector in the Heisenberg picture, called the Heisenberg state vector $|\psi(t)\rangle_H$, by the equation

$$|\psi(t)\rangle_H = U^{-1}(t, t_0)|\psi(t)\rangle_S. \tag{12.60}$$

Substituting the expression for $|\psi(t)\rangle_S$ from (12.49′) in (12.60), we obtain

$$|\psi(t)\rangle_H = U^{-1}(t, t_0)U(t, t_0)|\psi(t_0)\rangle_S = |\psi(t_0)\rangle_S, \tag{12.61}$$

as $U^{-1}(t, t_0)U(t, t_0) = I$. The above equation shows that the Heisenberg state vector $|\psi(t)\rangle_H$ does not change with time as at all times it is equal to the Schrödinger state vector $|\psi(t_0)\rangle_S$ at time t_0. Moreover, as the same physical system is to be represented by the Schrödinger and the Heisenberg pictures, these must be connected by a unitary transformation. We have shown that for such a transformation, the operators and the state vectors are connected by the equations

$$\hat{A} = U^{-1}AU \tag{12.32a'}$$

and

$$|\hat{\psi}\rangle = U^{-1}|\psi\rangle. \tag{12.32b'}$$

According to (12.60), the Heisenberg state vector is connected to the Schrödinger state vector by the equation

$$|\psi(t)\rangle_H = U^{-1}(t, t_0)|\psi(t_0)\rangle_S. \tag{12.60'}$$

where $U(t, t_0)$ is a unitary operator. Therefore, by (12.32a'), the Heisenberg and the Schrödinger operators must be related as

$$A_H(t) = U^{-1}(t, t_0)A_S U(t, t_0),$$

the operators in the Schrödinger picture being time-independent. Differentiating the last equation with respect to t, we obtain

$$\frac{dA_H(t)}{dt} = \frac{\partial U^{-1}(t, t_0)}{\partial t}A_S\, U(t, t_0) + U^{-1}(t, t_0)A_S\frac{\partial U(t, t_0)}{\partial t}. \tag{12.62}$$

Now, from (12.58), we have

$$\mathrm{i}\hbar\frac{\partial U}{\partial t} = H_S U. \tag{12.58'}$$

Taking the Hermitian conjugate of both sides, we obtain

$$-\mathrm{i}\hbar\frac{\partial U^\dagger}{\partial t} = U^\dagger H_S^\dagger.$$

Since H is Hermitian, the above equation may be written as

$$-\mathrm{i}\hbar\frac{\partial U^\dagger}{\partial t} = U^\dagger H_S.$$

Moreover, as U is unitary, $U^\dagger = U^{-1}$, the above equation reduces to

$$-\mathrm{i}\hbar\frac{\partial U^{-1}}{\partial t} = U^{-1}H_S. \tag{12.63}$$

Substituting the expressions for $i\hbar\partial U^{-1}/\partial t$ and $\mathrm{i}h\partial U/\partial t$ from (12.63) and (12.58′) in (12.62), we obtain

$$\begin{aligned}\mathrm{i}\hbar\frac{\mathrm{d}A_H(t)}{\mathrm{d}t} &= -U^{-1}H_S A_S U + U^{-1}A_S H_S U \\ &= -U^{-1}H_S U U^{-1}A_S U + U^{-1}A_S U U^{-1}H_S U \\ &= -H_H(t)A_H(t) + A_H(t)H_H(t) \\ &= [A_H(t), H_H(t)],\end{aligned} \tag{12.64}$$

where

$$H_H(t) = U^{-1}H_S U$$

and

$$A_H(t) = U^{-1}A_S U.$$

For the particular case $A_H(t) = H(t)$, (12.64) gives $\mathrm{d}H/\mathrm{d}t = 0$. This yields $H(t) =$ constant in time $= H$, say. Therefore, (12.64) becomes

$$\mathrm{i}\hbar\frac{\mathrm{d}A_H(t)}{\mathrm{d}t} = [A_H(t), H]. \tag{12.65}$$

This is *Heisenberg's equation of motion* for an operator. Note that the time-dependence has been transferred to the operators as the Heisenberg state vectors are independent of time. This picture is analogous to the classical description and it may be expected that the operators $A(t)$ are closely related to the corresponding time-dependent classical dynamical variables.

Example 12.4.

Problem

Show that the expectation values $\langle A_S \rangle$ and $\langle A_H \rangle$ of any operator A in the Schrödinger and the Heisenberg representations are the same.

Solution

We have

$$\begin{aligned}\langle A_S \rangle &= \langle \psi_S | A_S | \psi_S \rangle = \left\langle \psi_S \right| U A_H U^{-1} \left| \psi_S \right\rangle \\ &= \langle \psi_H | A_H | \psi_H \rangle \\ &= \langle A_H \rangle. \end{aligned} \tag{12.66}$$

That is, the average value of an observable is the same in both pictures as it must be because whatever picture we use, the results for physical measurements must be the same.

12.5.3 The interaction picture

Another type of representation called the *interaction* or *Dirac picture* is also significant. In this picture, *both the operators and the state vectors are time-dependent.* It is well-known that there are a few physical systems for which the state vectors and eigenvalues of their Hamiltonians can be determined exactly. It then becomes imperative to use approximation techniques such as perturbation theory to determine the time evolution of the system. Consider a physical system for which its Hamiltonian H can be split into two parts: a free (unperturbed) Hamiltonian $H^0(r)$, for which state vectors and eigenvalues can be exactly determined, and an interaction (perturbation) Hamiltonian $H^{\text{int}}(\mathbf{r}, t)$ which in general is time-dependent. Therefore, we may write

$$H = H^0(\mathbf{r}) + H^{\text{int}}(t). \tag{12.67}$$

Let $|\psi(t)\rangle_S$ denote the time-dependent vector in the Schrödinger picture. Then, the equation of motion in this picture may be expressed as

$$\begin{aligned} i\hbar \frac{\partial |\psi(t)\rangle_S}{\partial t} &= H \, |\psi(t)\rangle_S \\ &= [H^0 + H^{\text{int}}] |\psi(t)\rangle_S. \end{aligned} \tag{12.68}$$

Let us now *define* a new state vector $|\psi(t)\rangle_I$ such that

$$|\psi(t)\rangle_I = e^{iH^0(t-t_0)/\hbar} \, |\psi(t)\rangle_S. \tag{12.69a}$$

To make the calculations simple, let us take the initial time $t_0 = 0$. Then the above equation reduces to

$$|\psi(t)\rangle_I = e^{-iH^0 t/\hbar} \, |\psi(t)\rangle_S. \tag{12.69b}$$

This definition can be used to express the Schrödinger state vector $|\psi(t)\rangle_S$ in terms of the interaction vector:

$$|\psi(t)\rangle_S = e^{-iH^0 t/\hbar}\,|\psi(t)\rangle_I. \tag{12.70}$$

To find the equation of motion for $|\psi(t)\rangle_I$, put the above expression for $|\psi(t)\rangle_S$ in terms of $|\psi(t)\rangle_I$ in (12.68). This gives

$$\begin{aligned} i\hbar\frac{\partial}{\partial t}[e^{-iH^0 t/\hbar}|\psi(t)\rangle_I] &= [H^0 + H^{\text{int}}]e^{-iH^0 t/\hbar}\,|\psi(t)\rangle_I \\ &= H^0 e^{-iH^0 t/\hbar}\,|\psi(t)\rangle_I + H^{\text{int}} e^{-iH^0 t/\hbar}\,|\psi(t)\rangle_I, \end{aligned}$$

or

$$\begin{aligned} i\hbar\left[-\frac{iH^0}{\hbar}\right]e^{-iH^0 t/\hbar}\,|\psi(t)\rangle_I + i\hbar e^{-iH^0 t/\hbar}\frac{\partial}{\partial t}\,|\psi(t)\rangle_I \\ = H^0 e^{-iH^0 t/\hbar}\,|\psi\rangle_I + H^{\text{int}} e^{-iH^0 t/\hbar}\,|\psi\rangle_I, \end{aligned}$$

or

$$\begin{aligned} H^0 e^{-iH^0 t/\hbar}\,|\psi(t)\rangle_I + i\hbar e^{-iH^0 t/\hbar}\frac{\partial}{\partial t}\,|\psi(t)\rangle_I \\ = H^0 e^{-iH^0 t/\hbar}\,|\psi(t)\rangle_I + H^{\text{int}} e^{-iH^0 t/\hbar}\,|\psi(t)\rangle_I. \end{aligned}$$

Cancelling the term $H^0\, e^{-iH^0 t/\hbar}\,|\psi(t)\rangle_I$ on the two sides of the above equation, we obtain

$$i\hbar e^{-iH^0 t/\hbar}\frac{\partial}{\partial t}\,|\psi(t)\rangle_I = H^{\text{int}} e^{-iH^0 t/\hbar}\,|\psi(t)\rangle_I. \tag{12.71}$$

Multiplying both sides of the above equation from the left by $e^{-iH^0 t/\hbar}$, we obtain

$$i\hbar\frac{\partial}{\partial t}\,|\psi(t)\rangle_I = e^{-iH^0 t/\hbar} H^{\text{int}} e^{-iH^0 t/\hbar}\,|\psi(t)\rangle_I.$$

or

$$i\hbar\frac{\partial}{\partial t}\,|\psi(t)\rangle_I = H_I^{\text{int}}\,|\psi(t)\rangle_I, \tag{12.72}$$

where

$$H_I^{\text{int}} = e^{iH^0 t/\hbar} H^{\text{int}} e^{-iH^0 t/\hbar}. \tag{12.73}$$

Equation (12.72) shows that the temporal development of the state vectors is determined *only* by the interaction Hamiltonian.

If an observable A is represented by the operators A_S and A_I in the Schrödinger and the interaction pictures, then the expectation values must be equal:

$$\psi(t)_S|A_S|\psi(t)_S\rangle = \langle\psi(t)_I|A_I|\psi(t)_I\rangle, \tag{12.74}$$

but

$$|\psi(t)\rangle_I = e^{iH^0t/\hbar}\,|\psi(t)\rangle_S. \tag{12.69'}$$

Taking the Hermitian conjugate of the above equation, we obtain

$$\langle\psi(t)|_I = \langle\psi(t)|_S\; e^{-iH^0t/\hbar} \tag{12.75}$$

Substituting the expressions for $|\psi(t)\rangle_I$ and $\langle\psi(t)|_I$ from (12.69′) and (12.75) in (12.74), we obtain

$$\langle\psi_S(t)|A_S|\,\psi_S(t)\rangle = \left\langle\psi_S e^{-iH^0t/\hbar}\,|A_I|\,e^{iH^0t/\hbar}\psi_S\right\rangle.$$

Since $|\psi_S(t)\rangle$ is an arbitrary vector in the Schrödinger picture, the above equation will hold only if

$$A_S = e^{-iH^0t/\hbar}A_I(t)e^{iH^0t/\hbar}, \tag{12.76a}$$

or

$$A_I(t) = e^{iH^0t/\hbar}A_S(t)e^{-iH^0t/\hbar}. \tag{12.76b}$$

This equation shows that the time-dependence of operators in the interaction picture is determined by the unperturbed Hamiltonian H^0. Differentiating (12.76*b*) with respect to t, we obtain

$$\frac{d}{dt}(A_I(t)) = \frac{i}{\hbar}H^0e^{iH^0t/\hbar}A_Se^{-iH^0t/\hbar} + e^{iH^0t/\hbar}A_S\left(\frac{-i}{\hbar}H^0e^{-iH^0t/\hbar}\right).$$

By virtue of (12.76*b*), we obtain

$$\frac{d}{dt}(A_I(t)) = -\frac{i}{\hbar}[A_I(t),\, H^0]. \tag{12.77}$$

This differential equation determines the time development of operators in the interaction picture. In this picture, the temporal change of a state vector is entirely due to the interaction (perturbation) Hamiltonian H^{int} while the operators evolve in time due to the unperturbed part H^0 of the system.

12.6 A one-dimensional oscillator

The classical equation of motion for a one-dimensional harmonic oscillator of mass m is given by

$$F = m\,\frac{d^2q}{dt^2}$$

where m is the mass, q is the displacement of the oscillator from its position of equilibrium and F is the force acting on the oscillator. Since for a harmonic oscillator, the force is $-kq$, we have

$$-kq = m\frac{d^2q}{dt^2},$$

or

$$\frac{d^2q}{dt^2} + \frac{ka}{m} = 0,$$

or

$$\ddot{q} + \omega^2 q = 0, \tag{12.78}$$

where $\omega = \sqrt{k/m}$ is the angular frequency of the oscillator and for convenience we have written $\ddot{q}$ for d^2q/dt^2. The total energy H is given by

$$H = \frac{p^2}{2m} + \frac{1}{2}kq^2 = \frac{p^2}{2m} + \frac{1}{2}m\omega^2 q^2, \quad p = m\dot{q}.$$

For a particle of unit mass, this reduces to

$$H = \frac{1}{2}(p^2 + \omega^2 q^2), \quad p = \dot{q}. \tag{12.79}$$

Let us now quantise the system with the condition

$$[Q, P] = i\hbar, \tag{12.80}$$

where Q and P are the position and momentum operators. By taking $A_H = Q$ in (12.77), we obtain

$$\begin{aligned}
i\hbar\frac{dQ}{dt} &= [Q, H] \\
&= \left[Q, \frac{1}{2}P^2 + \frac{1}{2}\omega^2 Q^2\right] \\
&= \left[Q, \frac{1}{2}P^2\right] + \left[Q, \frac{1}{2}\omega^2 Q^2\right] \\
&= \frac{1}{2}[Q, P^2] + \frac{1}{2}\omega^2[Q, Q^2] \\
&= \frac{1}{2}[Q, P^2] \\
&= \frac{1}{2}[QP^2 - P^2Q] \\
&= \frac{1}{2}[QP^2 - P \cdot PQ].
\end{aligned}$$

By virtue of (12.80), it reduces to

$$\begin{aligned}
i\hbar\frac{dQ}{dt} &= \frac{1}{2}[QP^2 - P.(QP - i\hbar)] \\
&= \frac{1}{2}[QP^2 - PQP + i\hbar P] \\
&= \frac{1}{2}[(QP - PQ)P + i\hbar P] \\
&= \frac{1}{2}[i\hbar P + i\hbar P] = i\hbar P,
\end{aligned}$$

or

$$\frac{dQ}{dt} = P,$$

or

$$\dot{Q} = P.$$

Now, by taking $A_H = P$ in (12.77), we have

$$\begin{aligned}
i\hbar\frac{dP}{dt} &= [P, H] \\
&= \left[P, \frac{1}{2}P^2 + \frac{1}{2}\omega^2 Q^2\right] \\
&= \left[P, \frac{1}{2}P^2\right] + \left[P, \frac{1}{2}\omega^2 Q^2\right] \\
&= \frac{1}{2}\omega^2[P, Q^2] \\
&= \frac{1}{2}\omega^2[PQ^2 - Q^2P] \\
&= \frac{1}{2}\omega^2[PQ \cdot Q - Q^2P].
\end{aligned}$$

By using (12.80), we obtain

$$\begin{aligned}
i\hbar\frac{dP}{dt} &= \frac{1}{2}\omega^2[QPQ - Q^2P - i\hbar Q] \\
&= \frac{1}{2}\omega^2[Q(PQ - QP) - i\hbar Q] \\
&= \frac{1}{2}\omega^2[Q(-i\hbar) - i\hbar Q] \\
&= \frac{1}{2}\omega^2[-2i\hbar Q] \\
&= -i\hbar\omega^2 Q,
\end{aligned}$$

or

$$\frac{\mathrm{d}P}{\mathrm{d}t} = -\omega^2 Q,$$

or

$$\dot{P} = -\omega^2 Q. \tag{12.82}$$

Differentiating (12.81) with respect to time, we obtain

$$\ddot{Q} = \dot{P}. \tag{12.83}$$

Substituting the expression for $\dot{P}$ from (12.82) in the above equation, we obtain

$$\ddot{Q} = -\omega^2 Q,$$

or

$$\ddot{Q} + \omega^2 Q = 0. \tag{12.84}$$

This is identical to the classical equation of motion (12.78), though it is an operator equation. This is the advantage of the Heisenberg picture.

12.7 The relation between state vectors and wave functions

To find a relation between a wave function and a vector in a Hilbert space, consider a Hermitian operator α_{operator} representing a dynamical variable α and having a continuous spectrum of eigenvalues α', α'', α''', ... in the interval $-\infty$ to $+\infty$. Let $|\alpha'\rangle$, $|\alpha''\rangle$, $|\alpha'''\rangle$,... denote the corresponding complete orthonormal set of eigenvectors in the Hilbert space. Then for an arbitrary vector $|\psi\rangle$, the scalar products $\langle\alpha'|\psi\rangle$, $\langle\alpha''|\psi\rangle$, $\langle\alpha'''|\psi\rangle$,... are the components of the vector $|\psi\rangle$ along the orthonormal set of vectors $|\alpha'\rangle$, $|\alpha''\rangle$, $|\alpha'''\rangle$,.... These are infinite in number and each one of them is, by definition, a complex number. Let us denote them by $\psi(\alpha')$, $\psi(\alpha'')$, $\psi(\alpha''')$,.... These complex numbers may be regarded as the values of a function $\psi(\alpha)$ for $\alpha = \alpha'$, α'', α''',.... Hence, the components $\psi(\alpha')$, $\psi(\alpha'')$, $\psi(\alpha''')$,... and consequently the vector $|\psi\rangle$ itself represents the entire range of values of $\psi(\alpha)$; a component of $|\psi\rangle$ along, say, $|\alpha'\rangle$ representing the value of the wave function ψ for $\alpha = \alpha'$. This shows how the state vectors are related to the wave functions.

12.8 A free particle

Consider a physical system. Suppose that the states of the system are represented by vectors in a Hilbert space. Instead of considering the vectors in the Hilbert space, it is often convenient for the description of the physical system to introduce some *coordinate system* in the Hilbert space, i.e. to choose some *representation.*

Let us consider the motion of a free particle in a one-dimensional space. Let x' be the eigenvalue of its position operator x corresponding to the eigenvector $|x'\rangle$ specifying its state. Then, the eigenvalue equation is

$$x|x'\rangle = x'|x'\rangle. \tag{12.85}$$

A free particle, by definition, may occupy any position. Therefore, the position eigenvalue spectrum will be continuous. The result of a measurement of its position can be any number between $-\infty$ and $+\infty$. The normalisation condition can then be written as

$$\langle x'|x''\rangle = \delta(x' - x''). \qquad (12.86)$$

The eigenvectors of a Hermitian operator form a complete set of vectors. Therefore, these vectors can be used as a basis of representation and the completeness relation is expressed as

$$\int |x'\rangle\langle x'|\mathrm{d}x' = 1. \qquad (12.87)$$

We will now find the representation of momentum operator p_x in this coordinate representation. This is done as follows. We have

$$\langle x''|[x, p_x]|x'\rangle = \mathrm{i}\hbar\langle x''|x'\rangle, \qquad (12.88)$$

where we have used the relation $[x, p_x] = \mathrm{i}\hbar$. But the left-hand side of the above equation can also be expressed as

$$\langle x''|[x, p_x]|x'\rangle = \langle x''|(xp_x - p_x x)|x'\rangle. \qquad (12.89)$$

By taking the Hermitian conjugate of (12.85), we obtain

$$\langle x'|x^{\dagger} = x'^{*}\langle x'|.$$

Since the eigenvalues of the Hermitian operator x, for which $x^{\dagger} = x$, must be real, therefore for $x^{\dagger} = x$, we have $x'^{*} = x'$. Consequently, the above equation reduces to

$$\langle x'|\, x = x'\langle x'|.$$

Changing $'$ to $''$, we obtain

$$\langle x''|\, x = x''\langle x''| = \langle x''|\, x''. \qquad (12.90)$$

Substituting the expressions for $x|x'\rangle$ and $\langle x''|x$ from (12.85) and (12.90) in (12.89), we obtain

$$\begin{aligned}\langle x''|[x, p_x]|x'\rangle &= \left\langle x''\middle|x''p_x - p_x x'\middle|x'\right\rangle \\ &= \left\langle x''\middle|(x'' - x')p_x\middle|x'\right\rangle \\ &= (x'' - x')\left\langle x''\middle|p_x\middle|x'\right\rangle. \end{aligned} \qquad (12.91)$$

From (12.88) and (12.86), we have

$$\langle x''|[x, p_x]|x'\rangle = \mathrm{i}\hbar\langle x''|x\rangle = \mathrm{i}\hbar\delta(x'' - x'). \qquad (12.92)$$

By comparing (12.91) and (12.92), we find that the left-hand sides being the same, the right-hand sides of the two equations should be equal:

$$(x'' - x')\langle x''|p_x|x'\rangle = i\hbar\delta(x'' - x'). \tag{12.93}$$

By using the relation

$$x\frac{d}{dx}\delta(x) = -\delta(x), \tag{12.94}$$

we can write the right-hand side of (12.93) as

$$i\hbar\delta(x'' - x') = -i\hbar(x'' - x')\frac{\partial}{\partial x''}\delta(x'' - x'). \tag{12.95}$$

Substituting the expression $\langle x''|x'\rangle$ for $\delta(x'' - x')$ from (12.92) in (12.95) and by the use of (12.93), we obtain

$$\langle x''|p_x|x'\rangle = -i\hbar\frac{\partial}{\partial x''}\langle x''|x'\rangle. \tag{12.96}$$

Dirac's simple and elegant but abstract formalism which associates the vectors in an abstract space with the states of a physical system has transcended the classical limits of vision. The approach is novel and its ingenuity has to be appreciated. No doubt, this formalism has not only generalised the entire structure of quantum mechanics but has also made it more charming.

We close this chapter by quoting the exquisite remarks of Dirac: 'The steady progress of physics requires for its theoretical formulation a mathematics that gets continually more advanced. This is only natural and to be expected. What, however, was not expected by the scientific workers of the last century was that...the modern physical developments [would] require a mathematics that continually shifts its foundations and gets more abstract. ...It seems likely that this process...will continue in the future.'

Additional problem

12.6. Prove that

$$\left\langle p_x''\,|x|\,p_x'\right\rangle = -i\hbar\frac{\partial}{\partial p_x''}\left\langle p_x''|p_x'\right\rangle.$$

Lightning Source UK Ltd.
Milton Keynes UK
UKOW07n1045170116

266525UK00020B/478/P

9 780750 3120